Applications and Challenges of Maintenance and Safety Engineering in Industry 4.0

Alberto Martinetti
University of Twente, The Netherlands

Micaela Demichela
Politecnico di Torino, Italy

Sarbjeet Singh
Luleå University of Technology, Sweden

IGI Global
DISSEMINATOR OF KNOWLEDGE

A volume in the Advances in Civil and Industrial Engineering (ACIE) Book Series

Published in the United States of America by
IGI Global
Engineering Science Reference (an imprint of IGI Global)
701 E. Chocolate Avenue
Hershey PA, USA 17033
Tel: 717-533-8845
Fax: 717-533-8661
E-mail: cust@igi-global.com
Web site: http://www.igi-global.com

Library of Congress Cataloging-in-Publication Data

Library of Congress Cataloging-in-Publication Data

Names: Martinetti, Alberto, editor. | Demichela, Micaela, editor. | Singh, Sarbjeet, 1976- editor.
Title: Applications and challenges of maintenance and safety engineering in industry 4.0 / Alberto Martinetti, Micaela Demichela, Sarbjeet Singh, editors.
Description: Hershey, PA : Engineering Science Reference, 2020. | Includes bibliographical references and index. | Summary: "This book addresses safety and design for maintenance and reducing the factors that influence and degrade human performance and that provides technological advancements and emergent technologies that reduce the dependence on operator capabilities"-- Provided by publisher.
Identifiers: LCCN 2020000295 (print) | LCCN 2020000296 (ebook) | ISBN 9781799839040 (h/c) | ISBN 9781799852360 (s/c) | ISBN 9781799839057 (eISBN)
Subjects: LCSH: Industrial safety.
Classification: LCC T55 .A67 2020 (print) | LCC T55 (ebook) | DDC 620.8/6--dc23
LC record available at https://lccn.loc.gov/2020000295
LC ebook record available at https://lccn.loc.gov/2020000296

This book is published in the IGI Global book series Advances in Civil and Industrial Engineering (ACIE) (ISSN: 2326-6139; eISSN: 2326-6155)

British Cataloguing in Publication Data
A Cataloguing in Publication record for this book is available from the British Library.

For electronic access to this publication, please contact: eresources@igi-global.com.

Advances in Civil and Industrial Engineering (ACIE) Book Series

Ioan Constantin Dima
University Valahia of Târgovişte, Romania

ISSN:2326-6139
EISSN:2326-6155

Mission

Private and public sector infrastructures begin to age, or require change in the face of developing technologies, the fields of civil and industrial engineering have become increasingly important as a method to mitigate and manage these changes. As governments and the public at large begin to grapple with climate change and growing populations, civil engineering has become more interdisciplinary and the need for publications that discuss the rapid changes and advancements in the field have become more in-demand. Additionally, private corporations and companies are facing similar changes and challenges, with the pressure for new and innovative methods being placed on those involved in industrial engineering.

The **Advances in Civil and Industrial Engineering (ACIE) Book Series** aims to present research and methodology that will provide solutions and discussions to meet such needs. The latest methodologies, applications, tools, and analysis will be published through the books included in **ACIE** in order to keep the available research in civil and industrial engineering as current and timely as possible.

Coverage

- Materials Management
- Optimization Techniques
- Engineering Economics
- Structural Engineering
- Production Planning and Control
- Coastal Engineering
- Productivity
- Ergonomics
- Operations Research
- Transportation Engineering

IGI Global is currently accepting manuscripts for publication within this series. To submit a proposal for a volume in this series, please contact our Acquisition Editors at Acquisitions@igi-global.com or visit: http://www.igi-global.com/publish/.

The Advances in Civil and Industrial Engineering (ACIE) Book Series (ISSN 2326-6139) is published by IGI Global, 701 E. Chocolate Avenue, Hershey, PA 17033-1240, USA, www.igi-global.com. This series is composed of titles available for purchase individually; each title is edited to be contextually exclusive from any other title within the series. For pricing and ordering information please visit http://www.igi-global.com/book-series/advances-civil-industrial-engineering/73673. Postmaster: Send all address changes to above address.

Titles in this Series

For a list of additional titles in this series, please visit:
http://www.igi-global.com/book-series/advances-civil-industrial-engineering/73673

Multi-Objective Optimization of Industrial Power Generation Systems Emerging Research and Opportunities
Timothy Ganesan (Royal Bank of Canada, Canada)
Engineering Science Reference • © 2020 • 233pp • H/C (ISBN: 9781799817109) • US $195.00

Impact of Industry 4.0 on Architecture and Cultural Heritage
Cecilia Maria Bolognesi (Politecnico of Milano, Italy) and Cettina Santagati (Università di Catania, Italy)
Engineering Science Reference • © 2020 • 422pp • H/C (ISBN: 9781799812340) • US $225.00

Handbook of Research on Urban-Rural Synergy Development Through Housing, Landscape, and Tourism
Aleksandra Krstić-Furundžić (University of Belgrade, Serbia) and Aleksandra Djukić (University of Belgrade, Serbia)
Engineering Science Reference • © 2020 • 437pp • H/C (ISBN: 9781522599326) • US $275.00

Re-Coding Homes Through Flexible Interiors Emerging Research and Opportunities
Nilüfer Saglar Onay (Istanbul Technical University, Turkey) S. Banu Garip (Istanbul Technical University, Turkey) and Ervin Garip (Istanbul Technical University, Turkey)
Engineering Science Reference • © 2020 • 165pp • H/C (ISBN: 9781522589587) • US $175.00

Green Building Management and Smart Automation
Arun Solanki (Gautam Buddha University, India) and Anand Nayyar (Duy Tan University, Vietnam)
Engineering Science Reference • © 2020 • 312pp • H/C (ISBN: 9781522597544) • US $215.00

Handbook of Research on Implementation and Deployment of IoT Projects in Smart Cities
Krishnan Saravanan (Anna University Chennai – Regional Office Tirunelveli, India) Golden Julie (Anna University, India) and Harold Robinson (SCAD College of Engineering and Technology, India)
Engineering Science Reference • © 2019 • 415pp • H/C (ISBN: 9781522591993) • US $295.00

Handbook of Research on Digital Research Methods and Architectural Tools in Urban Planning and Design
Hisham Abusaada (Housing and Building National Research Center, Egypt) Carsten Vellguth (German Academic Exchange Service, Germany) and Abeer Elshater (Ain Shams University, Egypt)
Engineering Science Reference • © 2019 • 445pp • H/C (ISBN: 9781522592389) • US $265.00

Recycled Waste Materials in Concrete Construction Emerging Research and Opportunities
Jahangir Mirza (Research Institute of Hydro-Québec, Canada) Mohd Warid Hussin (Universiti Teknologi Malaysia, Malaysia) and Mohamed A. Ismail (Miami College of Henan University, China)
Engineering Science Reference • © 2019 • 169pp • H/C (ISBN: 9781522583257) • US $165.00

IGI Global
DISSEMINATOR OF KNOWLEDGE

701 East Chocolate Avenue, Hershey, PA 17033, USA
Tel: 717-533-8845 x100 • Fax: 717-533-8661
E-Mail: cust@igi-global.com • www.igi-global.com

Editorial Advisory Board

Table of Contents

Section 3

Detailed Table of Contents

Aiming to remain competitive, companies from diverse industries are paying more attention to Industry 4.0 concept and its benefits. Maintenance is seen as a specific area of action to successfully sustain a competitive leverage, and its fusion with Industry 4.0 is perceived to revolutionize the whole maintenance concept. Maintenance 4.0 emerges as a subset of Industry 4.0 in the form of self-learning and smart system that predicts failures, makes diagnoses, and establishes maintenance actions. This chapter presents a systematic literature review (SLR) on Maintenance 4.0, with the aim of outlining the current achievements as well as limitations of maintenance meeting Industry 4.0 demands. The analysis included 90 papers selected as being the most suitable to reach the proposed goal. A state of the art on Maintenance 4.0 is performed, followed by an analysis ambitioning the delineation of what future holds on this topic.

Industry 4.0 has received a massive amount of attention worldwide in the past few years as a technological infrastructure to provide efficient operations in existing production systems as well as fast-tracking the implementation of internet-connected technologies across various industries. Industry 4.0 technologies have been considered as a strategy and implemented successfully in various developed countries. However, in emerging economies (or developing countries), the implementation of Industry 4.0 is not as successful as developed nations because of various challenges. However, fast-moving economies can take advantage of Industry 4.0 techniques as their requirement to operate at faster rates, capitalizing on new technologies that can drive efficiencies. This chapter examines the sustainability issues of Industry 4.0 in developing or emerging economies countries. These sustainability issues are related to scientific, technological, and societal issues.

Several industries are looking for smart methods to increase their production throughput and operational efficiency at the lowest cost, reduced risk, and reduced spending of resources considering demands from stakeholders, governments, and competitors. To achieve this, industries are looking for possible solutions to the above problems by adopting emerging technologies. A foremost concept that is setting the pace and direction for many sectors and services is Industry 4.0. The focus is on augmenting machines and infrastructure with wireless connectivity, sensors, and intelligent systems to monitor, visualize, and communicate incidences between different entities for decision making. An aspect of physical asset management that has been enormously influenced by the new industrial set-up is the maintenance process. This chapter highlights the issues and challenges of Industry 4.0 from maintenance process viewpoint according to EN 60300-3-14. Further, a conceptual model on how maintenance process can be integrated into Industrial 4.0 architecture is proposed to enhance its value.

Sharing of tacit knowledge is a key topic of research within the knowledge management community. Considering its embodied nature, organizations have always struggled with embedding it into their processes. Proper execution of complex processes such as system integration asks for an adequate sharing of tacit knowledge. Acknowledging the importance of lessons learned for system integration and their presence in tacit and explicit form, a case study was conducted within the Netherlands Railways. It was determined that non-sensitivity to the tacit dimension of lessons learned has resulted in their lack of utilization. Consequently, LEAF framework was developed, where LEAF stands for learnability, embraceability, applicability, and findability. The framework suggests that addressing these four features collectively can eventually lead to an adequate knowledge-sharing strategy for lessons learned. Lastly, the chapter presents an example from the Netherlands Railways to emphasize the key role technological solutions of Industry 4.0 can play in facilitating tacit knowledge sharing.

The convergence of information technology (IT) and operational technology (OT) and the associated paradigm shift toward fourth industrial revolution (aka Industry 4.0) in companies has brought tremendous changes in technology vision with innovative technologies such as robotics, big data, cloud computing, online monitoring, internet of things (IoT), cyber-physical systems (CPS), cognitive computing, and artificial intelligence (AI). However, this transition towards the fourth industrial revolution has many benefits in productivity, efficiency, revenues, customer experience, and profitability, but also imposes many challenges. One of the challenges is to manage and secure large amount of data generated from internet of things (IoT) devices that provide many entry points for hackers in the form of a threat to exploit new and existing vulnerabilities within the network. This chapter investigates various cybersecurity issues and challenges in Industry 4.0 with more focus on three industrial case studies.

Section 2

In recent years augmented reality has begun to be a presence in various industrial sectors. In augmented reality the operator's perception of reality is enriched through virtual information useful to help him in his working activity. Augmented reality can be generated through various technical solutions. A

The main idea of this chapter is to promote maintenance centered to safety, in accordance to adaptive fuzzy inference model, which has online adjustment to working conditions. Input data for this model are quality of service indicators of analyzed engineering system: reliability, maintainability, failure consequence, and severity and detectability. Indicators in final form are obtained with permanent monitoring of the engineering system and statistical processing. Level of safety is established by composition and ranking of indicators according to fuzzy inference engine. The problem of monitoring and processing of indicators comprising safety is solved by using the features that Industry4.0 provides. Maintenance centered to safety is important for complex, multi-hierarchy engineering systems. Sudden failures on such systems could have significant financial and environmental effect. Developed model will be tested in the final part of the chapter, in the case study of bucket wheel excavator.

Dealing with maintenance activities in complex systems often configures the so-called management of change (MoC). MoC is a process for evaluating and controlling modifications to facility design, operation, organization, or activities—prior to implementation—to be sure no new hazards are introduced. Traditionally, MoC is related to technical changes. Safety implications from organizational changes have recently led to proposed integrated management of both types. An inadequate MoC is recognized as a recurring cause of accidents, often resulting in major accidents, mainly in the process industry. Despite this recognised criticality, the MoC workflow in many companies is still far from being mature and there are still evident shortcomings in its application that have to be compensated. The technical solutions composed by the enabling technologies within Industry 4.0 can offer a valid support to overcome the MoC shortcomings, as will be discussed within this chapter.

Safety is an important concern for critical assets, such as aircrafts. E-maintenance strategies have long been explored for maintenance decision support and optimizing the operational availability of aircraft assets. Data-driven tools are an important influencer of day-to-day maintenance processes, which if optimally used may support practitioners to design more effective maintenance strategies. Recent trends show a correlation between e-maintenance strategies and enhanced use of data-driven tools for optimally managing technical assets. However, using data-driven tools for designing e-maintenance strategies is challenging because of aspects such as data-readiness and modelling-related challenges. This chapter presents a data-exploration approach for aiding root cause analysis of aircraft systems. The approach embeds algorithms for data preparation, text mining, and association rule mining and is validated in a use-case of maintenance of aircraft equipment, discussed in this chapter.

The demand for increased efficiency of production processes, while maintaining quality and safety in the operating environment, are permanent requirements of industrial revolutions. In the information age, data acquisition and its use to affect business strategies are being carried out by sensing production lines, tracking processes, and the product itself throughout its life cycle. Industry 4.0 requires an organizational transformation in terms of culture, process, and technology for the organization to be able to harness the potential of information. This chapter seeks to establish the difficulties and challenges of organizational transformation from the analysis of an aviation MRO company in light of integrated logistics support (ILS). The discussion will lead to the points to be taken into account from all elements of the ILS that will produce a roadmap for decision-makers to follow.

The chapter describes the role of additive manufacturing (AM) in Industry 4.0 (I4.0) for maintenance engineering. A brief introduction of the fourth industrial revolution and related technologies has been included. The different AM processes with significant contributions in the relevant industry sectors have been discussed along with suitable examples. Difference between the manufacturing capabilities of conventional and AM technologies has also been presented. Owing to its high degree of design freedom, AM helps to reduce the spare parts inventory cost, component assembly cost, and can replace the discontinued parts easily. A case study presenting these key distinctive features of AM, which make it an indispensable technology for I4.0, are also discussed. Furthermore, the barriers to the adoption of AM technology by manufacturers and possible remedial actions are also discussed in brief. The knowledge gaps in terms of materials and design tools for AM have been identified and a probable road ahead has been discussed.

In this chapter, the authors present an overview of methods that can be used to evaluate risks and opportunities for deferred maintenance interventions on aging equipment, and underline the importance to include monetised risk considerations and timeline considerations, to evaluate different scenarios connected with the possible options. Asset managers are compelled to continue operating aging assets while deferring maintenance and investment due to the constant pressure to reduce maintenance costs as well as short-term budget constraints in a changing market environment. Monetised risk values offer the opportunity to support risk-based decision-making using the data collected from the field. The chapter presents examples of two different methods and their practical applicability in two case studies in the energy sector for a company managing power stations. The use of the existing and the new proposed solutions are discussed on the basis of their applicability to the concrete examples.

In this chapter, the authors present a critical analysis about the current maintenance and inspection process carried out on hazardous lifting equipment. In Italy, a mandatory audit schema is working requesting a periodic interaction between owners of the lifting equipment and inspectors. The current condition has been analyzed aiming to evaluate potential points of criticalities. A smart platform integrating physical devices—based on internet of things technologies, mobile, and cloud applications—has been developed in order to provide companies and inspectors with a reliable and modular tool to organize, certify, and trace maintenance activities developed on the specific equipment. The final purpose is to guarantee a high level of safety for this type of hazardous equipment.

Preface

OVERVIEW

Why are Safety and Maintenance pulled together in this manuscript? Which is the reason behind this choice?

Those probably could be the first questions a reader could rise starting reading *Applications and Challenges of Safety and Maintenance in Industry 4.0.*

The motivation that fueled us, as editors, to discuss those two topics in one opera lies on the deep characteristic of them.

Safety and Maintenance are often considered not as primarily research sciences. Mainly, due to their intrinsic nature. They are "invisible". They act on the backstage. They become relevant, necessary and "hot" only when their absence steps in a play.

However, it is becoming clear how underestimating their importance usually lets us save little money at the beginning and oblige us to pay back (with high interest rate) at the end. Lack of Safety and Maintenance is responsible everywhere of industrial and occupational accidents, infrastructures - road, bridge and buildings - collapses or simply of production stops in factories and powerplants.

Therefore, researchers, engineers, politicians and business leaders are starting to realize how Safety and Maintenance can play a key role in the overall economy.

As in other domains, the introduction of Industry 4.0 solutions represents a game changer for Safety and Maintenance, bringing multiple applications but several challenges as well in terms of feasibility, cost-effectiveness and complexity.

Applications and Challenges of Safety and Maintenance in Industry 4.0 wants to reflect and offers case studies regarding how Industry 4.0, Safety and Maintenance have been combined so far, highlighting pros and cons.

Consequently, the reader will be able to find inspirations, results and tests on different topics such as IoT, augmented reality, additive manufacturing, fuzzy logic and sensing applications deployed in different industrial sectors from mining to aviation and manufacturing for Safety or Maintenance goals. Not only from a scientific angle, but also from industrial perspective.

As editors, our intention is to create a collection of experiences and knowledge for paving the way to a better and more effective deployment of Industry 4.0 solutions that can contribute to increase both Safety and Maintenance.

SUMMARY OF TOPICS

Safety, Maintenance and Industry 4.0 are quite broad areas of interest by definition, characterized not only by technical disciplines but also by social, organizational and managerial aspects. Therefore, the spectrum of topics that the reader will encounter is wide. In this manuscript interesting discussions about how Industry 4.0 is penetrating into Safety and Maintenance domains are offered.

The main topics presented in the chapters are related to both theoretical and practical problems in relation to Safety and Maintenance and they drive the reader through aspects of Industry 4.0 of great relevance nowadays:

- Safety 4.0
- Maintenance 4.0
- Cost Benefit for Aging Equipment in Industry 4.0
- Digital Technologies for Inspections
- Issues and Challenges of Cybersecurity
- Operator 4.0
- Additive Manufacturing in Maintenance
- Societal and Sustainability Challenges of Industry 4.0
- Augmented Technology for Safety and Maintenance
- Tacit Knowledge Sharing
- Management of Change (MOC)
- E-Maintenance
- Maintenance, Repair and Overhaul (MRO) 4.0
- Applying fuzzy inference model

The topics mentioned above are connected to different industrial domains highlighting even more the relevance that Safety and Maintenance play in all the working domains.

In *Handbook of Research on Applications and Challenges of Safety and Maintenance in Industry 4.0* the reader will be able to find examples applied in:

- Aviation Industry
- Manufacturing Industry
- Mining Industry
- Lifting Equipment Industry
- Railway Industry
- Consumer Product Industry
- IT Industry

TARGET AUDIENCE

Applications and Challenges of Safety and Maintenance in Industry 4.0 looks at a broad worldwide audience of researchers, engineers, professionals and technological experts. This book gives the opportunity to get in contact with some of the last solutions and novelties of Industry 4.0 applied to vital sectors such as Safety and Maintenance.

The readers can get inspired and learn important lessons on difficulties and challenges of introducing new technologies to improve the accident prevention or to move the Maintenance strategies towards a dynamic and flexible system based on innovative applications.

IMPORTANCE OF EACH CHAPTER

As already mentioned, *Applications and Challenges of Safety and Maintenance in Industry 4.0* is mix of different topics disciplines with three common "leitmotif, Safety, Maintenance and Industry 4.0. The manuscript is structured in three different sections without, however, rigid boundaries and with several common points. Therefore, the reader can recognize chapters with lines strongly connected to a different section.

As editors, we decided to divide our work as such:

The first section sets the base line for understanding state of the art, evolution, knowledge sharing, challenges and issues of Safety and Maintenance in Industry 4.0 dealing with their impacts on society and their sustainability on long-term planning.

The second section focuses on analyzing the impacts of Industry 4.0 on Safety. Discussions on how to empower operators with digital technologies such as Augmented Reality to support them in their tasks and on how the Management of Change can drive the introductions of innovative solutions without compromising the level of Safety.

The third section explores the concept of Maintenance 4.0 and E-Maintenance trying to point out the advantages of adopting Additive Manufacturing, Big Data and IoT in connection with monetized risk values.

Section 1

Chapter 1, "Maintenance 4.0: Where Are We? A Systematic Literature Review", provides a literature review with the objective to analyse the current state and future progresses of Maintenance 4.0 in a systematic manner. The authors claim that this chapter can be used for anyone approaching Maintenance 4.0 at an industrial level or to perform academic researches.

Chapter 2, "Industry 4.0 in Emerging Economies: Technological and Societal Challenges for Sustainability", discusses the status of Industry 4.0 in developing and advanced economies and presented some significant sustainability challenges of Industry 4.0 that are related to technological and societal problems. Technological Challenges, such as digital culture, scalability, integration & interoperability, standardization, information privacy & security real-time analysis; and Societal challenges such as energy intensity technologies, jobs and inequality, skills for new technologies, resistance to change and policy and regulatory environment are discussed.

Chapter 3, "Evolution of Maintenance Process in Industry 4.0", addresses several issues and challenges in the Industry 4.0 framework with respect to Maintenance processes that need attention to improve its value adding capability and productivity. A conceptual integration of the different elements of the

Maintenance process according to EN standard with five levels (5Cs) based on Industry 4.0 architecture have been proposed in this chapter to address some of the issues raised.

Chapter 4, "Tacit Knowledge Sharing for System Integration: A Case of Netherlands Railways in Industry 4.0", presents the case for tacit knowledge sharing in the era of Industry 4.0. The work presented recognizes the significance of lessons learned for the proper execution of system integration processes and propose to pay attention to not just the explicit but also the tacit dimension of lessons learned. It presents an example to demonstrate ways in which technological solutions of Industry 4.0 can facilitate in effective knowledge sharing and developing a knowledge sharing strategy which can bridge the gap between tacit expectation and developed explicit models.

Chapter 5, "Cybersecurity Issues and Challenges in Industry 4.0", focuses on various cybersecurity issues and challenges in Industry 4.0 along with three industrial case studies. This chapter suggests significance of three security elements i.e. Confidentiality, Integrity, and Availability for the components of information system along with risks, impacts and countermeasures.

Section 2

Chapter 6, "Safety 4.0 Analyzing the Impact of Digital Technologies", presents an expeditious index methodology aimed at analyzing new risks introduced by the adoption of Augmented Reality systems. The presented methodology verified by comparing three Augmented Reality alternatives applied to a Maintenance process. The chapter conclusions advocate the use of Augmented Reality systems applied to Maintenance can bring advantages both in terms of production and Safety but after carefully evaluation with respect to the pros and cons.

Chapter 7, "Operator 4.0 Within the Framework of Industry 4.0", presents the concept of Operator 4.0, integration of operators in smart factories, human-in-the-loop in Cyber-Physical Systems. This chapter also describes the knowledge and understanding for the development of Human Cyber-Physical Systems for new generation automated systems and design principles of Industry 4.0 for the perspective of operator 4.0.

Chapter 8, "Augmented Technology for Safety and Maintenance in Industry 4.0", highlights the processes used in industry 4.0 for the Safety and Maintenance and the related physical system with the cyber world. The chapter focuses on the various challenges existing in the highly sophisticated and automated processes involved in Industry 4.0.

Chapter 9, "Applying Fuzzy Inference Model in Maintenance Centered to Safety: Case Study Bucket Wheel Excavator", undergoes the analysis of Bucket Wheel Excavator. This engineering system is typical representative of complex engineering systems, operating in difficult working conditions. This chapter provides two case studies of catastrophic failures of this machine, with analysis of reasons for reduced Safety.

Chapter 10, "The Support From Industry 4.0 to the Management of Change (MOC)", shows how Management of Change is a powerful tool for Maintenance activities in complex systems in order to evaluate and control modifications to facility design, operation, organization, or activities-prior to implementation-to ensure no new hazards – for the operator, the quality, the environment, the assets – will be introduced.

Section 3

Chapter 11, "Towards E-Maintenance: An Exploration Approach for Aircraft Maintenance Data", presents a data mining approach for supporting Maintenance decisions of aircraft systems. The approach integrates varying data and text mining methods, and embeds a classification algorithm for predicting critical failure events of aircraft equipment based on their repair lead-times. Moreover, the proposed approach provides a robust method for improving Maintenance auditing processes and identifying violations of audit procedures.

Chapter 12, "MRO 4.0: Mapping Challenges Through the ILS Approach", provides an overview of Maintenance, Repair, and Overhaul (MRO) processes that could be enhanced by technologies from the Industry 4.0 paradigm like Machine Learning, IoT, and Big Data Analytics. The chapter aims to identify challenges, business opportunities, and technology gaps regarding MRO 4.0 processes, by analyzing and scrutinizing each MRO process against ILS Elements framework. Finally, it describes Integrated Logistics Support (ILS) framework, and the relationship between ILS elements and an overall MRO process.

Chapter 13, "Role of Additive Manufacturing in Industry 4.0 for Maintenance Engineering", briefly discusses the technologies related to Industry 4.0. This chapter presents a case study to discuss the advantages of Additive Manufacturing (AM) over the traditional manufacturing process. A framework (Product-Service-System) to overcome the high investment barrier has also been included in this chapter. Lastly, a road map has been presented for the future research work in the field of material and design tools for AM in the framework of Industry 4.0.

Chapter 14, "Monetized Risk Values and Cost Benefit Evaluation of Maintenance Options for Aging Equipment", presents the fact that most of industrial equipment used beyond the useful life foreseen at the design stage and therefore need a more frequent Maintenance for optimum productivity and Safety. This chapter describes three case studies where this issue is addressed through qualitative and quantitative methodologies able to support the operational optimization in the industrial domain.

Chapter 15, "Supporting Maintenance and Mandatory Inspections Through Digital Technologies on Lifting Equipment", proposes a smart tool to support companies in tracing all Maintenance activities developed on hazardous lifting equipment, which requires mandatory periodic inspection under an audit schema. The proposed tool aims to overcome two current criticalities outlined for the mandatory inspection process in hazardous equipment: uncertainty in equipment identification and lack of information during mandatory inspection activities.

CONCLUSION

The chapters mentioned in the previous sections have been written by 42 authors from 11 countries (the Netherlands, Italy, Sweden, Serbia, Brazil, Belgium, Iran, India, Taiwan, Portugal and Ireland) with a broad spectrum of experiences and working both in academia and industries.

Similarly, the editorial board of the book *Applications and Challenges of Safety and Maintenance in Industry 4.0* is composed by experts coming from different knowledge fields with different experiences with a strong vision on where Safety and Maintenance should be driven with the help of Industry 4.0 solutions. This multidisciplinary approach was able to create a vibrant and working sharing platform where theoretical approaches and industrial cases have been discussed in order to highlight possible applications and challenges.

The present book has also stimulated (and hopefully will continue to stimulate) the cooperation and connection between industries and academia and between experts of different domains in order on one hand to evaluate the quality of research on Safety and Maintenance and on the other hand to accelerate the deployment of innovative solutions in industries.

Overall, through the three sections the following main points have been brought to the spotlight:

- The possibility of having a fully-defined Safety 4.0 domain where Industry 4.0 solutions can contribute to risk reductions.
- The possibility of having a fully-defined Maintenance 4.0 domain where Industry 4.0 solutions can contribute to improve reliability and asset management.
- Interaction of different fields and industrial domains can stimulate synergies and new decision-making strategy.
- Knowledge sharing and Management of Change need to be carefully addressed when new technologies are deployed in environments not ready yet for embed them.

To conclude, this manuscript in your hand is only the starting point of a discussion regarding the pervasiveness of Industry 4.0 tools in traditional domains such as Safety and Maintenance. The hope of the editors is to just have opened an eye on future solutions that can play an important role in the sustainability of our society.

The editors want to express their gratitude to all those have spent time and efforts in writing, editing, reviewing or simply in providing valuable suggestions for producing a useful reference for these important topics.

Alberto Martinetti
University of Twente, The Netherlands

Micaela Demichela
Politecnico di Torino, Italy

Sarbjeet Singh
Luleå University of Technology, Sweden

Section 1

Chapter 1
Maintenance 4.0:
Where Are We? A Systematic Literature Review

Alberto Martinetti
https://orcid.org/0000-0002-9633-1431
University of Twente, The Netherlands

Micaela Demichela
https://orcid.org/0000-0001-5247-7634
Polytechnic of Turin, Italy

Sarbjeet Singh
https://orcid.org/0000-0001-7229-4050
Luleå University of Technology, Sweden

Gonçalo Matias Soares
University of Aveiro, Portugal

João Castro Silva
University of Aveiro, Portugal

ABSTRACT

Aiming to remain competitive, companies from diverse industries are paying more attention to Industry 4.0 concept and its benefits. Maintenance is seen as a specific area of action to successfully sustain a competitive leverage, and its fusion with Industry 4.0 is perceived to revolutionize the whole maintenance concept. Maintenance 4.0 emerges as a subset of Industry 4.0 in the form of self-learning and smart system that predicts failures, makes diagnoses, and establishes maintenance actions. This chapter presents a systematic literature review (SLR) on Maintenance 4.0, with the aim of outlining the current achievements as well as limitations of maintenance meeting Industry 4.0 demands. The analysis included 90 papers selected as being the most suitable to reach the proposed goal. A state of the art on Maintenance 4.0 is performed, followed by an analysis ambitioning the delineation of what future holds on this topic.

DOI: 10.4018/978-1-7998-3904-0.ch001

INTRODUCTION

Maintenance plays an important role in every company or industry within the engineering field. By maintenance it is meant the necessary actions to retain or restore any functionality of a product to the specified operable condition, in order to achieve its maximum useful life. These actions can be technical, administrative and managerial (Komonen, 2002). In order to fulfil customer's needs, maintenance had several developments, as it also is a required service that influences a company effectiveness. Therefore, over the years, technological developments allowed the change of maintenance service practices and its whole paradigm (Wan, Gao & Li, 2019).

In the recent decades, maintenance recognition as an effective part of a company competitiveness has grown. Several maintenance techniques were widely implemented to enhance production performance and to accomplish customers' requirements (Waeyenbergh & Pintelon, 2002). Recent technology advances revolutionized the concept of maintenance in its whole sense, for instance, shortening time to market and customized mass production. Industry 4.0 (I4.0) is the term given to the revolution that allowed a huge shift in maintenance and industry it-self. Therefore, the comprehension of this subject is mandatory to fully understand the aim of this paper.

According to (Peres et al., 2018) Industry 4.0 aims to introduce and take advantage of the interconnected world along the entire value chain, allowing the sharing and processing of the data avail-able in all of its actors to generate relevant knowledge and optimize the overall process (Peres et al., 2018). The term was carried by the German Government in 2012 and was also considered as the fourth industrial revolution. Governments and industries world-wide have noticed this trend and acted to benefit from what this new industrial revolution wave could provide (Yongxin Liao et al., 2017).

The four design principles of I4.0 that allow its implementation on maintenance are described by (Ashraf, 2018) as:

1. Cyber-Physical Systems (CPS) concern the integration of physical processes and computational systems. CPS focuses on the digital part of a manufacturing system, more specifically, in real-time data acquisition, interaction and communication between physical and digital world. At the same time, these systems trans-late intelligent computations and cognitive decision from the digital to the physical world;
2. Internet of Things (IoT) refers to the wireless communication between sensors and computing devices through an inter-net network. IoT is the technical infrastructure of CPS, i.e., through CPS, IoT allows human and machine to be connected in a whole manufacturing system and enables the automation of maintenance decision-making;
3. Cloud computing provides, through the inter-net, high-speed data access and complex computing power for large-scale engineering problems. The cloud includes hardware storage, operating systems, program execution environment, testing, application development, databases, etc;
4. Big data covers the huge amount of data that is generated by enterprise resources like sensors or production and control systems and that is further stored on the cloud servers. Data from all these sources is complex, decentralized and fast moving, which makes human analytical capabilities insufficient to deal with the big volume of data. The key enablers of I4.0 are reasonably described through literature (pre-sent papers).

I4.0 aims at improving production efficiency and decreasing the related cost(Mohamed et al., 2019). Therefore, with I4.0 emergence, maintenance is a particular area of action that is required to successfully sustain a competitive advantage (Rødseth et al., 2017). As (Mohamed et al., 2019) mentioned, Maintenance 4.0 is a subgroup of I4.0 as it possesses self-learning and smart systems that predict failures, make diagnosis and trigger maintenance actions. A smart maintenance environment is enabled, mainly due to the deployment of CPS, which allows a high degree of networking, digitization, de-centralization, efficiency and availability. I4.0 is then expected to bring along several advantages to maintenance.

MAIN FOCUS OF THE CHAPTER

Several studies presented the inclusion of I4.0 principles in maintenance, such as Ravna, (2016), Al-Najjar et al., (2018), Cao & Zhang, (2016), Donovan et al., (2015) and Yan et al., (2017). However, a clear document that addresses the state of art of Maintenance 4.0 (M4.0), concerning what has been done so far in terms of principles used, applications and limitations is still missing.

In order to provide an appropriate view on this topic, the objective of our work is to review and analyse the progress of M4.0 through a Systematic Literature Review (SLR), defined as a rigorous literature review which ensures the re-producibility and scalability of the study as well as the objectivity of the results (Kitchenham et al., 2009).

Possible future developments and implementations within this subject will also be pointed out. Consequently, all the relevant evidence to clarify these matters will be considered, and reliable judgements about its validity and implications will be conducted. This paper tries to provide indications for further developments on Maintenance 4.0, both academical and practical, offering a current overall insight of the topic.

The paper is organised as follows. In the Methodology section the methods used to conduct the literature reviews are explained. Section 3 reports on the main results of the SLR, providing an overview of the state of the art of M4.0 and analysing its future advancements and applications. Finally, Section 4 concludes this paper and points out future works.

METHODOLOGY

In order to assess the state of the art of M4.0 and its future developments and implementations, a SLR approach has been used. According to (Boothet al., 2016), by and large the best evidence for many decisions comes from a systematic review of all the evidence. With this standpoint, this SLR aims to search, appraise, synthetize and analyse all the studies relevant for the M4.0 research field.

The methodology utilised in this review is described by Booth et al., (2016) and its main aim is to synthetize the available literature on the topic to provide evidence of future fields of research. The different phases to conduct this SLR are planning, defining the scope, searching, assessing, synthetizing, analysing and writing and within each step, different procedures are followed. To better understand the methodology adopted, Figure 1 outlines the steps (blue) and its outcomes (grey); a description of these will be provided below.

Planning

Planning is the very first step to conduct a SLR, as shown in Figure 1. The timing indicates the available time to carry out this SLR, which in the case presented consisted of a five-month timescale. This helped to define the level of complexity and exhaustiveness the work could include. Relating the identification of databases, this systematic search used five electronic databases, based on (Kitchenham et al., 2009).

- IEEE Xplore (www.ieeexplore.com)
- Google Scholar (www.google-scholar.com)
- Science Direct (www.sciencedirect.com)
- Scopus (www.scopus.com)
- Web of Science (www.webofscience.com)

Defining Scope

Defining the scope aims at the proper formulation of literature study. Part of this process was made in the introduction section, which consequently lead to a literature search concerning the related works on M4.0 (Kans et al., 2016; Qian et al., 2019). This created the space to formulate suitable questions to focus our research on. Consequently, two research questions were defined:

Q1: What is the state of art of Maintenance 4.0?
Q2: Which are the potential future developments and implementations of Maintenance 4.0?

Searching

Searching consists in obtaining a comprehensive set of papers through the electronic databases that might contribute to answer the predefined questions. To this end, searching string and fields were set within each database, utilising a pre-selected set of keywords and Boolean operators to provide a completer and more detailed first screening: "Maintenance 4.0" AND "Industry 4.0" AND "Maintenance.

The findings of this searching process updated at the May 2019 results in the collection of 908 documents, as shown in Table 1.

The first column reports the databases utilised. The second column reports the "search fields" where the search string has been applied. The third column reports the number of documents returned by the database.

Since this step was carried out separately for each of the databases, the total number includes duplicated papers, that will be ahead excluded, as detailed in Figure 2. It is worth to notice that this phase does not involve any reading, once only electronic database tools were used.

ASSESSING

The narrowing of the hundreds of documents previously found is the aim of this assessing step. The reduction of the total number of papers is intended to only retain the essential ones that help to answer the research questions.

Figure 1. Methodology employed to conduct SLR

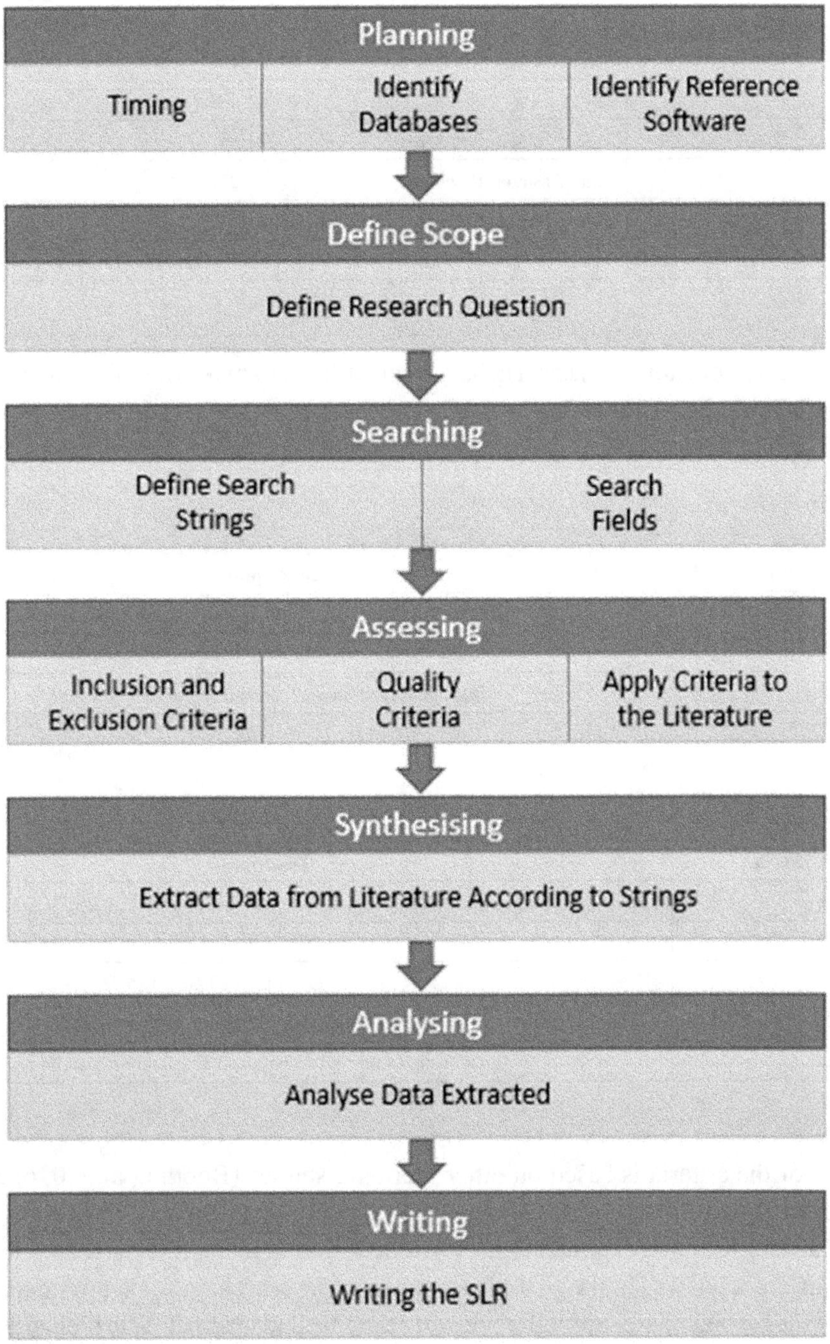

Table 1. Outcome of the searching phase

Electronic Database	Search fields	Documents returned
IEEE Xplore	Metadata	126
Google Scholar	Title	433
Science Direct	Title; Abstract; Keywords	69
Scopus	Title; Abstract; Keywords	237
Web of Science	Title; Abstract; Keywords	43
		Total products = 908

Inclusion (IC) and Exclusion (EC) criteria have been utilized to make the first screening of the documents as shown in Table 2 and Table 3:

Table 2. Inclusion criteria

Inclusion Criteria	Description
IC1	Study that concerns the use of Industry 4.0 in maintenance
IC2	Study that concerns the Maintenance 4.0 state of art
IC3	Document type: journal, conference or review article

Table 3. Exclusion criteria

Exclusion Criteria	Description
EC1	Not open access
EC2	Not in English
EC3	Older than 2000
EC4	Not engineering or computer science field
EC5	Maintenance 4.0 is only mentioned as a future research topic

The selection of the criteria is based on other literature studies (Booth et al., 2016; Kitchenham et al., 2009; Corrêa et al., 2013).

These criteria were separately applied to the documents found in the five databases listed in Section 2.1 and in two different phases. In the first phase, through the searching tools provided by each of the databases and in the second phase, through reviewing the title and abstract. When elucidative outcomes about the suitability of the document could not be taken, the re-viewing of the introduction and conclusions was also done is this phase.

The results of the IC and EC are the collection of 90 documents, as outlined in Figure 2.

The next step was defining quality criteria (QC) in order to strengthen the extraction of quantitative and quality data for the synthesis and results analysis, as it is shown in Table 2. Quality criteria were based on Palmarini et al., (2018) and on the authors point of view:

Figure 2. Paper selection process

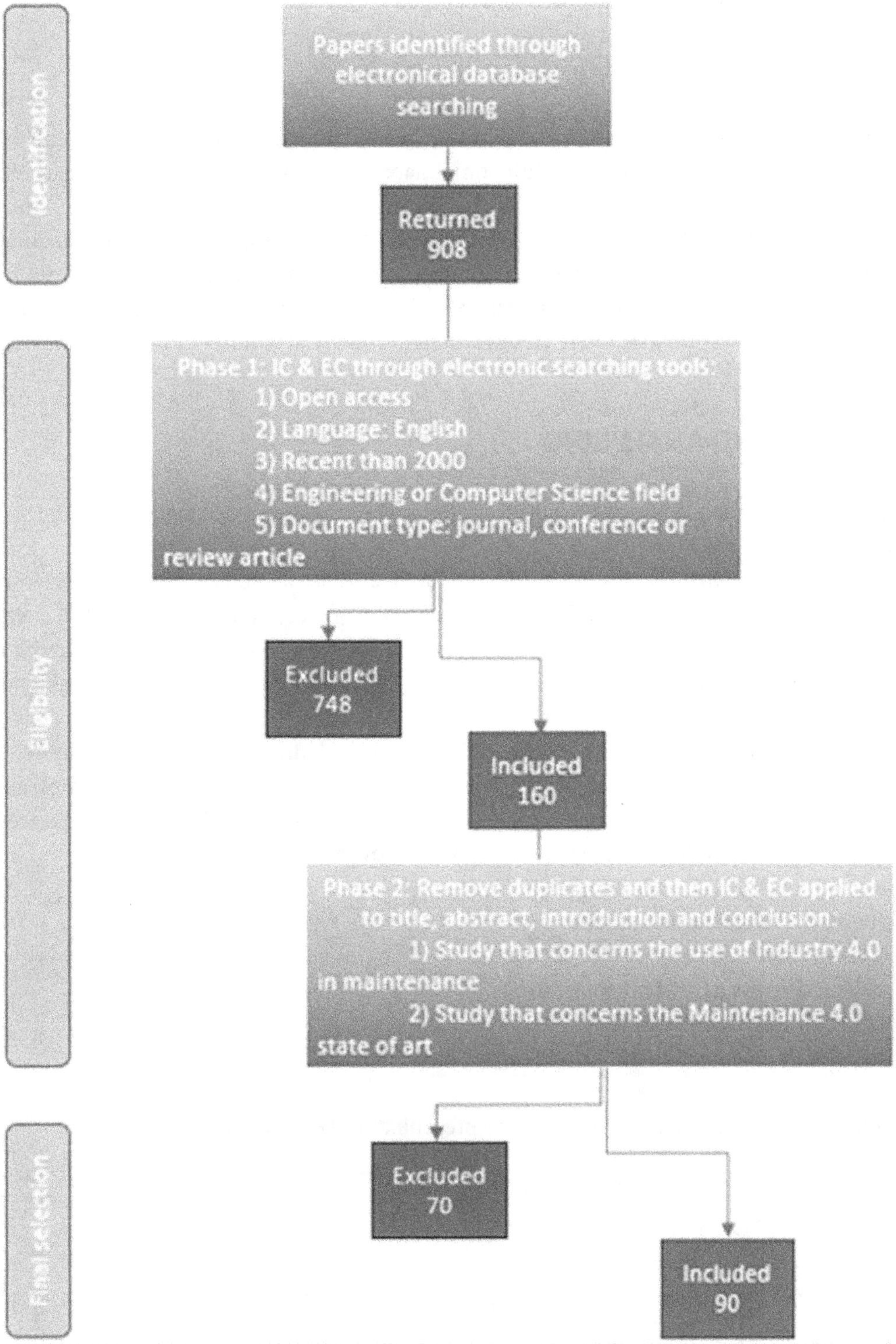

- QC1 The document is clear
- QC2 The methodology is well exposed and detailed
- QC3 The document is updated
- QC4 Analytical results were provided

For each one of the 90 papers selected, a score from 0 to 4 has been calculated summing up the scores assigned for each of the QC. 1 point means full compliance of the QC; 0.5 points for the partial compliance; and 0 points for the non-compliance. Annex 1 shows the results of the QC. For QC3, the authors considered the papers from 2016 till the current date as updated and "fresh" source of information, due to the intense researches that have been made on both I4.0 and M4.0 However, these criteria did not result in the exclusion of any paper, since it is a subjective analysis and every paper contained useful information for the SLR.. Annex 1 provides the reader raw data to check the validity of this quality assessment.

SYNTHESIZING AND ANALYSING

In order to answer the proposed research questions, the authors analysed and synthetized the 90 articles through the systematic search.

It is relevant to clarify that the results of this SLR are based on the 90 articles selected. However, as guidance for the description of the results and its presentation, papers in the form of SLR were analysed (Liao et al., 2017; Palmarini et al., 2018). Furthermore, the authors found it valuable to check the bibliographies of the papers found through the electronic database search to increase the extent and value of the SLR.

This step consisted of finding patterns and correlations through the different studies. The detection of common features and trends made possible the selection of main characteristics of I4.0 implementation in maintenance, that will onward be analysed. The main characteristics of M4.0, defined according to the authors' perspective and to the domains of the papers reviewed, are:

- Maintenance techniques evolution;
- M4.0 framework;
- Augmented Reality as a M4.0 application;
- Smart Factories in Industry;
- Costs of M4.0 implementation.

The explanation of each characteristic will be presented in the following section.

RESULTS

The aim of this paper, and consequently of the SLR, was to answer the research questions:

- **Q1:** What is the state of art of Maintenance 4.0?
- **Q2:** Which are the potential future developments and implementations of Maintenance 4.0?

These questions are answered separately in the following subsections.

Answer to Q1

In order to describe the state of the art of M4.0, an analysis covering 90 papers was conducted as said. The analysis is divided in categories: Maintenance techniques evolution in M4.0 context; M4.0 framework; Augmented Reality as a M4.0 application; Smart Factories in Industry; Costs as M4.0 consequence. A description of each of the classes is provided below.

Maintenance Techniques Evolution

In line with the technological evolution and the introduction of I4.0, maintenance techniques are evolving as well fairly rapidly. This subsection presents a brief description of these concepts and describes why they are gaining or losing importance. Also, a quantification of the papers is done, in order to show which maintenance techniques are mentioned in the analysed papers.

Figure 3. Maintenance techniques presented in the papers reviewed

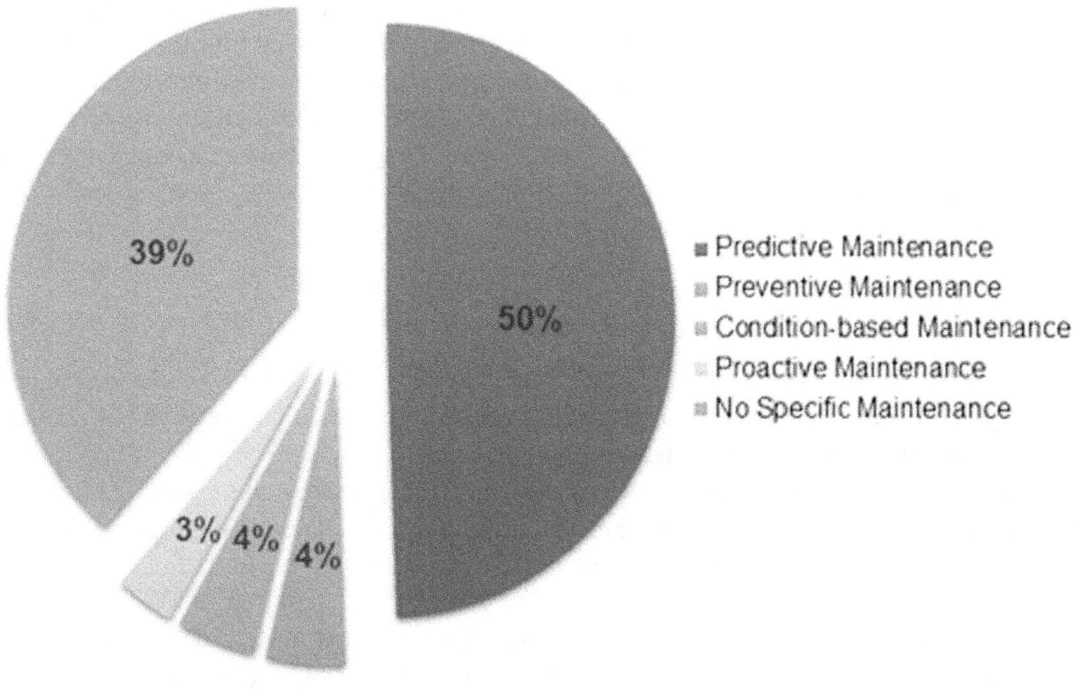

Promoting the explanation of the most used and implemented techniques regarding maintenance actions in the I4.0 era, the authors considered that an analysis concerning the maintenance procedures before this time would be helpful for the readers.

Failure-based maintenance defines the maintenance actions according to the assets break down, which implies the need for additional spares parts, redundancy equipment and labour, which makes this technique very costly (Wan et al., 2019). This maintenance technique was not given relevance in the papers analysed. The logical evolution led to the adoption of maintenance techniques, focusing on the scheduling of maintenance actions, based on the experts' opinions, in order to reduce the probability of failures (preventive maintenance). These actions were taken regardless the system condition, which caused unnecessary maintenance actions and loss of production time (Al-Najjar et al., 2018). Only 5% of the analysed papers mentioned preventive maintenance in I4.0 era, (A Y Alqahtani et al., 2019) a technique to find the optimal preventive maintenance and warranty policy, and the reasons for this maintenance technique to be losing relevance are discussed in the other papers (Huang et al., 2018) (Roy, R. et al., 2018) (Koulali et al., 2018). Condition-based maintenance (CBM) takes the data regarding symptoms/indicators of failure based on Condition Monitoring parameters (temperature, pressure, vibration, etc) and acts just before failure occurrence (Al-Najjar et al., 2018). CBM can be used as proactive when high quality information is available, and predictive otherwise. Proactive maintenance is a modern maintenance approach, which is being more and more discussed, and it is present in 3% of the papers analysed. This technique is based on a continuous activity with the purpose of rooting the failures' causes. The maintenance technique which gets more relevance, along the analysed papers is Predictive Maintenance (PdM), since it is discussed in 50% of the 90 papers analysed. PdM concerns the application of condition-based indicators and alerts to define when maintenance actions must be performed, optimizing this way the maintenance cadences and maximizing the asset availability (Klathae, 2019).

With the introduction of I4.0, the entire value chain can now be integrated and share digitalized information to cooperate and execute tasks, which will generate a big data mass from different the various elements over the network (Al-Najjar et al., 2018). A company with this data mass from different systems could achieve a better detection of problems and their root-causes, as well as diagnosis and prediction of damage development, assessment of effects, and reliable planning of maintenance activities, avoiding unplanned downtime (Lee et al., 2014). Given this, it can be said that data coverage, quality and its utilization are important factors for maintenance in I4.0.

M4.0 Framework

An implementation model and reviewing of established concepts is a key factor for companies intending to adapt to M4.0 vision. This subsection presents a framework based on the collection of the various design concepts, either implemented or not.

After analysing all the selected papers, it was noticed that various frameworks regarding Maintenance 4.0 have been established. Most of these frameworks are based or related to predictive maintenance techniques, once they apply condition monitoring of assets/systems to predict its failure and define a maintenance strategy.

Different design concepts were conceived in the last few years, and most of them were based on a specific industry or were even conceived for a specific company. In order to understand what was already conceived and sometimes implemented, the authors found it interesting to propose a framework (Figure 5), combining the ones in study.

Figure 4. Complexity of the frameworks presented in the analysed paper

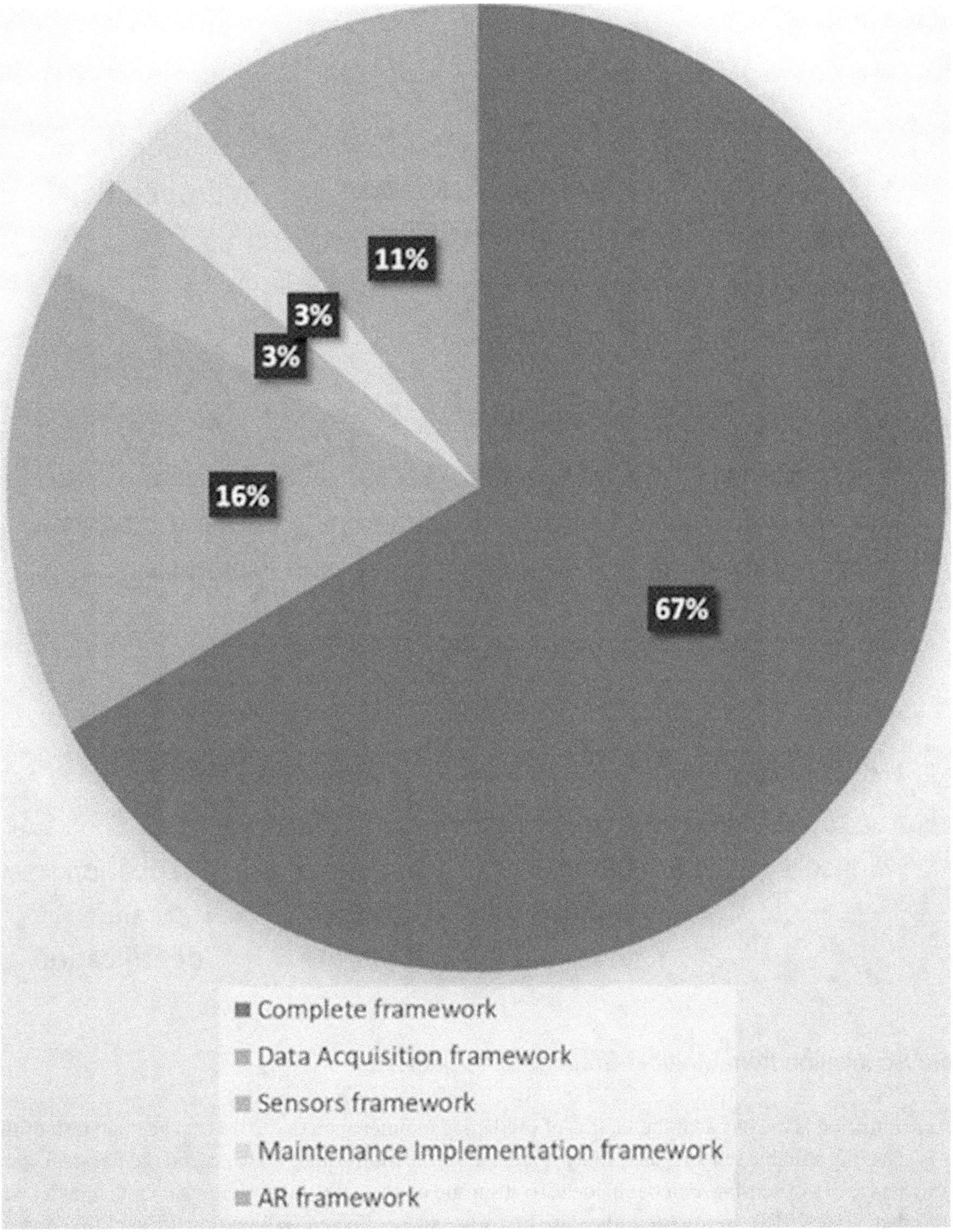

Figure 5. Proposed framework

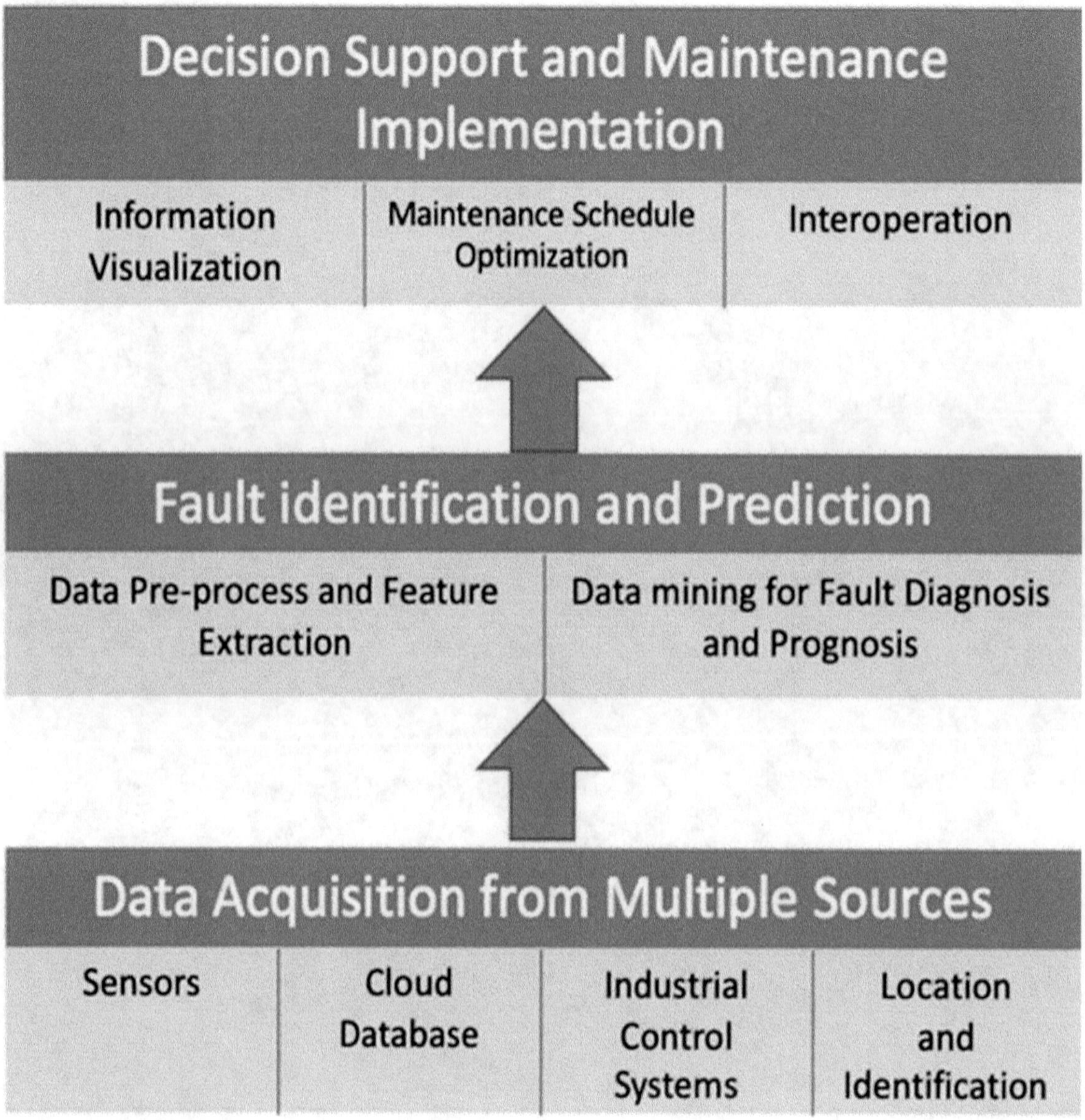

Data Acquisition from Multiple Sources

Data acquisition is the physical foundation of predictive maintenance (Li, 2018). "The main task of this tier is selecting suitable sensors, data sources and data collection strategies to extend the physical world through a variety of sensing, detection, identification and connect the objects or make them interact with each other." (Li, 2018). In this step, the data that was once collected will be transformed to a domain which contains maximum information about the condition of the equipment (Ashraf, 2018). The ability of virtualization, which means the monitoring of the physical processes and the creation of a virtual copy of the physical world by linking the collected data to virtual models (also called Digital Twin), is considered fundamental (Li, 2018). The four following common types of data sources are helpful to achieve PdM in I4.0.

Sensors

With the fast advancements in sensor technology, streaming real-time data is now very uncomplicated (Li, 2018). Various types of sensors can now be designed to get different kinds of data. The selection of sensors determines the representation of the machine health by the collected data, considering specifications and cost-effectiveness (Li et al., 2017). Obtaining the right smart sensor is a key factor for obtaining a complete data acquisition and improve the cost-effectiveness. According to (Li, 2018), the following sensors and transducers are widely applied in data collection:

- Mechanical sensor systems to monitor parameters such as acceleration, displacement, velocity, torque, location, strain, and cutting forces (static and dynamic)
- Optical transducers, such as photo detectors and lasers
- Thermal transducers, such as thermocouples, and thermography
- Audible sensors, such as ultrasonic sensors, and acoustic emission sensors
- Environment sensors systems, such as spectrometer, PH indicators and temperature sensors.

Cloud Database

The possibility of storing data collected by the sensors in a cloud, revealed itself to be a great perk. "An enabling factor in becoming an agile manufacturer has been the development of manufacturing support technology that allows the marketers, the designers and the production personnel to share a common database of parts and products and to share data on production capacities and problems, particularly where small initial problems may have larger downstream effects." (Li, 2018).

Industrial Control Systems

In the I4.0 era, Supervisory Control and Data Source (SCADA) systems are an important source that collect data. These systems are often incorporated into the production process, giving the state of the production in real time (Simon et al., 2018). They can also be used to detect faults and condition degradation, through the develop of an algorithm for monitoring (Welte et al., 2018).

Location and Identification

Data regarding location and identification may be brought by various types of devices/systems. Inventory Management model gives information regarding quantities and location (Ashraf, 2018), Computerized Maintenance Management Systems (CMMS) may report the number of assets at a specific location and the master list report of these assets (Ashraf, 2018), and Radio Frequency Identification (RFID) provides accurate data on product attributes, such as location and characteristics (Ammar Y. Alqahtani et al., 2019), which can be from distance (Kans et al., 2016).

"RFID can enhance the forecasting accuracy of the demand of the spare parts and significantly improve the efficiency of the storage, distribution and marketing, and significantly decrease costs after analysing the use of these data. Based on the data generated by the sensors of equipment and products, they can know the risk and the location of the failures, so that they can also predict where and when they need spare parts, which will greatly reduce inventory and optimize the supply chain."(Cao & Zhang, 2016).

The data acquired from the different categories of data sources, allows to identify degradations or failures in components/systems. This identification enables the prediction of the behaviour of the assets. In the following section, the fault identification and prediction part of the PdM framework will be described.

Fault Identification and Prediction

Through the various papers collected, a thorough description of the fault identification and prediction was conceived. Its function is to discover relevant knowledge or information regarding faults or degradation data. This step is the core processor for data mining in the framework, which formulation will be done after all the data is stored in a data warehouse (Li, 2018). However, during the discovery process, much irrelevant data and redundant information, such as noise, unreliable data (e.g., incorrect sensors monitoring), may be collected, causing a more demanding training phase (Li et al., 2017). Therefore, a data manipulation model should be applied, which includes two steps, (1) pre-processing and conditioning (2) feature extraction. In the first step signal characteristics and quality are improved (Ashraf, 2018). Furthermore, diagnosis model can be established for detection, isolation and identification of fault, and also a prognosis model will focus on the prediction of occurring a fault in the future (Ashraf, 2018). Figure 6 demonstrates the data analysis process flow, both for fault diagnosis and prognosis.

Data Pre-Process and Feature Extraction

Generally, the major steps involved in data pre-processing are data cleaning, data integration, data transformation and data reduction. Data cleaning is the process of detecting and correcting corrupt or inaccurate records deriving out of the database by filling in missing values, smoothing noisy data, identifying or removing outliers, and resolving inconsistences (Li et al., 2017). "Data integration is the process of merging data from multiple data stores." (Li et al., 2017). Wary integration can help to reduce and avoid redundancies and inconsistencies in the resulting data set (Ashraf, 2018). Data transformation regards the data transformed or consolidated into appropriate forms for knowledge discovery, so that data mining process may be more efficient, and the patterns found may be more straightforward to understand. Data reduction attain a reduced representation of the data set that is much smaller in volume and can produce the same (or almost the same) analytical results (Li, 2018). There are various dimensionality reduction methods (Li et al., 2017). Among them, a forthright approach is to apply feature extraction methods to the data set, which extracts features from pre-processed signals that are characteristic of an incipient failure or fault (Li et al., 2017). Commonly, the features can be extracted from three domains: time domain, frequency domain and time-frequency domain (Li, 2018). In data transformation, the data are transformed or consolidated into appropriate forms for Data Mining (DM), in a way that DM process may be more efficient, and the patterns obtained may be easier to understand (Li et al., 2017). The development of the storage media and computation ability produces massive data during the data acquisition process (Li, 2018). Data pre-processing can effectively clean raw data, reduce data dimension, and store it back to the warehouse for knowledge discovery (Li et al., 2017).

Figure 6. Data analysis for process flow

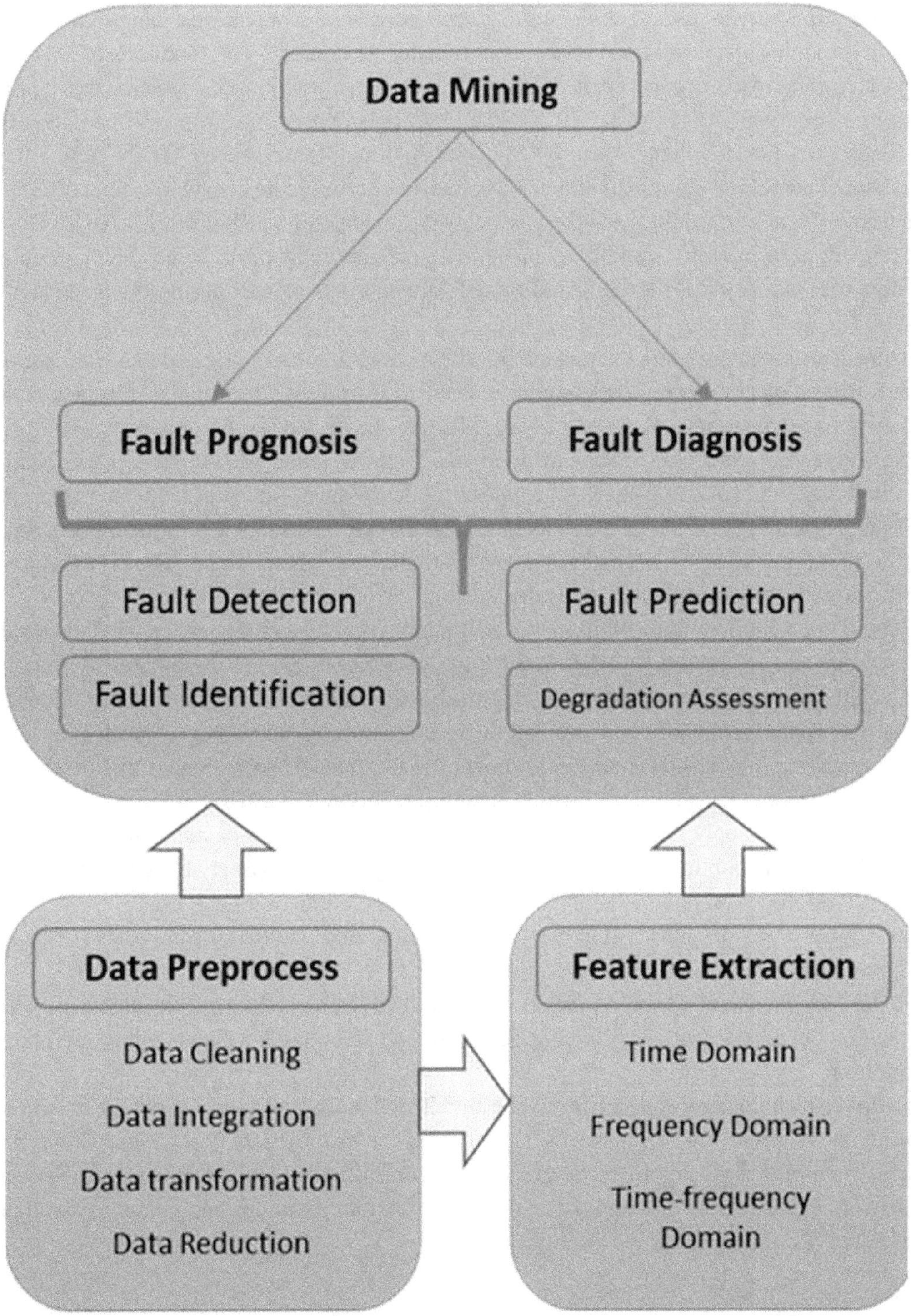

Data Mining for Fault Diagnosis and Prognosis

"DM has the capability to discover hidden links, recognize unknown patterns, and predict future trends by digging through and analysing enormous sets of data." (Li, 2018). The functions of DM can be categorized according to the task performed, such as clustering, classification, decision trees, predication, regression, association, etc (Li et al., 2017). Data mining's analysis methods can be categorized into three groups: statistics, Machine Learning (ML) and Artificial Intelligence (AI) (Li et al., 2017).

A statistical model is a set of mathematical functions describing the behaviour of the objects in target class in terms of random variables and their associated probability distributions (Li, 2018). Statistics has an inherent connection with data mining during data collection, analysis, interpretation and presentation, which means it is widely leveraged to model data and data classes during the process of DM (Li et al., 2017).

Machine learning investigates the method by which computers can study and make predictions based on data. It is employed in a range of computing tasks to learn how to recognize complex patterns and make decisions automatically (Li et al., 2017). ML may be divided in two main types: (1) predictive or supervised learning process, which aims to form a map from inputs to outputs, give a labelled set of input-output data; and (2) descriptive or unsupervised learning, which goal is to determine interesting patterns and knowledge from large amounts of data, with no classes labelled in the input samples (Li et al., 2017). This type is not as well defined as the first one, since there are no patterns to search for, and there is no error metric to evaluate the results.

AI is built upon heuristic algorithm (Li, 2018). Includes several techniques, such as genetic algorithm (GA), artificial neural network (ANN), fuzzy logic systems (FLS), and case-based reasoning (CBR) (Li et al., 2017). The intention is to apply human-thought-like processing to solve problems. "It uses techniques for writing computer code to represent and manipulate knowledge, which is exactly apt for computer process in modern business environment." (Li, 2018). AI solves one of the most challenging problems in the research area of PdM, which is fault prediction (Li, 2018).

It is appropriate to apply DM methods to fault diagnosis and prognosis in machine centres, due to its high complexity and coupling features among a wide range of faults (Li et al., 2017). The DM module in the framework focus on fault detection, classification and prediction for PdM.

Fault diagnosis and prognosis strategies can be divided into two major categories (Li et al., 2017), model based and data-driven.

A model-based technique leans on the accuracy of the dynamic system model and utilizes the actual system and model to generate the differentiation between two outputs, which is indicative of a potential fault condition.

Data-driven techniques normally address the anticipated fault condition, and only that, which must be taken from known prototype fault patterns (Li, 2018). If the historic data is obtained easily, data-driven techniques are useful, if not, hybrid techniques, which combine model based and data-driven techniques, can be used to evaluate the condition (Li et al., 2017). Various other common diagnosis and prognosis algorithms are also listed in the shown framework.

Decision Support and Maintenance Implementation

The function of this tier is to manipulate the analysis results in the data mining tier, transform them into meaningful information for maintenance strategy, share and publish this information on a common cloud-based network, and eventually provide optimal schedule for maintenance implementation (Li, 2018). Its main functions are visualization, maintenance scheduling and the interoperation.

Information Visualization

With visualization techniques, users will be able to timely grasp the current operating status of the system without being onsite (Cao & Zhang, 2016).

The aim is to help users understanding and analysing data through progressive, iterative visual examination (Li, 2018). In PdM, the application of key performance indicators (KPI) is one of most efficient methods to illustrate result of data mining (Li, 2018). Afterwards, KPI tracking can be achieved through publishing on a digital dashboard (also called a spider chart), and offers insights into the current condition, performance or degradation (Li, 2018). An optimized schedule can be provided according to the evaluation's result (Li et al., 2017).

Maintenance Scheduling Optimization

In this step decisions regarding maintenance activities schedule are optimized. The goal is making these decisions sustaining on the available information in order to optimize certain objectives such as maintenance costs, and development of potential failures (Li, 2018). In PdM, what is sought is the maximization of the equipment's up-time with no failures, minimizing the repair time and decreasing total maintenance costs according to the prediction from data mining (Li, 2018).

Attention should be paid to the relationship between production and maintenance, which has been considered a conflict in management decision (Li, 2018), which makes it a nondeterministic polynomial time (NP) problem (Li et al., 2017). To solve complex scheduling problems for PdM, heuristic algorithms such as, genetic algorithm, particle swarm intelligence (SI) optimization, genetic simulated annealing algorithm, artificial bee colony algorithm and ant colony algorithm, are very good options (Li, 2018). Although these techniques are not global solutions, they are not normally restricted to size or structure of the problem, which means they all are a reliable solution for maintenance scheduling optimization.

Interoperation

The term interoperation entails that a system is performing an operation for another system. In PdM, interoperability stands for the ability of interaction in data, services and processed between enterprise results (Li et al., 2017).

After acquiring information and knowledge regarding the failures, based on the result of data analysis in the cyber world, it should be applied to interact with the physical world, to implement maintenance and evaluate the influence of failures or degradation and solutions to the system (Li et al., 2017). Accordingly, it is important to have in mind coordination among those systems.

Smart Factories in Industry

Smart Factories represent the implementation of the previous PdM framework in a factory or company, being considered a M4.0 scenario. As described, this framework allows a global interconnection of a factory, as the different components communicate with each other to enable intelligent decision making i.e., when and how to intervene in the most cost-effective way. This concept still requires further developments due to its complexity, since it depends on organizational scale, deployed technological assets, and type of production system (Ansari et al., 2018). However, smart factories are already implemented in specific areas of industry. This section intends to expose the different industries that have explored M4.0 as smart factories.

Within these industry fields, Marine industry is the one getting more attention (6 papers were analysed). These papers regard not only potential advantages of M4.0 integration, but also design concepts based on case studies.

The shipbuilding industry is also addressed in (O. Blanco-Novoa et al., 2018) (Fernández-Caramés, T.M. et al., 2018), as the necessary technologies to design an industrial system for developing applications for an Industry 4.0 shipyard are described. (Hadjina & Matulja, 2018) reviews academic and industrial progresses of I4.0 in the shipbuilding sector.

Some papers refer to specific sector applications in M4.0 context. Machine centres take large attention nowadays, once there are 14 papers regarding CNC, milling machines, drilling machines, etc.

Augmented Reality and Digital Twin

Augmentation Reality (AR) has been gaining a notable relevance in the last years, quite as much as I4.0. The relation between these concepts is possible through Digital Twins (DT), once DT helps in the creation of an illustrative virtual imagery of the equipment for further maintenance applications, which is explained in this subsection.

As presented in Figure 4, AR is widely discussed in the analysed papers (11% of the analysed frameworks), and a few frameworks regarding M4.0 applied to AR are presented

As well as the term Industry 4.0, AR is a buzzword in today's industry. They can be related, as some of the papers collected prove. One of the most common ways to connect them is through DT.

DT is one of the main concepts related to Industry 4.0, and it refers to a digital replica of physical assets, processes and systems that can be used in real-time for control and decision purposes (Vatn, 2018). The digital twin depiction is seen as a requirement for effective synchronization of operation and maintenance within industries.

AR technology enables the users to visualize and interact with 3D objects, in a real environment (Rabah et al., 2018). It combines the real world with the virtual one, to provide a live form of computer-generated graphics, which has been comprehended as a technology that uses virtual imagery illustrations to overlay the real environment that digitally enhance user's performance (Wang et al., 2018).

Figure 7. M4.0 in different industries

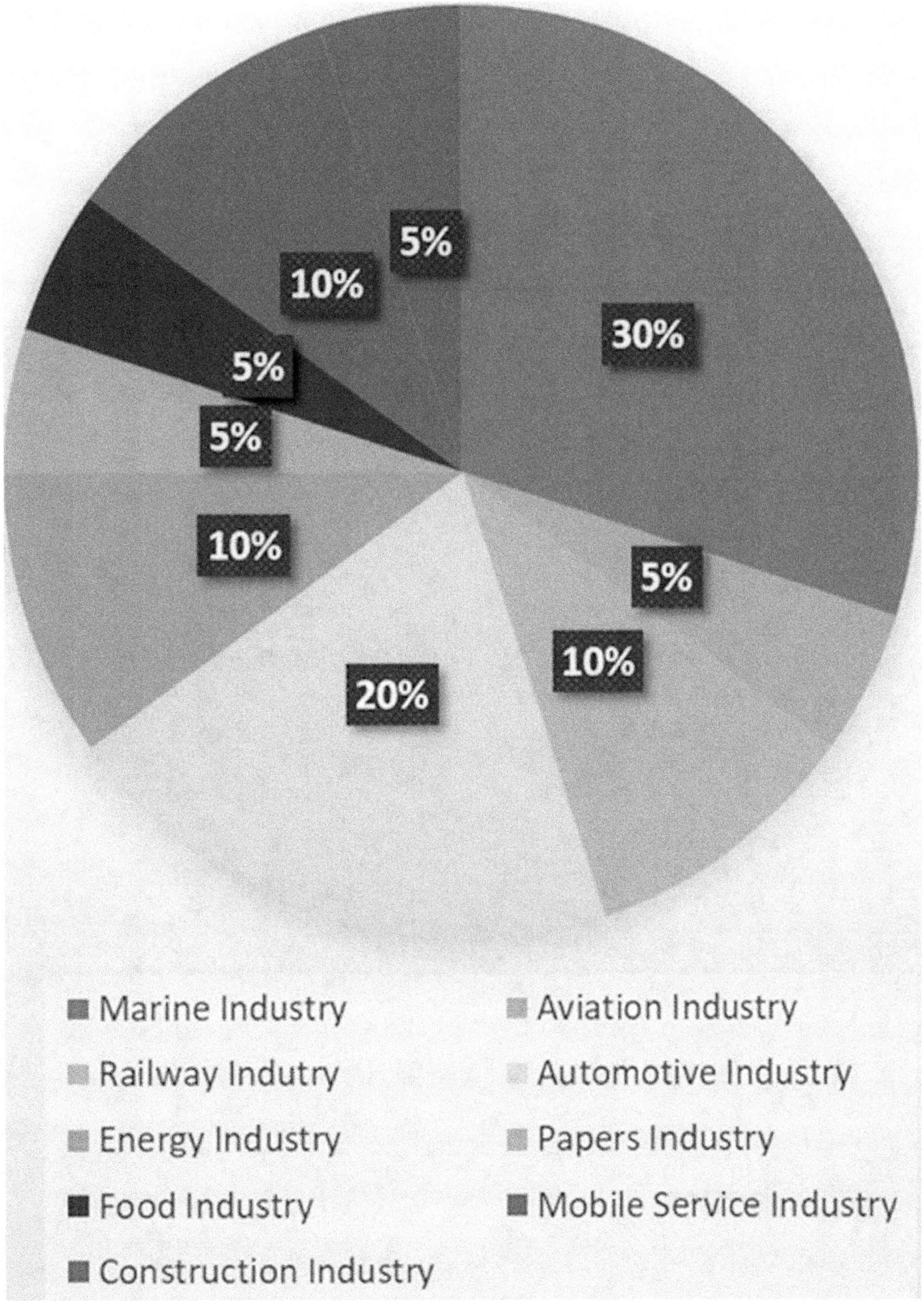

Figure 8. Specific sector applications with M4.0 concepts through the reviewed papers

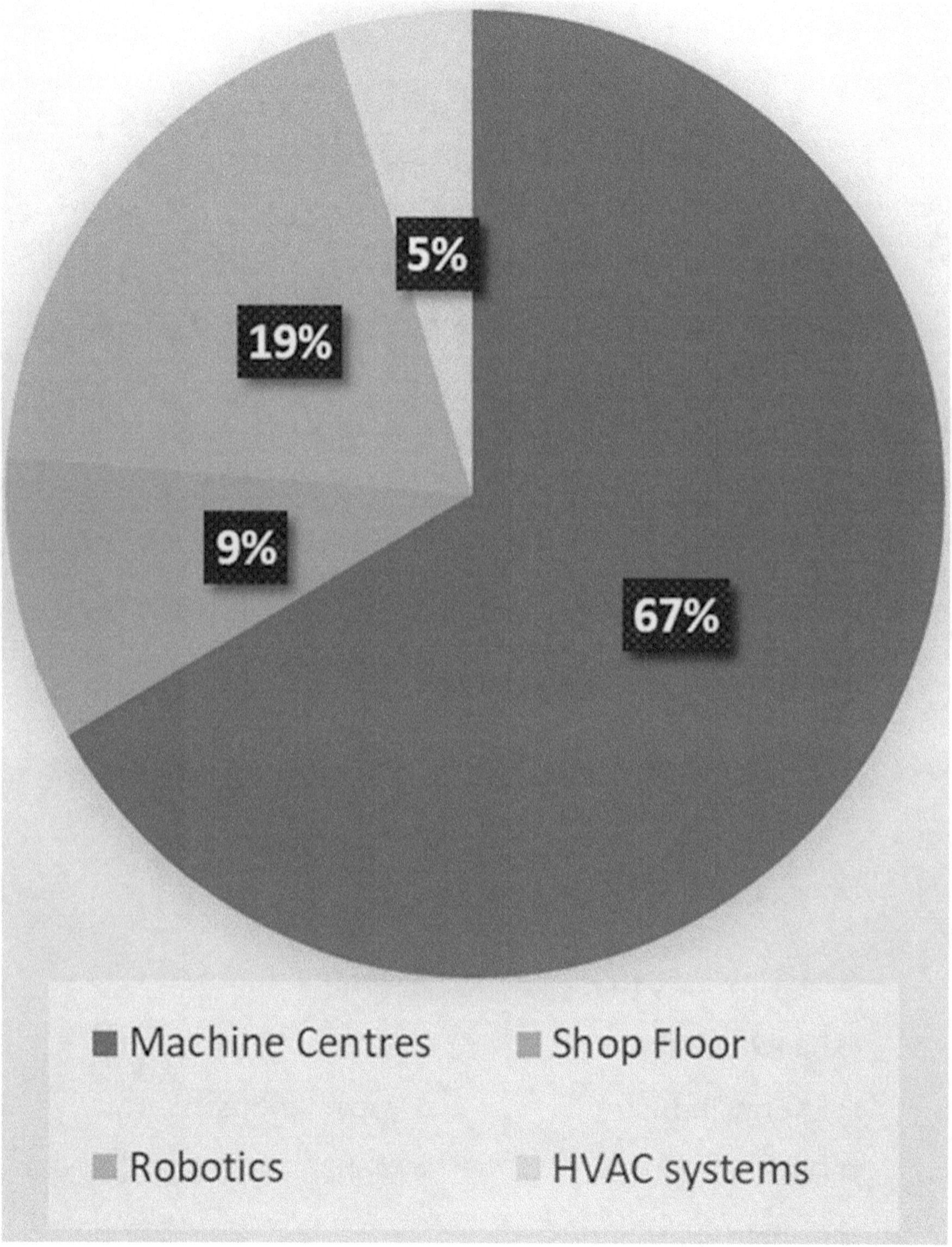

(Rabah et al., 2018) presents a method of monitoring and detecting forces and failures in rolling bearings. This method improves the flow of the process thanks to AR technologies which offer a full DT integration in its environmental context. It presents a primary proof-of-concept and a first step towards the development of a PdM strategy.

A PdM approach in this kind of system is presented in (Rabah et al., 2018), to detect anomalies. The anomalies could be given by quality controls and sensors installed in the devices (Blanco-Novoa et al., 2018). These systems could evaluate the state of the process, make an interpretation of the data and establish maintenance scenarios. The resulting maintenance procedure would be presented through the DT. Information would be transported via a platform to operators in real time through AR. AR would then interact in real time, be in contact with all the process components and perform the maintenance phase.

Costs as M4.0 Consequence

Through the papers reviewed it is perceptible that the implementation of I4.0 in maintenance brought numerous advantages. Costs are a fundamental parameter within any project or idea a company might want to develop. As (Bokrantz et al., 2017) predicted, M4.0 enables more cost-effective maintenance with fewer resources, reducing the maintenance related costs.

From the 90 papers checked, 14 of them referred the cost/financial parameter in manifold forms. 6 of them presented a case study that allowed to validate the study proposal. The case studies were carried in different industries and maintenance fields, such as: mechanical industry (engine cylinder head) (Yihai He, Xiao Han, Changchao Gu, 2018), railway sector (Vatn, 2018), food processing plant (Ashraf, 2018) and aquaculture industry (Steinsland, 2018). To better visualize how costs are addressed in the 14 papers mentioned, the pie chart Figure 9 is provided.

Models to assess the costs involved in a M4.0 context are presented (Tedeschi et al., 2018), (Rødseth et al., 2017) and (Vatn, 2018). These papers formulate models that aim to estimate the cost of implementing M4.0 components, considering important parameters associated to its implementation, such as the level of vulnerability of the machines or data loss caused by cyber threats.

Models to improve the cost and availability of M4.0 systems or strategies are also presented (Felsberger et al., 2019), (Yihai He, Xiao Han, Changchao Gu, 2018), (Ashraf, 2018), (Ammar Y. Alqahtani et al., 2019), (Huang et al., 2018). In (Felsberger et al., 2019) an approach to choose the load-sharing strategy of maintenance that achieves the highest availability and better maintenance costs is discussed. It is concluded that the ratio between repair and downtime costs directly influences the selection of the most economic load-sharing policy. However, for each system load, a different load-sharing policy is optimal, and the choosing of the right one, can reach a reduction of system costs by almost an order of magnitude. A cost oriented PdM model based on mission reliability state is proposed in (Yihai He, Xiao Han, Changchao Gu, 2018). Mission reliability is obtained to characterize the production status of the equipment. This model includes the modelling of CM, PdM, production capacity, indirect losses and product quality costs, and the sum of each one them results in the cumulative comprehensive cost. With the reduction of mission reliability threshold, CM, indirect losses and product quality loss costs gradually increase. This is due to the increasing of equipment failure, which consequently leads to the declining

Figure 9. Cost models in the reviewed papers

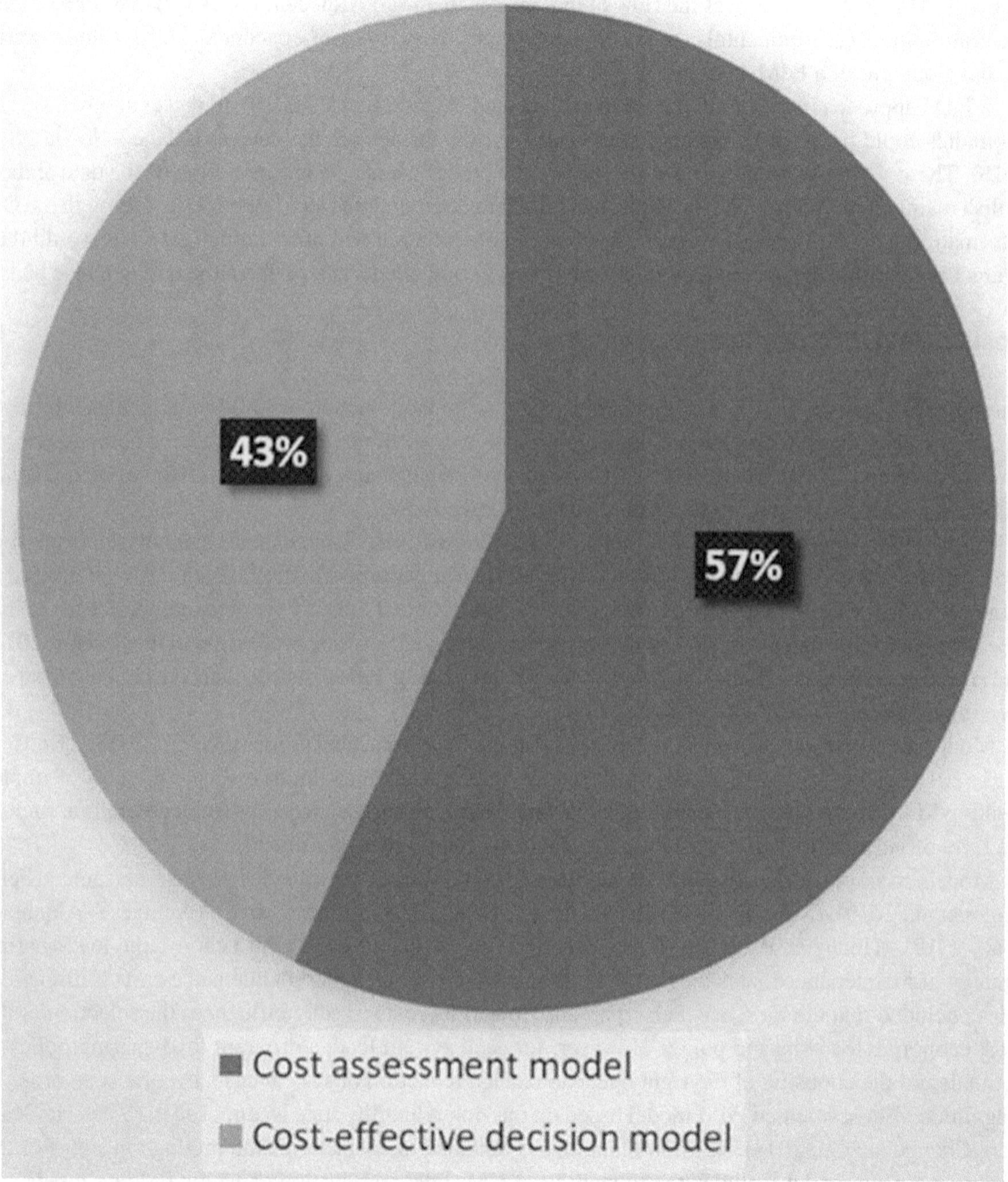

of qualified rate of work in process and task completion probability. In addition, PdM and production capacity loss costs gradually decrease because of the reduced number of PdM activities.

Al-Najjar et al., 2018 reports how to identify the most suitable maintenance technique to be further developed in an M4.0 context, performing, among others, a cost-effective analysis of each one of them. The assessment showed that CM is ranked low in cost effectiveness due to the impossibility of reducing the probability of failures. PM has tools and methods to reduce the probability of production stoppages. However, they are not always early enough and therefore are graded middle in cost effectiveness. PdM detects damages initiation before this impact the production, which consequently rate PdM as the most cost-effective technique.

The outsourcing of maintenance activities is reported to bring great financial benefits to companies. According to Vacík & Špaček, 2018, in a Maintenance 4.0 environment, the outsourcing of maintenance activities allows 39% of cost savings, herewith higher quality of services. Bigger cost savings are in the trading, service and production sector.

Applying M4.0, may mean increased maintenance cost, however, does not matter how much maintenance budget will increase, as long as maintenance cost per high quality product is decreasing (Al-Najjar et al., 2018), which is feasible with M4.0.

Answer to Q2

The answer to this question must be found in the discussions, conclusions and future works of the papers reviewed in this SLR. The answer to the Q1 has partially answered this question.

Important to notice that developments in the I4.0 will have a great impact on maintenance as technology is the driving force of industry related activities these days. Even though the advantages of I4.0 implementation in maintenance have been identified and proven at an academic level, improvements on I4.0 and maintenance process organisation are required in order to provide robust, reliable and adaptable solutions for practical implementations (Qi & Tao, 2018), (Vatn, 2018), (De Pace et al., 2019), (Roy et al., 2016).

Therefore, to answer this research question, future developments and implementations on I4.0 and maintenance will be described in distinct subsections below.

I4.0 key Enablers Future Developments and Implementations

The key enablers of I4.0 were described in the introduction. These components allow the implementation of I4.0 within maintenance which impacts several areas such as cost reduction (Lee et al., 2014). However, these components still present space for improvements. The IoT acts as an enabler for more efficient maintenance, but the need for fundamental understanding of the degradation mechanisms and root causes for the failure modes remains unchanged. It is therefore required to develop adjustable architecture of the IoT based products (Roy et al., 2016).

Real time data capture, analysis and modelling of the "big data" from the products in use is vital within a "highly connected" manufacturing and use environment so that the maintenance efficiency can be improved (Roy et al., 2016). Hence, CPS will be implemented as these systems help companies to design the layout of sensor nets, to achieve coordination and controlling of smart machines and to extend the usage of information and communication technologies to the maintenance processes (Mueller et al., 2017).

Maintenance Paradigm, Future Developments and Implementations

In a time with high competition in the different industries, it is very important to have as effective and efficient production as possible. Since maintenance plays a crucial role to achieve this goal, its management is essential to avoid unexpected downtime and late deliveries (Ravna, 2016). Reliable inspection and maintenance methodologies incorporating new technologies would facilitate cost effective and efficient asset management (Seneviratne et al., 2018).

The maintenance department's typical workday includes highly technical and specialized tasks related to a broad range of advanced technologies. Until recently, information technology has not been considered relevant for maintenance staff (Fusko et al., 2018). It is necessary to adapt the direction of current maintenance since "Smart Things" are present to a large extent. The development of production, logistical and technical services will allow to achieve long-term sustainability.

Maintenance is believed to evolve into an even more sophisticated system (Fusko et al., 2018). To reach this refined system and success, it is recommended to start with a relatively small set of standardized models for critical processes in the value chain of the company (Vatn, 2018). One of those processes is the selection of the right maintenance technique or strategy. It is expected that PdM technique will be able to detect damage initiation (Steinsland, 2018). Total Quality Maintenance is mentioned to be the future of maintenance techniques as it detects deviations in the condition and performance of all essential elements involved in a production system (Al-Najjar et al., 2018). Personal competence and raw material quality are, among others, additionally concerned, as the whole process is covered and not only the equipment. Quality of maintenance programs is improved by the previously mentioned techniques, as they are strategic instead of reactive. I4.0 supports this view with the value that can be gained from generating data on whole processes to create smart machines and enable a prescriptive maintenance strategy. These fundamental characteristics will lead to the implementation of the previous mentioned maintenance techniques in M4.0 environments.

From the viewpoint of production, intelligent maintenance is basically a new concept, the transformation from classic maintenance to digitally based maintenance has to go through each step of process design and requires further implementations (Rakyta et al., 2016).

CONCLUSION AND FUTURE WORKS

The main objective of this research was to review and analyse the current state and future progresses of M4.0 in a systematic manner, answering two main research questions: Q1) What is the state of art of M4.0? Q2) Which are the potential future developments and implementations of M4.0? In the proposed SLR, special attention was given to different aspects that supported the answering of the research questions. The evolution of maintenance techniques within the I4.0 demands were described, as well as the different frameworks that allow the employment of such techniques. AR was outlined as being a key technology pointed out by I4.0 and with applicability on maintenance. Moreover. the cost factor within the M4.0 concept was analysed. The highlighted categories provided a legitimate state of art on M4.0 context. Besides, future directions concerning developments and implementations were offered based on the 90 articles reviewed in the SLR, which answered the second research question.

In general, I4.0 clearness of main concepts and further practical implementations are needed to convert maintenance to even more intelligent and autonomous processes. Standardized models of maintenance management are required in order to comply with I4.0 claims and to achieve long-term sustainability in companies. When considering the results of the present study, several limitations should be noted.

Regarding the validity and objectivity of the SLR, the authors provided a reproducible methodology, which is only subjective in the application of the quality criteria and in the selection of the main characteristics of M4.0. The authors´ knowledge on performing this type of paper and on the M4.0 topic might affect the objectivity of the SLR. From a completeness point of view, this review could be more comprehensive if more databases and more languages were also taken into account.

The data extraction process was explained and applied systematically therefore, the authors believe this SLR provides a contribution to M4.0. The paper can be used for anyone approaching M4.0 at an industrial level or to perform academic researches.

Future literature works could aim to find a correlation between M4.0 systems and their application in a systematic way, i.e., to point out common architectures that include all aspects of M4.0. Moreover, in the extensive and accelerated context of digital engineering, what will the role of human maintenance operators be? Will the maintenance sector become fully human-less and self-incorporated in companies?

M4.0 is on the edge of deploying its full potential, but as remarked by this paper, several areas require further improvements.

REFERENCES

Al-Najjar, B., Algabroun, H., & Jonsson, M. (2018). Maintenance 4.0 to fulfill the demands of Industry 4.0 and Factory of the Future. *Journal of Engineering Research and Application*, *8*(11), 20–31. Www.Ijera.Com. doi:10.9790/9622-0811022031

Alqahtani, A. Y., Gupta, S. M., & Nakashima, K. (2018, December). Warranty and maintenance analysis of sensor embedded products using internet of things in industry 4.0. *International Journal of Production Economics*, *208*, 483–499. doi:10.1016/j.ijpe.2018.12.022

Alqahtani, A. Y., Gupta, S. M., & Nakashima, K. (2019). Warranty and maintenance analysis of sensor embedded products using internet of things in industry 4.0. *International Journal of Production Economics*, *208*, 483–499. doi:10.1016/j.ijpe.2018.12.022

Ansari, F., Khobreh, M., Seidenberg, U., & Sihn, W. (2018). A problem-solving ontology for human-centered cyber physical production systems. *CIRP Journal of Manufacturing Science and Technology*, *22*, 91–106. doi:10.1016/j.cirpj.2018.06.002

Ashraf, W. (2018). *A model to assess and customize computerized maintenance management systems complies with industry 4.0 vision and requirements: A case study in food processing plant*. Academic Press.

Blanco-Novoa, O., Fernandez-Carames, T. M., Fraga-Lamas, P., & Vilar-Montesinos, M. A. (2018). A Practical Evaluation of Commercial Industrial Augmented Reality Systems in an Industry 4.0 Shipyard. *IEEE Access: Practical Innovations, Open Solutions*, *6*, 8201–8218. doi:10.1109/ACCESS.2018.2802699

Blanco-Novoa, Ó., Fernández-Caramés, T. M., Fraga-Lamas, P., & Vilar-Montesinos, M. A. (2018). A Practical Evaluation of Commercial Industrial Augmented Reality Systems in an Industry 4.0 Shipyard. *IEEE Access: Practical Innovations, Open Solutions*, *6*, 8201–8218. doi:10.1109/ACCESS.2018.2802699

Bokrantz, J., Skoogh, A., Berlin, C., & Stahre, J. (2017). Maintenance in digitalised manufacturing: Delphi-based scenarios for 2030. *International Journal of Production Economics, 191*, 154–169. doi:10.1016/j.ijpe.2017.06.010

Cao, J., & Zhang, S. (2016). *The Application and Design of Big Data in Operation and Maintenance of Industry 4.0.pdf*. Academic Press.

De Pace, F., Manuri, F., Sanna, A., & Zappia, D. (2019). A Comparison Between Two Different Approaches for a Collaborative Mixed-Virtual Environment in Industrial Maintenance. *Frontiers in Robotics and AI*, *6*(March), 1–14. doi:10.3389/frobt.2019.00018

Donovan, P. O., Leahy, K., Bruton, K., & Sullivan, D. T. J. O. (2015). An industrial big data pipeline for data - driven analytics maintenance applications in large - scale smart manufacturing facilities. *Journal of Big Data*. doi:10.118640537-015-0034-z

Felsberger, L., Todd, B., & Kranzlmüller, D. (2019). Cost and Availability Improvements for Fault-Tolerant Systems Through Optimal Load-Sharing Policies. *Procedia Computer Science, 151*(2018), 592–599. doi:10.1016/j.procs.2019.04.080

Fernández-Caramés, T.M., Fraga-Lamas, P., Suárez-Albela, M., & Vilar-Montesinos, M. (2018). *A Fog Computing and Cloudlet Based Augmented Reality System for the Industry 4.0 Shipyard.pdf*. Academic Press.

Fusko, M., Rakyta, M., Krajcovic, M., Dulina, L., Gaso, M., & Grznar, P. (2018). *Basics of designing maintenance processes in Industry 4.0*. Academic Press.

Hadjina, M., & Matulja, T. (2018). *Toward shipbuilding 4.0-an Industry 4.0 changing the face of the shipbuilding industry*. Academic Press.

He, Han, & Gu. (2018). *Cost-oriented predictive maintenance based on mission reliability state for cyber manufacturing systems*. Academic Press.

Huang, K., Blue, J., Weng, H., & Liu, S. (2018). *Single machine scheduling with consideration of preventive maintenance and machine health*. Academic Press.

Kans, M., Galar, D., & Thaduri, A. (2016). Maintenance 4.0 in railway transportation industry. In Lecture Notes in Mechanical Engineering, F4, 317–331. doi:10.1007/978-3-319-27064-7_30

Klathae, V. (2019). *The Predictable Maintenance 4.0 by Applying Digital Technology A Case Study of Heavy Construction Machinery.pdf*. Academic Press.

Komonen, K. (2002). *A cost model of industrial maintenance for profitability analysis and benchmarking*. Academic Press.

Koulali, M., Koulali, S., Tembine, H., & Kobbane, A. (2018). Industrial Internet of Things-Based Prognostic Health Management: A Mean-Field Stochastic Game Approach. *IEEE Access: Practical Innovations, Open Solutions*, *6*, 54388–54395. doi:10.1109/ACCESS.2018.2871859

Lee, J., Kao, H., & Yang, S. (2014). *Service innovation and smart analytics for Industry 4.0 and big data environment*. doi:10.1016/j.procir.2014.02.001

Li, Z. (2018). *Deep Learning Driven Approaches for Predictive Maintenance*. Academic Press.

Li, Z., Wang, Y., & Wang, K.-S. (2017). Intelligent predictive maintenance for fault diagnosis and prognosis in machine centers: Industry 4.0 scenario. *Advances in Manufacturing*, *5*(4), 377–387. doi:10.100740436-017-0203-8

Liao, Deschamps, Loures, & Ramos. (2017). Past, present and future of Industry 4.0 - a systematic literature review and research agenda proposal. International Journal of Production Research, 55(12), 3609–3629. doi:10.1080/00207543.2017.1308576

Mohamed, N., Al-jaroodi, J., & Lazarova-Molnar, S. (2019). Leveraging the Capabilities of Industry 4.0. *International Journal of Production Research*, *55*(12), 3609–3629. doi:10.1080/00207543.2017.1308576

Mueller, E., Chen, X.-L., & Riedel, R. (2017). Challenges and Requirements for the Application of Industry 4.0: A Special Insight with the Usage of Cyber-Physical System [English Edition]. *Chinese Journal of Mechanical Engineering*, *30*(5), 1050–1057. doi:10.100710033-017-0164-7

Palmarini, R., Erkoyuncu, J. A., Roy, R., & Torabmostaedi, H. (2018). A systematic review of augmented reality applications in maintenance. *Robotics and Computer-Integrated Manufacturing*, *49*(March 2017), 215–228. doi:10.1016/j.rcim.2017.06.002

Peres, R. S., Dionisio Rocha, A., Leitao, P., & Barata, J. (2018). IDARTS – Towards intelligent data analysis and real-time supervision for industry 4.0. *Computers in Industry*, *101*, 138–146. doi:10.1016/j.compind.2018.07.004

Qi, Q., & Tao, F. (2018). *Digital Twin and Big Data Towards Smart Manufacturing and Industry 4.0 360 Degree Comparison*. Academic Press.

Qian, Y., Dersch, M. S., Gao, Z., & Edwards, J. R. (2019). Railroad infrastructure 4.0: Development and application of an automatic ballast support condition assessment system. *Transportation Geotechnics*, *19*, 19–34. doi:10.1016/j.trgeo.2019.01.002

Rabah, S., Assila, A., Khouri, E., Maier, F., Ababsa, F., Bourny, V., ... Mérienne, F. (2018). Towards improving the future of manufacturing through digital twin and augmented reality technologies. *Procedia Manufacturing*, *17*, 460–467. doi:10.1016/j.promfg.2018.10.070

Rakyta, M., Fusko, M., Herčko, J., Závodská, Ľ., & Zrnić, N. (2016). Proactive approach to smart maintenance and logistics as a auxiliary and service processes in a company. *Journal of Applied Engineering Science*, 433–442.

Ravna, R. (2016). *Industry 4.0 and Maintenance*. Academic Press.

Rødseth, H., Schjølberg, P., & Marhaug, A. (2017). Deep digital maintenance. *Advances in Manufacturing*, *5*(4), 299–310. doi:10.100740436-017-0202-9

Roy, R., Stark, R., Tracht, K., Takata, S., & Mori, M. (2016). Continuous maintenance and the future – Foundations and technological challenges. *CIRP Annals - Manufacturing Technology*, *65*(2), 667–688. doi:10.1016/j.cirp.2016.06.006

Roy, R., Stark, R., Tracht, K., Takata, S., & Mori, M. (2018). *Continuous maintenance and the future – Foundations and technological challenges.pdf*. Academic Press.

Seneviratne, D., Ciani, L., Catelani, M., & Galar, D. (2018). Smart maintenance and inspection of linear assets: An Industry 4.0 approach. *Acta IMEKO*, *7*(1), 50–56. doi:10.21014/acta_imeko.v7i1.519

Simon, J., Trojanova, M., Zbihlej, J., & Sarosi, J. (2018). Mass customization model in food industry using industry 4.0 standard with fuzzy-based multi-criteria decision making methodology. *Advances in Mechanical Engineering*, *10*(3). doi:10.1177/1687814018766776

Steinsland, K. (2018). *A model to develop and implement a predictive maintenance programme complying with Industry 4.0 vision: A case study in Skretting AS*. Academic Press.

Tedeschi, S., Rodrigues, D., Emmanouilidis, C., Erkoyuncu, J., Roy, R., & Starr, A. (2018). A cost estimation approach for IoT modular architectures implementation in legacy systems. *Procedia Manufacturing*, *19*, 103–110. doi:10.1016/j.promfg.2018.01.015

Vacík, E., & Špaček, M. (2018). Process innovation in service sector matching industry 4.0 environment. *Annals of DAAAM and Proceedings of the International DAAAM Symposium*, *29*(1), 352–360. doi:10.2507/29th.daaam.proceedings.05

Vatn, J. (2018). Industry 4.0 and real-time synchronization of operation and maintenance. *Safety and Reliability - Safe Societies in a Changing World - Proceedings of the 28th International European Safety and Reliability Conference, ESREL 2018*, 681–686. https://www.scopus.com/inward/record.uri?eid=2-s2.0-85058076521&partnerID=40&md5=027b2998282b9231868068dfaf3e7000

Waeyenbergh, G., & Pintelon, L. (2002). A framework for maintenance concept development. *International Journal of Production Economics*, *77*(3), 299–313. doi:10.1016/S0925-5273(01)00156-6

Wan, S., Li, D., Gao, J., & Li, J. (2019). A knowledge based machine tool maintenance planning system using case-based reasoning techniques. *Robotics and Computer-Integrated Manufacturing*, *58*, 80–96. doi:10.1016/j.rcim.2019.01.012

Wang, J., Erkoyuncu, J., & Roy, R. (2018). A Conceptual Design for Smell Based Augmented Reality: Case Study in Maintenance Diagnosis. *Procedia CIRP*, *78*, 109–114. doi:10.1016/j.procir.2018.09.067

Welte, T. M., Foros, J., Nielsen, M. H., & Adsten, M. (2018). *MonitorX – Experience from a Norwegian-Swedish research project on industry 4.0 and digitalization applied to fault detection and maintenance of hydropower plants*. Academic Press.

Yan, J., Meng, Y., Lu, L., & Li, L. (2017). *Industrial Big Data in an Industry 4.0 Environment : Challenges, Schemes, and Applications for Predictive Maintenance. 5*. Academic Press.

APPENDIX

Quality Criteria Selected For this SLR

Table 4.

	QC1	QC2	QC3	QC4	SUM		QC1	QC2	QC3	QC4	SUM
Blanco-Nova, Ó et al., (2018)	1	1	1	1	4	He, Y. et al., (2018)	1	1	1	1	4
LIN, CHUN-CHENG et al., (2019)	1	1	1	1	4	Pagalday, G. et al., (2018)	0,5	1	1	1	3,5
Qi, Quinglin & Tao, Fei (2018)	0,5	0,5	1	0	2	Tantik, E. & Anderl, R. (2017)	1	1	1	0	3
Yan, J. et al., (2017)	1	1	1	1	4	Gregor, M. et al., (2016)	1	1	0,5	0	2,5
Bagheri, M. & Zollanvari, A. (2018)	1	1	1	1	4	Pelantova, V. & Cecak, P. (2018)	0,5	0,5	1	1	3
Vatn, J (2018)	1	1	0,5	1	3,5	Bengtsson, M. & Lundström, G. (2018)	1	1	1	0	3
Cao, J. & Zhang, S. (2016)	1	1	0,5	0	2,5	Rakyta, M. et al., (2016)	0,5	0,5	0,5	0	1,5
Ravna, R. (2016)	1	1	0,5	0	2,5	Huang, K. et al., (2018)	1	1	1	1	4
Al-najjar, B. et al., (2018)	1	1	1	1	4	Umer, M. et al., (2018)	1	1	1	0	3
Welte, T.M. et al., (2018)	1	1	1	1	4	Pace, F. et al., (2019)	1	1	1	1	4
Chukwuekwe, D. O. et al., (2016)	0,5	1	0,5	0	2	Fernández-Caramés, T. M. et al., (2018)	1	1	1	1	4
Marhaug, A. & Schjølberg, P. (2016)	1	1	0,5	0	2,5	Antosz, K. (2018)	1	1	1	1	4
Cao, J. & Zhang, S. (2016)	1	1	0,5	0	2,5	Peres, R. S. et al., (2018)	1	1	1	1	4
Ashraf, W. (2018)	1	1	1	1	4	Heynicke, R. et al., (2018)	1	1	1	1	4
Steinsland, K. (2018)	1	1	1	1	4	Roy, R. et al., (2016)	1	1	0,5	1	3,5
Vafeiadis, T. et al., (2018)	1	1	1	1	4	Reis, M. S. & Gins, G. (2017)	0,5	0,5	1	0	2
Li, Z (2018)	1	1	1	1	4	Bokrantz, J. et al., (2017)	0,5	1	0,5	0	2
Li, Z et al., (2017)	1	1	1	1	4	Silva, M. et al., (2018)	1	1	1	1	4
Ceruti, A. et al., (2019)	1	1	1	1	4	Stanić, V. et al., (2018)	1	1	1	1	4
Dang, Tuan et al., (2016)	1	1	0,5	0	2,5	He, Yihai et al., (2018)	1	1	1	1	4
Seneviratne, D. et al., (2018)	1	1	1	0	3	Rødseth, H. et al., (2017)	0,5	0,5	1	1	3
Alqahtani, A. Y. et al., (2019)	1	1	1	1	4	Simon, J. et al., (2018)	1	1	1	1	4
Tedeschi, S. et al., (2018)	1	1	1	0	3	Zhang, H. et al., (2017)	1	1	1	1	4
Felsberger, L. et al., (2019)	0,5	1	1	1	3,5	Tao, F. & Zhang, M. (2017)	1	1	1	1	4
Mourtzis, D. et al., (2018)	1	1	1	1	4	Mueller, E. et al., (2017)	1	1	1	1	4
Rabah, S. et al., (2018)	1	1	1	1	4	Tsai, W. & Lai, S. (2018)	1	1	1	1	4
Klein, M. et al., (2018)	1	1	1	0	3	Yan, J. et al., (2017)	1	1	1	1	4
Schimtt, R. et al., (2016)	1	0,5	0,5	1	3	Alonso, A. et al., (2018)	1	1	1	1	4
Welte, T.M. et al., (2018)	1	0,5	1	0	2,5	Koulali, M. et al., (2018)	1	1	1	0	3
Lee, W. J. et al., (2019)	1	1	1	1	4	Mohamed, N. et al., (2019)	1	1	1	1	4
Pinto, R. & Cerquitelli, T. (2019)	1	1	1	1	4	Batista, N. C. et al., (2017)	0,5	0,5	0,5	1	2,5
Lee, J. et al., (2014)	1	1	0,5	0	2,5	Mabkhot, M. M. et al., (2018)	1	1	1	0	3
Demminger, C. et al., (2016)	1	1	0,5	0	2,5	Jantunen, E. et al., (2018)	0,5	1	0	1	2,5
Wang, J. et al., (2018)	1	1	1	1	4	Uhlmann, E. et al., (2017)	1	1	1	1	4
Mocan, M. et al., (2018)	1	1	1	1	4	Celik, M. & Kandemir, C. (2017)	1	1	1	0	3
O'Donovan, P. et al., (2015)	1	1	0,5	0,5	3	Rimpault, X. et al., (2018)	1	1	1	0	3
Fusko, M. et al., (2018)	1	0,5	1	1	3,5	Cachada, A. et al., (2018)	1	1	1	1	4

continued on following page

Table 4. Continued

	QC1	QC2	QC3	QC4	SUM		QC1	QC2	QC3	QC4	SUM
Medojevic, M. et al., (2018)	1	1	1	1	4	Jantunen, E. et al., (2018)	1	1	1	0	3
Vacík, E. & Špaček, M. (2018)	1	1	1	1	4	Algabroun, H. M. et al., (2017)	1	1	1	0	3
Koch, P. J. et al., (2017)	1	1	0,5	0	2,5	Kaasinen, E. et al., (2018)	1	1	1	1	4
Zhou, P. et al., (2019)	1	1	1	0	3	INŽINIERSTVA, K. P. et al., (2019)	1	1	1	1	4
Katona, A. % Panfilov, P. (2018)	1	1	1	1	4	Galar, D. & Seneviratne, D. (2016)	1	1	0,5	1	3,5
Kans, M. & Ingwald, A. (2016)	1	1	0,5	0	2,5	Klathae, V. & Ruangchoengchum, P. (2019)	1	1	1	1	4
Åkerman, M. et al., (2018)	1	1	1	0	3	Ansari, F. et al., (2018)	1	1	1	1	4
Bierer, A. et al., (2016)	1	1	0,5	0	2,5	Koskinen, K. T. et al., (2015)	1	1	0,5	1	3,5

Chapter 2
Industry 4.0 in Emerging Economies:
Technological and Societal Challenges for Sustainability

Pratima Verma
https://orcid.org/0000-0001-7179-3878
Department of Information Management, Chaoyang University of Technology, Taiwan

Vimal Kumar
Department of Information Management, Chaoyang University of Technology, Taiwan

Priyanka C. Bhatt
https://orcid.org/0000-0001-5638-6844
Department of Information Management, Chaoyang University of Technology, Taiwan

Vinayak Arvind kumar Drave
https://orcid.org/0000-0002-3554-0766
Jindal Global Business School, O.P. Jindal Global University, India

Sung-Chi Hsu
Department and Graduate Institute of Construction Engineering, Chaoyang University of Technology, Taiwan

Kuei Kuei Lai
Department of Business Adminstration, Chaoyang University of Technology, Taiwan

Vijay Pal
Department of Mechanical Engineering, Indian Institute of Technology, Jammu, India

ABSTRACT

Industry 4.0 has received a massive amount of attention worldwide in the past few years as a technological infrastructure to provide efficient operations in existing production systems as well as fast-tracking the implementation of internet-connected technologies across various industries. Industry 4.0 technologies have been considered as a strategy and implemented successfully in various developed countries. However, in emerging economies (or developing countries), the implementation of Industry 4.0 is not as successful as developed nations because of various challenges. However, fast-moving economies can take advantage of Industry 4.0 techniques as their requirement to operate at faster rates, capitalizing on new technologies that can drive efficiencies. This chapter examines the sustainability issues of Industry 4.0 in developing or emerging economies countries. These sustainability issues are related to scientific, technological, and societal issues.

DOI: 10.4018/978-1-7998-3904-0.ch002

INTRODUCTION

Industry 4.0 (IR 4.0) has congregated massive attention from nations as well as organizations as a network of technologies for economic development within the past five years. Industry 4.0 offers information and communication technologies (ICT) and digital manufacturing technologies which enhance the operational processes of enterprises swiftly and efficiently. IR 4.0 has shaped a trend of information exchange among various networks, (including Cyber-Physical Systems (CPS), the Internet of Things (IoT), Cloud Computing, Cognitive Computing, and so on). The fourth revolution in manufacturing techniques following mechanization, and electronics; and non-manufacturing techniques following communication, security, has been widely publicized by myriad companies with an aim to combine automation and robotization with human expertise. Manufacturing dependent countries producing a variety of products/services for their economic development can leverage the most from Industry 4.0 (Arifiani & Arifiani, 2019). Emerging technologies, such as AI, connected machines, 3D printing, etc. provide a cost-effective, qualitative and flexible solutions to manufacturing industries. Additionally, the exponential level of connected networks and devices in IR 4.0, resulting in faster automation of processes can disruptively affect the development of wealth as well as the economy in emerging nations. In line with these statements, we can see that IR 4.0 can provide a solid competitive edge for organizations that create innovative business models enabled by advanced technologies. Certainly, we can say that the implementation of IR 4.0 technologies enhances or benefits the organization. However, the big question is - how is the IR 4.0 enhancing competitiveness in emerging economies?

A number of developing countries are struggling with a lot of issues viz., electricity, infrastructure, and government policy & legal system, etc. Even a few developing countries are not enough capable to successfully implement Industry 3.0 or they are not ready yet for introducing new paradigms for ensuring reliability and safety (Martinetti et al., 2017; Martinetti et al. 2018). Developing countries have the potential to get benefit from industry 4.0, but the issue is that "sustainability" in the long-term. According to Merlino et al. (2015), the latest technology is not enough but the organizations need the business model to go with it. Emerging economies can be explained to fulfill two criteria; first, increasing the development of the economy, and second, liberal government policies for the free-market system (Kumar et al., 2017). There are five major characteristics of the emerging market i.e. "business heterogeneity, sociopolitical governance, unbranded competition, a chronic shortage of resources, and inadequate infrastructure" (Verma et al., 2017). One of the important factors that help drive emerging economies is industrialization.

The importance of sustainability concepts has been rising both at the local and global levels. The extreme implementation of IR 4.0 in terms of automation and connectivity has huge potential to affect economic development, but it is still a long way for emerging economies to exploit completely the benefits of IR 4.0. Gibbons (2009) addresses the three sustainability issues – the economic, social and environmental impact on firms in emerging economies. When we try to implement any kind of strategy, process, technologies, etc. in emerging countries then we should focus on these issues.

The primary aim of this chapter is to recapitulate and evaluate the literature concerning sustainability issues of industry 4.0 in emerging economies. In the following sections, we present the above-mentioned concerns in a more systematic view based on the sustainability of Industry 4.0. The most important sustainability issues are technological and societal issues.

This chapter is further structured as follows: Section 2 provides detailed literature about the current development of emerging economies. This section also highlights the outline of industry 4.0 origins, principles and technologies, as well as its relationship with the organization and developing & advanced economies. In section 3, we discuss the sustainability factors and sustainability issues of industry 4.0 in developing countries. In section 4, we provide the discussion and conclusion with major findings that appeared from the literature and indicate directions for future research.

BACKGROUND

Industrial Revolution 4.0

Similar to the first three industrial revolutions, the origin of the fourth industrial revolution is rooted in new means of creating value from a manufacturing perspective (Reyes et al., 2019). The term 'Industry 4.0' was devised in 2011 which considers an industry with main features consisting of "connected machines, smart products and systems, and inter-related solutions" (Tortorella & Fettermann, 2018). Industry 4.0 was described as a novel model for decentralized and smart manufacturing systems (Hunsaker & Youngdahl, 2018). Main technological innovations in Industry 4.0 have been considered as the automation process, information and communication technologies, artificial intelligence, enterprise resource planning, etc. (Santos et al., 2017). Technologies connected in the system of IR 4.0 combine in the manufacturing domain to support the manufacturing of voluminous and highly customized products. Additionally, the term "Industry 4.0" relates to a predicted fourth industrial revolution through horizontal and vertical interconnection in real-time based on digital technologies (Müller, 2020).

In other words, IR 4.0 is also known as a smart manufacturing system (SFS), also considered as the most suitable term for IR 4.0 (Lasi et al., 2014). It combines the Internet of Things (Almada-Lobo, 2016; Lasi et al., 2014; Stock & Seliger, 2016), Cyber-Physical Systems, Internet of Services(Almada-Lobo, 2016). It also focuses on the establishment of intelligent products and production processes (Brettel et al., 2014).

Industry 4.0 technologies are used in various areas viz. production & operation management, supply chain, health care, education, marketing, etc. In the context of production and operation management, IR 4.0 acts as a strategy that helps upsurge product development, manufacturing, quality, flexibility and efficiency in the production process (Tortorella & Fettermann, 2018; Vyatkin et al., 2007). IR 4.0 acts as an enabler to real-time monitoring of various manufacturing systems, such as, "*production status, energy consumption, the flow of materials, customers' orders, and suppliers' data*" (Luthra & Mangla, 2018). Additionally, when we considered supply chain management then industry 4.0 provides the technologies that enhance the delivery process and sharing information, a strong security system, and speed up assembly lines. Based on these literature surveys, we can say that definite industry 4.0 is profitable for manufacturing companies. IR 4.0 is considered promising in terms of increasing income levels as well as much improved life quality worldwide. Two things need to be understood, firstly, is industry 4.0 really an important aspect? And secondly, if it is so important, then which countries (developed or developing) are taking more advantage from industrial revolution 4.0. There is a number of developed countries in which the industry 4.0 implementation is successful. Nkamnebe (2011) emphasizes that, with the globalization convention, the global market is promising great opportunities for firms both in the advanced and emerging markets. Additionally, numerous studies are claiming that there are various factors in emerging economies which are more favorable to IR 4.0 application.

Figure 1. Key technologies of Industry 4.0

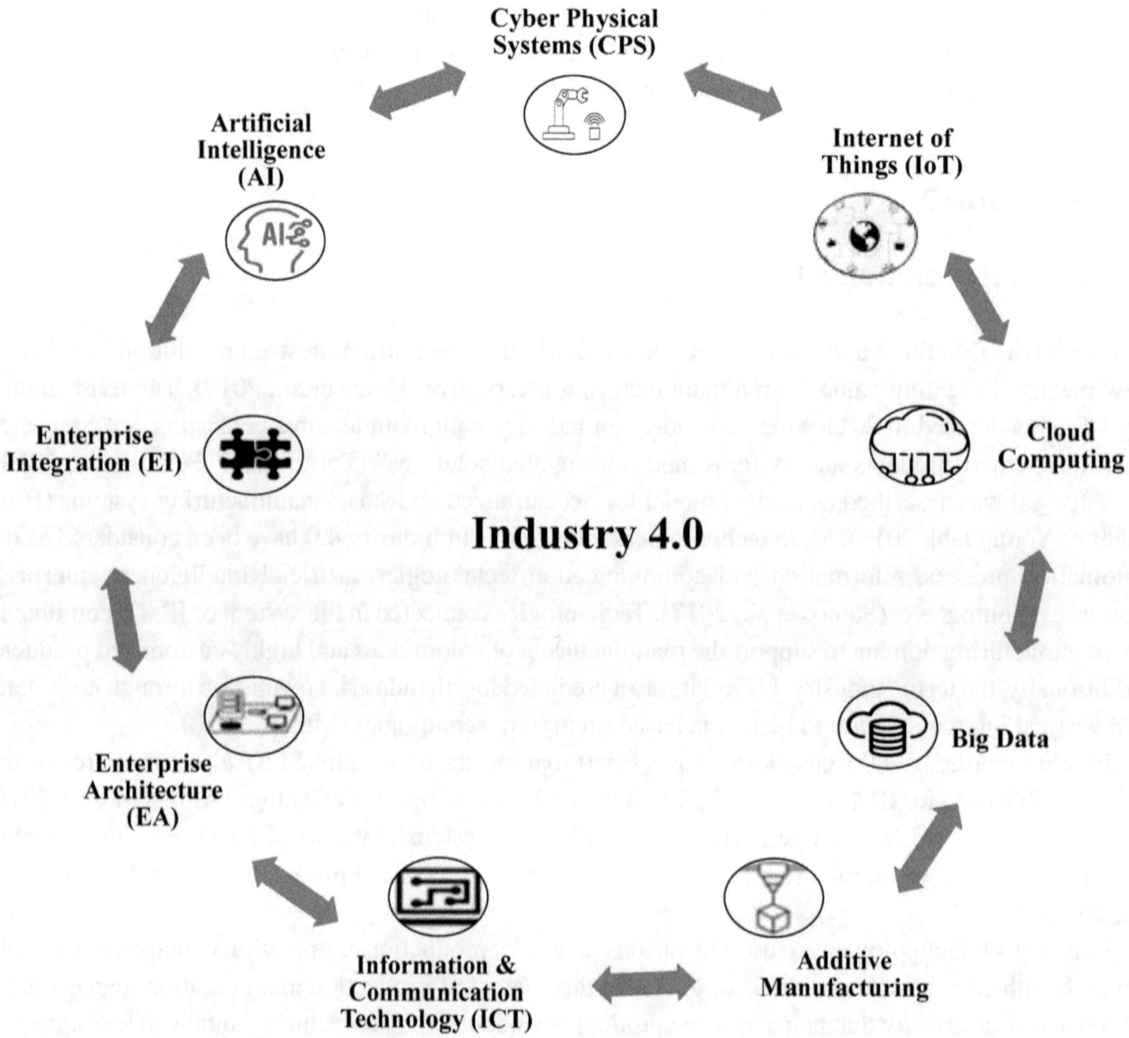

Industry 4.0 in Advanced Economies

Industry 4.0 implementation aims to build the current business economy and utilizing its resources in an effective and efficient way, as advanced-economy countries such as the United States, Finland, Switzerland, Germany etc. get benefitted from it. According to a recent report (World Economic Forum, 2017), only 25 countries (Western, European, and Asian countries like Japan, Republic of Korea, and Malaysia) are considered to leverage the maximum benefits from implementing Industry 4.0. For instance, the United Kingdom recently announced its industrial strategy focused on leveraging the combined benefits of AI and renewable energy technologies. A huge amount, approximately around £400 million is invested in electrical automobiles infrastructure followed by more than £100 million investment in 5G technologies. With keeping above in mind, no doubt that industrial revolution 4.0 is enough successful in developed economies. On the contrary to this, some estimates suggest that large shares of United States operations will be automated in future decades which will most probably result in technologies replacing humans at the workplace (Pfeffer, 2010). Moreover, according to one study between 1990 and 2007, focusing on local labor markets in the United States, found that the increase in industrial robot use resulted in large and robust negative effects on employment and wages. The same study found that there are positive effects of automation on wages and no changes in total employment in Germany.

Industry 4.0 in Emerging Economies

Emerging economies or emerging markets are also known as developing economies, that are nations in which growing economies and growing consuming population both are exits. They are rapidly industrializing and adopting a free market or mixed economy. In other words, emerging markets have a huge potential to contribute internationally to economic growth through trade and finance (Wang et al., 2016). According to Kumar et al. (2017), unpredictable requirements and extreme competition combined with innumerable prospects in addition to the dynamic business ecosystem form the major enablers in emerging. There is a number of factors defining emerging markets or emerging economies. According to Sheth (2011), five characteristics of emerging markets are "*market heterogeneity, sociopolitical governance, unbranded competition, a chronic shortage of resources, and inadequate infrastructure*". Lower-than-average per capita income, slow economic growth, High volatility, currency Swings, and potential for growth, these are the five characteristics which mainly describe the emerging economies. The emerging market involves countries such as "*Brazil, Chile, China, Colombia, Hungary, Indonesia, India, Malaysia, Mexico, Peru, Philippines, Russia, South Africa, Thailand, and Turkey*", according to Morgan Stanley Capital International (MSCI). There are many ways to take advantage of the high growth rates and opportunities in emerging markets. Industry 4.0 is one of the ways to take advantage of such growth and opportunities for emerging economies or emerging markets. The fast-moving economies are approaching to exploit IR 4.0 benefits to reach faster capitalization rates based on new technologies for efficient operations. After the financial-economic crisis of 2008, various emerging economies focused on rising commodity prices rather than investing in their technological infrastructure for their growing economies.

THEORETICAL FRAMEWORK

Sustainability Issues of Industry 4.0 in Emerging Economies

Although Industry 4.0 seems to be capturing the attention everywhere, it is yet to reach its complete potential, which may take more than a decade from now as well. There are many challenges and issues for implementation of industry 4.0, such as, *"scientific challenges, technological challenges, economic challenges, social problems, and political issues"* (Zhou et al., 2015). According to Arifiani & Arifiani (2019) lack of proper infrastructure, cheap labor, high-cost technologies, government indecisiveness and a dearth of knowledge sources, are some of the challenges of industry 4.0 in emerging economies. In addition to this, IR 4.0 is still considered to be in the early stages (Roblek et al., 2016). Oztemel & Gursev (2020) also outlined for the 4th industrial revolution is not mature enough for most of the reel life implementations. According to Kovacs (2018), the sustainable development of IR 4.0 with emerging digital market space is possible only by the inclusion of growth measures that can provide a way for political as well as social stability via trust.

Figure 2. Factors of emerging economies

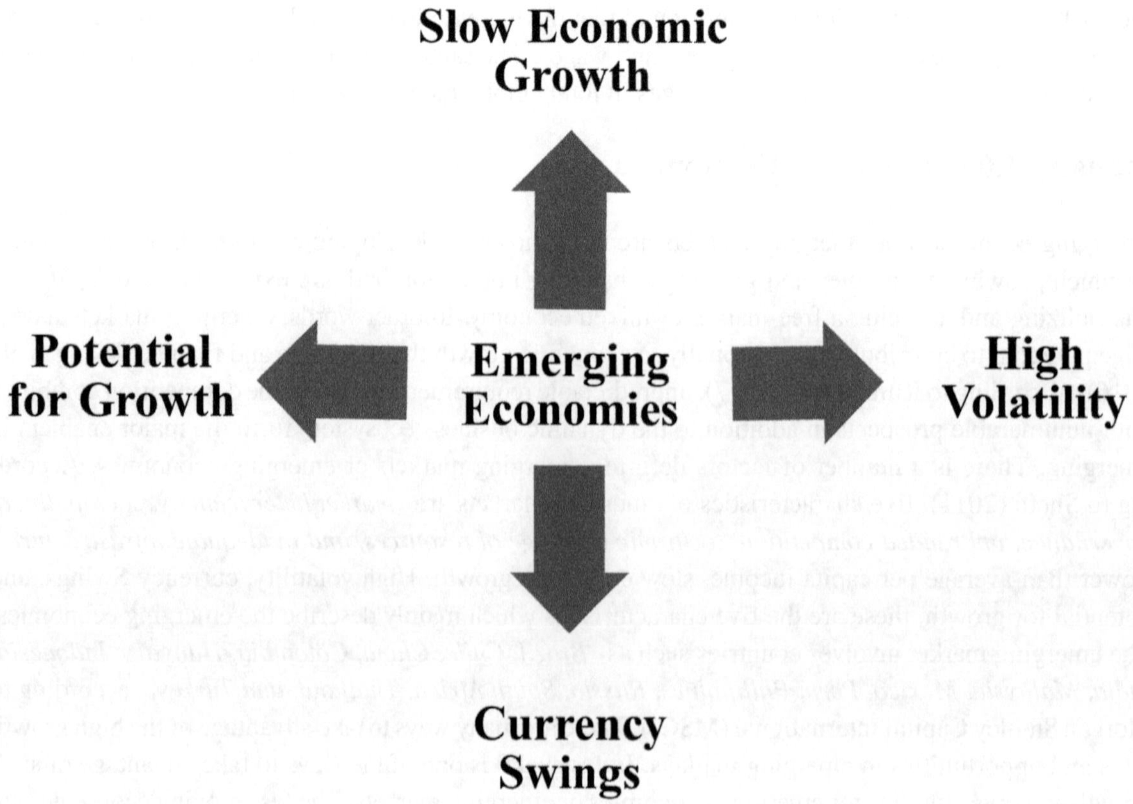

Table 1. BRICS Industrial revolution 4.0

Country	Strategy	Focus
Brazil	New national strategy on industry 4.0	Comprise a new framework for competitiveness, innovation, and entrepreneurship.
Russia	National Technology Initiative	NTI is a long term program of the public-private partnership in the development of promising new markets based on high-tech solutions that will determine the development of the global and Russian economy in the next 15-20 years.
India	Make in India	Aims to transform India into a global design and manufacturing hub
China	Made in China 2015	Aims to transform China into a leading manufacturing hub and by taking advantage of technology advances in manufacturing
South Africa	National E-strategy	Aims to position South Africa as a significant player in the development of ICTs throughout the value chain of the sector as well as accelerate the uptake and usage of ICTs in other social and economic sectors

Source: Manda and Ben Dhaou (Manda & Ben Dhaou, 2019). *Responding to the challenges and opportunities in the 4th Industrial revolution in developing countries*

The concept of sustainability is vast and is found in various areas of business. It is a broad concept consisting of various constructs such as governance, industry ethics, societal justice, green policies, etc. It is now considered a prominent component of any organizations' core strategy. Verma et al. (2017) address the three sustainability issues shown in Figure 3. Emerging economies are usually much more familiar to the sustainability challenges as compared to developed economies. Challenges related to the industry development and suburbanization, such as lack of energy-efficient resources, pollution, poor physical and technological infrastructure, lack of waste management, corrupt government, lack of skilled manpower.

Figure 3. Sustainable issue in emerging economies
Source: Verma et al. (Verma et al., 2017)

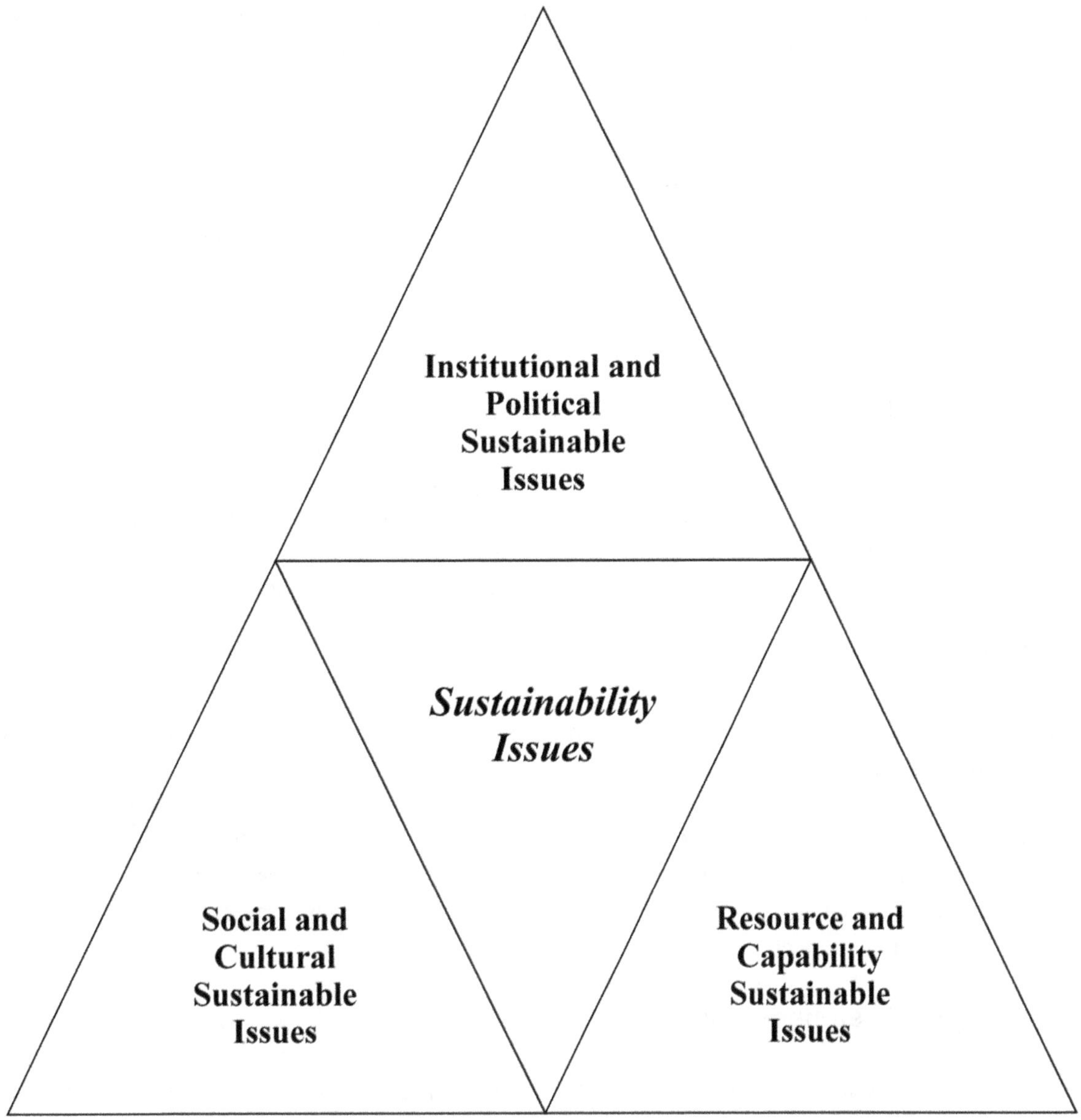

Technological Issues for Sustainability of Industry 4.0

Basic pillars of IR 4.0 are varied range of technologies, that when put together with the help of a wide range of business models, from small to complex levels of technological infrastructure. The technological landscape has changed exponentially over the last decade, and so have made changing businesses more ready for the volatile nature of technology.

Lee (2015) emphasizes the overall goal of industry 4.0 as a flexible, highly efficient system to create high volume processes/production with the major goal of cost reduction.

Industry 4.0 has given a way to Economy 4.0, which "*encompasses a full digital value chain from suppliers of materials and components required for production, through all the brokers and providers of necessary services, up to the final recipients of the entire production, i.e., end-customers, regardless of who they are*" (Cellary, 2019). However, emerging economies are yet far behind becoming economy 4.0. Although, industry 4.0 has the major implication in Smart Factory systems (SFS) (Preuveneers & Ilie-Zudor, 2017). However, with the complex architecture of technology, come various issues related to the implementation and application of the technology. Technological diversity often leads to challenges related to different types. For example, an agriculture sector implementing IoT services might have to deal with low and high-frequency bandwidth sensors; or high-tech companies have to deal with security and privacy of huge amounts of data related to their business operations as well as customers. In a nutshell, industry 4.0 poses technical challenges related to both hardware and software. The related issues are discussed further. Emerging economies have to deal with various issues related to implementation of SFS.

Digitization

A digital culture is the most important aspect of laying the ground for industry 4.0. And to connect heterogeneous networks together to form a complete ecosystem needs a digitally ready ecosystem. Emerging economies still have a long way for a 100 percent digital environment, which prevents the use of Industry 4.0 to its full effect. This results in the efficient implementation of I4.0 in certain geographical areas and excluding rural or remote areas from the success of the technology. An economy where every user is connected to the digital ecosystem will be able to leverage the benefits of I4.0 (Xu et al., 2018).

Scalability

I4.0 environment most necessary includes real-time high-speed data transfer between large scale heterogeneous devices or networks. This feature is defined as scalability. The number of increasing physical objects connected to the network makes it a real challenge to enable an efficient transaction among the devices (Bi & Cochran, 2014; Lin et al., 2016). The significant challenges related to scalability are the sustainability among device naming and addresses, the new connecting devices entering into the existing network, and the management of services to address the data flow among heterogeneous device networks (Díaz et al., 2016; Palattella et al., 2016).

Integration and Interoperability

The most discussed issue (Datta, 2017; Grody, 2018; Lu, 2017; Luthra & Mangla, 2018; Melnyk et al., 2018; Saweros & Song, 2019; Sisinni et al., 2018; Zhang et al., 2018) for implementing I4.0 is the use of heterogeneous devices across the network. The use of such heterogeneous devices involves the capability of the system to ensure efficient communication between them to perform operations in real-time. Interoperability means the ability of various devices (sensors, devices, machines, etc.) to connect with each other and provide meaningful output for human/machine interpretation.

Standardization

For efficient communication and identification between heterogeneous networks, global communication protocols are the essential requirements. Uniform industry standards are required In case of communication standards, both established and emerging economies have to establish the industry open standard protocols (e.g., SOS over CoAP, OGC Sensor Things API, and Tiny SOS, etc.) (Hopali & Vayvay, 2018; Lu, 2017; Preuveneers & Ilie-Zudor, 2017). Various researchers (Kim et al., 2017; McNamara, 2014; Palattella et al., 2016; Singh et al., 2019; Xu et al., 2018) address the issue of standardization when various technologies come into play together, such as IoT, CPS, Cloud Computing, etc.

Information Privacy and Security

When there is a network of huge devices communicating with each other, the most challenging and most prioritized concern becomes the privacy and security of the data interchange between these devices. Various researchers (Fernández-Caramés et al., 2019; Huang et al., 2019; Laszka et al., 2017; Onik et al., 2019; Sadeghi et al., 2015; Schulz & Freund, 2019; Sisinni et al., 2018; Ulz et al., 2017) have discussed the concerns and solutions related to the privacy and security of data in the context of industry 4.0. Sadeghi et al. (Sadeghi et al., 2015) discuss in detail the different surfaces of attacks (physical, electronic, networks, human, etc.) and security concerns from the perspective of IR 4.0. Sisinni et al. (Sisinni et al., 2018) provide a three-tier Industrial internet of things architecture for security at different levels of the connected network. However, the extent to which emerging economies can adopt the measures for tackling such issues is yet to be studied.

Real-Time Data Analysis

The interconnectivity among a variety of networks is the most prominent feature of Industry 4.0. However, it is not just the interaction or the flow of data among different devices, it is the capability of organizations to transform that data into value, what matters most in the industry 4.0 landscape. The huge and exponentially increasing number of data is generated in the I4.0 context, and it is important to interpret, analyze and derive accurate meaning from such voluminous real-time data efficiently (Cezarino et al., 2019; Hopali & Vayvay, 2018; Kunst et al., 2019; Webster et al., 2018; Xu et al., 2018). The real-time data analytics have their own set of challenges (such as, cloud data storage, data management from the different formats in, data extraction, creating data structures supporting the technology implementation, big data analytics, and so on).

Societal Issues for Sustainability of Industry 4.0

According to the world economic forum (WEF), industry 4.0 is both a social and technological revolution. Due to the technological changes in the workplace have serious social implications such as temporary and casual work risks, unemployment, lesser wages, worker's exploitation, and health & safety. These issues can lead to social disruption. The technological advancements or digitization lead to technological advantages and, in turn, give rise to competitive gaps and act as a driver of social inequality. IR 4.0 has a considerable impact on the labor market structure, demand particular set of skills, and the policy issues in emerging economies (Kergroach, 2017). In this section, we discuss these societal issues such as energy intensity technologies, jobs and inequality, skills & resistance of people, and policy and regulatory environment for industry 4.0.

Energy Intensity Technologies

All the technologies of the fourth industrial revolution need incessant consumption of energy. Various emerging technologies, viz. Blockchain, 3D printing, Artificial intelligence, Robotics, Drones & autonomous vehicles, biotechnologies, neuro-technologies, IoT, Geo-engineering, etc. require a continuous as well as enormous power input (Tronchoni & Brennan, 2017). A considerable number of emerging economies struggle with the electricity supply to their daily households. For example, India ranks third in energy consumption worldwide as well as electricity production worldwide up to 1272 TWh in FY 2014–15, but approximately more than 30 million households still don't have any access to the electricity (World Economic Forum, 2017). In line with this, the energy requirement is more in emerging economies such as India, China, Brazil and South Africa is much higher than the developed countries (Mangla et al., 2020). Brazil is also one of the developing countries which have challenges in electricity production and security. Likewise, India, a few communities in remote rural areas of Brazil are still lacking electricity. Consequently, we can say that the complete electrification of many developing countries is still a challenge for implementation of the IR 4.0 energy consuming technologies.

Jobs and Inequality

This is a second and most critical factor for the sustainability of industry 4.0 in emerging and developed countries as well. Due to increased automation, the world of work is changing. Job creation, job destruction, increase productivity, and production process are few challenges or benefits from digitalization.

For example, approximately more than 40% and more than 50% of occupations can be replaced by automation and computerization in the United States and Europe, respectively (Kovacs, 2018). According to UNCTAD, jobs are positively and negatively affected by Industry 4.0 technologies (see in Table 2). Countries, such as South Africa are already being affected by the unemployment rate of more than 25% (Manda & Ben Dhaou, 2019) (Davos, 2016). Major transformations in industrial landscape occurring due to technological changes and upgrades are expected to affect jobs significantly worldwide, resulting in both creations as well as displacement of various job domains (World Economic Forum, 2017).

New jobs require new kinds of skills, and labor productivity is also influenced by the industry 4.0 technologies, e.g., fast and efficient operations between manufacturing steps as well as enhanced Research & Development process (for e.g., using technology such as 3D-printing) (Arthur, 2017). As a result, there is an increasing need by governments to ensure that the IR 4.0 benefits are reaped among cities and it should not aid the unemployment of youth especially in the low- or middle-income economies (World Economic Forum, 2017). Job inequality does not only emerge from the job destruction, but also from the digital divide in a country. An incessant need to keep up with the changing pace of technology is required by the emerging economies to tackle this issue of the digital divide resulting in job inequality.

Table 2. Estimated impact of Industry 4.0 technologies on jobs Source: UNCTAD

Estimate	Time frame	Technology	Study
47 percent of total United States employment at high risk of being automated	10-20 years	Artificial intelligence and robotics	Frey and Osborne, 2017
9 percent of total employment in the United States and 21 countries of the Organization for Economic Cooperation (OCED) and Development at high risk of being automated	10-20 years	Artificial intelligence and robotics	Arntz et al., 2016, 2017
50 percent of today's work activities worldwide could be automated	by 2055	Artificial intelligence and robotics	McKinsey Global Institute, 2017
8.5 percent of the global manufacturing workforce, mostly in lower-income regions of major economies, could become redundant	20 years	Industrial robotics	Oxford Economics, 2019

Skills for New Technologies

IR 4.0 requires a highly skilled workforce, but a number of developing countries are struggling to produce a highly skilled labor force. For example, in South Africa, the work-force consists of approximately 30% unskilled, more than 40% semi-skilled and approximately 23% of skilled workers (Davos, 2016). Additionally, this is one of the key aspects of industry 4.0 to put more focus on the people/employee who works in the industry. As we already discuss in the previous issue that if new jobs are created then people have to learn new skill sets and competencies. Additionally, IR 4.0 technologies and their implementation require specialist skills as well as basic digital awareness or knowledge. According to a research by PwC Russia, Singapore ranked first among the countries to implement and use new technological initiatives, but only approximately more than 40% of its citizens showed readiness or acceptance to technological use. Technological innovation is one of the primary factors to effect workspaces, skilled workforce and their demands, and the occupational structure (Kergroach, 2017). It is very hard to learn a new skill set for employees who are already working in industries in many years. Additionally, if people are ready to learn technologies or skills then this may also lead to a slow process in manufacturing or non-manufacturing companies.

According to World Economic Forum (2017), skills mismatch and unemployment occurred due to the fluctuating environment of employments resulting from the developments made in technological and manufacturing processes.

Resistance of People

Resistance from the personnel or people is one of the most common problems in organizations or economies while implementing or introducing new technological innovations. Technological innovation is the underlying foundation of most of the industries for their business model development. It doesn't matter, how efficient the technology is, what matters is the acceptance of the employees to work on that technology. Sometimes industries have to face reduced efficiency of its employees, or outright denial to implement new technologies. It is human tendency to work in their comfort zone, which aids the employee's expectation of a failed new system rather than learning and gaining a new skill (Fontaine et al.,

2016). Organizations or employees both need to continuously advance their skills, tools and techniques and processes to sustain in the dynamic market environment.

Implementing new technology involves implementing new strategies or plans of action, which would demand a change in working condition, style, supervision personnel, position, structure as well as resources (Verma & Sharma, 2016). Such changes are difficult for employees to accept and adapt and hence comes the resistance, one of the great sustainability challenges for industry 4.0 technologies.

Regulatory Environment

This is the last and critically important factor for the implementation of industry 4.0 in emerging economies. There is a number of researchers and executives who argue that they don't need a new industry policy but demand a better regulatory framework. Governments need to establish evident, and flexible policies, protocols, and principles (World Economic Forum, 2017). Supported industry policy and regulatory environment to industry 4.0 this is a way to sustain automation in emerging economies. Recently, there is a number of developing countries, such as Vietnam, that have created their own industrial policy. Various sectors are forecasted to be affected by the impact of IR 4.0, which demand new policies for transforming the traditional industrial landscape in emerging economies.

DISCUSSION AND CONCLUSION

Implementation of IR 4.0 technologies in the industrial sector has increased the structural as well as non-structural challenges in emerging economies. Therefore, it is aided in the delayed implementation and adoption of IR 4.0 among such economies to leverage its technological infrastructure (Arifiani & Arifiani, 2019). However, various emerging countries, specifically South-west (Brazil, Nigeria) as well as Asian (India, Indonesia, Malaysia), despite its challenges have proved to be the front-runners in leveraging IR 4.0 (Berawi, 2018; Ezenwa, 2018; Kamarul Bahrin et al., 2016).

Based on the literature survey, we found that developed and emerging economies have their own opportunities and benefit. But in this chapter, we considered the emerging economies because it has a fast-economic growth rate. Recent reviews about industry 4.0 show that, even though a plethora of literature is available on the basic concepts, challenges, issues, implementation of the IR 4.0 concept, the sustainability issues of IR 4.0 in emerging economies are not well researched upon. This chapter has made an attempt to fill this gap, firstly by discussing the current status of industry 4.0 in developing and advanced economies and their challenges. The success of the IR 4.0 will not depend on single factors but also needs a collective effort of each and every player that plays a critical role in its implementation. The chapter has presented some significant sustainability challenges of IR 4.0 that are related to technological and societal. Technological Challenges, such as digital culture, scalability, integration & interoperability, standardization, information privacy & security real-time analysis; and Societal challenges such as energy intensity technologies, jobs and inequality, skills for new technologies, resistance to change and policy and regulatory environment are discussed. Technological and societal challenges discussed are intertwined together in the context of IR 4.0. Each issue is in some manner the result of another one.

IR 4.0 is an innovative concept, which is accepted and implemented by developed countries to a much higher extent when compared to the emerging (or developing) economies. However, to leverage the benefits of IR 4.0, based on Prevention through Design (PtD) approach (Borchiellini et al., 2013;

Labagnara et al., 2013), emerging economies need to define and design regulatory and industry policies for implementation of IR 4.0 in their business landscape. Societal acceptance of industry 4.0 technologies is the only source of sustainability in developing countries.

REFERENCES

Almada-Lobo, F. (2016). The Industry 4.0 revolution and the future of Manufacturing Execution Systems (MES). *Journal of Innovation Management*, *3*(4), 16–21. doi:10.24840/2183-0606_003.004_0003

Arifiani, L., & Arifiani, L. (2019). The effect of disruption technology, opportunities and challenges of telecommunication industry 4.0 in Indonesia. *International Journal of Recent Technology and Engineering*, *7*(6), 808–819. doi:10.35940/ijeat.F1040.0986S319

Arthur, W. B. (2017). Where is technology taking the economy? *McKinsey Quarterly*. https://usblearn.belpark.sun.ac.za/pluginfile.php/44861/mod_resource/content/0/7b - Arthur %282017%29 Where is technology taking the economy.pdf

Berawi, M. A. (2018). Utilizing Big Data in Industry 4.0: Managing Competitive Advantages and Business Ethics. *International Journal of Technology*, *9*(3), 430. doi:10.14716/ijtech.v9i3.1948

Bi, Z., & Cochran, D. (2014). Big data analytics with applications. *Journal of Management Analytics*, *1*(4), 249–265. doi:10.1080/23270012.2014.992985

Borchellini, R., Cardu, M., Colella, F., Labagnara, D., Martinetti, A., Patrucco, M., . . . Verda, V. (2013). A Prevention through Design Approach for the Environmental S&H Conditions and the Ventilation System at an Italian Underground Quarry. *Chemical Engineering Transactions, AIDIC - The Italian Association of Chemical Engineering, 32*, 181-186. doi:10.3303/CET1332031

Brettel, M., Friederichsen, N., Keller, M., & Rosenberg, M. (2014). How Virtualization, Decentralization and Network Building Change the Manufacturing Landscape: An Industry 4.0 Perspective. *International Journal of Science, Engineering and Technology*, *8*(1), 37–44.

Cellary, W. (2019). Non-technical Challenges of Industry 4.0. In Advances in Information and Communication Technology (Vol. 568, pp. 245–258). Springer International Publishing. doi:10.1007/978-3-030-28464-0_1

Cezarino, L. O., Liboni, L. B., Oliveira Stefanelli, N., Oliveira, B. G., & Stocco, L. C. (2019). Diving into emerging economies bottleneck: Industry 4.0 and implications for circular economy. *Management Decision*, MD-10-2018-1084. doi:10.1108/MD-10-2018-1084

Da Xu, L., Xu, E. L., & Li, L. (2018). Industry 4.0: State of the art and future trends. *International Journal of Production Research*, *56*(8), 2941–2962. doi:10.1080/00207543.2018.1444806

Datta, S. (2017). Emergence of Digital Twins Is this the march of reason? *Journal of Innovation Management*, *5*(3), 14–33. doi:10.24840/2183-0606_005.003_0003

Davos. (2016). Emerging economies are less likely to benefit from fourth industrial revolution. *The Economic Times*.

Díaz, M., Martín, C., & Rubio, B. (2016). State-of-the-art, challenges, and open issues in the integration of Internet of things and cloud computing. *Journal of Network and Computer Applications*, *67*, 99–117. doi:10.1016/j.jnca.2016.01.010

Ezenwa, O. (2018). Effect of competitive intelligence on competitive advantage in Innoson technical and industry limited, Enugu state, Nigeria. *International Journal of Business, Economics & Management*. doi:10.31295/ijbem.v1n1.25

Fernández-Caramés, T. M., Blanco-Novoa, O., Froiz-Míguez, I., & Fraga-Lamas, P. (2019). Towards an Autonomous Industry 4.0 Warehouse: A UAV and Blockchain-Based System for Inventory and Traceability Applications in Big Data-Driven Supply Chain Management. *Sensors (Basel)*, *19*(10), 2394. doi:10.339019102394 PMID:31130644

Fontaine, O., Martinetti, A., & Michaelides-Mateou, S. (2016). Remote pilot aircraft system (RPAS): Just culture, human factors and learnt lessons. In *Chemical engineering transactions, AIDIC* (pp. 205–210). The Italian Association of Chemical Engineering. doi:10.3303/CET1653035

Gibbons, J. (2009). Debunking the myths surrounding sustainability. *The Irish Times*.

Grody, A. D. (2018). Rebuilding Financial Industry Infrastructure. *Journal of Risk Management in Financial Institutions*, *11*(1), 34–46.

Hopali, E., & Vayvay, Ö. (2018). Industry 4.0 as the last industrial revolution and its opportunities for developing countries. *Analyzing the Impacts of Industry 4.0 in Modern Business Environments*, 65–80. doi:10.4018/978-1-5225-3468-6.ch004

Huang, J., Kong, L., Chen, G., Wu, M. Y., Liu, X., & Zeng, P. (2019). Towards secure industrial iot: Blockchain system with credit-based consensus mechanism. *IEEE Transactions on Industrial Informatics*, *15*(6), 3680–3689. doi:10.1109/TII.2019.2903342

Hunsaker, B. T., & Youngdahl, W. E. (2018). *Intel: AI and Industry 4.0 Strategy*. Thunderbird School of Global Management, HBR.

Kamarul Bahrin, M. A., Othman, M. F., Nor Azli, N. H., & Talib, M. F. (2016). Industry 4.0: A review on industrial automation and robotic. *Jurnal Teknologi*, *78*(6-13), 6–13. doi:10.11113/jt.v78.9285

Kergroach, S. (2017). Industry 4.0: New Challenges and Opportunities for the Labour Market. *Foresight and STI Governance*, *11*(4), 6–8. doi:10.17323/2500-2597.2017.4.6.8

Kim, D., Lee, H., & Kwak, J. (2017). Standards as a driving force that influences emerging technological trajectories in the converging world of the Internet and things: An investigation of the M2M/IoT patent network. *Research Policy*, *46*(7), 1234–1254. doi:10.1016/j.respol.2017.05.008

Kovacs, O. (2018). The dark corners of industry 4.0 – Grounding economic governance 2.0. *Technology in Society*, *55*, 140–145. doi:10.1016/j.techsoc.2018.07.009

Kumar, V., Verma, P., Sharma, R. R. K., & Khan, A. F. (2017). Conquering in emerging markets: Critical success factors to enhance supply chain performance. *Benchmarking: An International Journal*, *24*(3), 570–593. doi:10.1108/BIJ-05-2016-0078

Kunst, R., Avila, L., Binotto, A., Pignaton, E., Bampi, S., & Rochol, J. (2019). Improving devices communication in Industry 4.0 wireless networks. *Engineering Applications of Artificial Intelligence, 83*, 1–12. doi:10.1016/j.engappai.2019.04.014

Labagnara, D., Martinetti, A., & Patrucco, M. (2013). Tunneling operations, occupational S&H and environmental protection: A Prevention through Design approach. *American Journal of Applied Sciences, 11*(11), 1371–1377. doi:10.3844/ajassp.2013.1371.1377

Lasi, H., Fettke, P., Kemper, H.-G., Feld, T., & Hoffmann, M. (2014). Industry 4.0. *Business & Information Systems Engineering*, *6*(4), 239–242. doi:10.100712599-014-0334-4

Laszka, A., Dubey, A., Walker, M., & Schmidt, D. (2017). Providing privacy, safety, and security in IoT-based transactive energy systems using distributed ledgers. *7th International Conference on the Internet of Things, IoT 2017*. 10.1145/3131542.3131562

Lee, E. (2015). The Past, Present and Future of Cyber-Physical Systems: A Focus on Models. *Sensors (Basel), 15*(3), 4837–4869. doi:10.3390150304837 PMID:25730486

Lin, C. C., Deng, D. J., Chen, Z. Y., & Chen, K. C. (2016). Key design of driving industry 4.0: Joint energy-efficient deployment and scheduling in group-based industrial wireless sensor networks. *IEEE Communications Magazine*, *54*(10), 46–52. doi:10.1109/MCOM.2016.7588228

Lu, Y. (2017). Industry 4.0: A survey on technologies, applications and open research issues. *Journal of Industrial Information Integration*, *6*, 1–10. doi:10.1016/j.jii.2017.04.005

Luthra, S., & Mangla, S. K. (2018). Evaluating challenges to Industry 4.0 initiatives for supply chain sustainability in emerging economies. *Process Safety and Environmental Protection, 117*, 168–179. doi:10.1016/j.psep.2018.04.018

Manda, M. I., & Ben Dhaou, S. (2019). Responding to the challenges and opportunities in the 4th Industrial revolution in developing countries. *Proceedings of the 12th International Conference on Theory and Practice of Electronic Governance - ICEGOV2019*, 244–253. 10.1145/3326365.3326398

Mangla, S. K., Luthra, S., Jakhar, S., Gandhi, S., Muduli, K., & Kumar, A. (2020). A step to clean energy - Sustainability in energy system management in an emerging economy context. *Journal of Cleaner Production*, 242.

Martinetti, A., Braaksma, A. J. J., & van Dongen, L. A. M. (2017). Beyond RAMS Design: Towards an Integral Asset and Process Approach. In L. Redding, R. Roy, & A. Shaw (Eds.), Advances in Through-life Engineering Services (pp. 417-428). Springer. doi:10.1007/978-3-319-49938-3_25

Martinetti, A., Chatzimichailidou, M., Maida, L., & van Dongen, L. A. M. (2018). Safety I-II, Resilience and Antifragility Engineering: A debate explained through an accident occurred on a Mobile Elevating Work Platform. *International Journal of Occupational Safety and Ergonomics*, *25*(1), 66–75. doi:10.1080/10803548.2018.1444724 PMID:29473459

McNamara, J. E. (2014). Interface Standards. *Technical Aspects of Data Communication*, 17–36. doi:10.1016/b978-1-4831-8400-5.50007-3

Melnyk, S. A., Flynn, B. B., & Awaysheh, A. (2018). The best of times and the worst of times: Empirical operations and supply chain management research. *International Journal of Production Research, 56*(1–2), 164–192. doi:10.1080/00207543.2017.1391423

Merlino, G., Bruneo, D., Distefano, S., Longo, F., Puliafito, A., & Al-Anbuky, A. (2015). A Smart City Lighting Case Study on an OpenStack-Powered Infrastructure. *Sensors (Basel), 15*(7), 16314–16335. doi:10.3390150716314 PMID:26153775

Müller, J. M. (2020). Industry 4.0 in the Context of the Triple Bottom Line of Sustainability. In Customer Satisfaction and Sustainability Initiatives in the Fourth Industrial Revolution (pp. 1–20). IGI Global.

Nkamnebe, A. D. (2011). Sustainability marketing in the emerging markets: Imperatives, challenges, and agenda setting. *International Journal of Emerging Markets, 6*(3), 217–232. doi:10.1108/17468801111144058

Onik, M. M. H., Kim, C.-S., & Yang, J. (2019). Personal Data Privacy Challenges of the Fourth Industrial Revolution. *21st International Conference on Advanced Communication Technology, ICACT 2019,* 635–638. 10.23919/ICACT.2019.8701932

Oztemel, E., & Gursev, S. (2020). Literature review of Industry 4.0 and related technologies. *Journal of Intelligent Manufacturing, 31*(1), 127–182. doi:10.100710845-018-1433-8

Palattella, M. R., Dohler, M., Grieco, A., Rizzo, G., Torsner, J., Engel, T., & Ladid, L. (2016). Internet of Things in the 5G Era: Enablers, Architecture, and Business Models. *IEEE Journal on Selected Areas in Communications, 34*(3), 510–527. doi:10.1109/JSAC.2016.2525418

Pfeffer, J. (2010). Building Sustainable Organizations: The Human Factor Executive Overview. *The Academy of Management Perspectives, 24*(1), 34–45.

Preuveneers, D., & Ilie-Zudor, E. (2017). The intelligent industry of the future: A survey on emerging trends, research challenges and opportunities in Industry 4.0. *Journal of Ambient Intelligence and Smart Environments, 9*(3), 287–298. doi:10.3233/AIS-170432

Reyes Garcia, J. R., Martinetti, A., Jauregui Becker, J. M., Singh, S., & van Dongen, L. A. M. (2019). Towards an Industry 4.0-Based Maintenance Approach in the Manufacturing Processes. In V. González-Prida Diaz & J. P. Zamora Bonilla (Eds.), *Handbook of Research on Industrial Advancement in Scientific Knowledge* (pp. 135–159). IGI Global. doi:10.4018/978-1-5225-7152-0.ch008

Roblek, V., Meško, M., & Krapež, A. (2016). A Complex View of Industry 4.0. *SAGE Open, 6*(2), 215824401665398. doi:10.1177/2158244016653987

Sadeghi, A. R., Wachsmann, C., & Waidner, M. (2015). Security and privacy challenges in industrial Internet of Things. *Proceedings - Design Automation Conference,* 1–6. 10.1145/2744769.2747942

Santos, M. Y., Oliveira e Sá, J., Andrade, C., Vale Lima, F., Costa, E., Costa, C., ... Galvão, J. (2017). A Big Data system supporting Bosch Braga Industry 4.0 strategy. *International Journal of Information Management, 37*(6), 750–760. doi:10.1016/j.ijinfomgt.2017.07.012

Saweros, E., & Song, Y.-T. (2019). Connecting heterogeneous electronic health record systems using tangle. In *13th International Conference on Ubiquitous Information Management and Communication, IMCOM 2019* (Vol. 935, pp. 858–869). 10.1007/978-3-030-19063-7_68

Schulz, K. F., & Freund, D. (2019). A multichain architecture for distributed supply chain design in industry 4.0. In *21st International Conference on Business Information Systems, BIS 2018* (*Vol. 339*, pp. 277–288). 10.1007/978-3-030-04849-5_25

Sheth, J. N. (2011). Impact of Emerging Markets on Marketing: Rethinking Existing Perspectives and Practices. *Journal of Marketing*, *75*(4), 166–182. doi:10.1509/jmkg.75.4.166

Singh, A., Payal, A., & Bharti, S. (2019). A walkthrough of the emerging IoT paradigm: Visualizing inside functionalities, key features, and open issues. *Journal of Network and Computer Applications*, *143*(February), 111–151. doi:10.1016/j.jnca.2019.06.013

Sisinni, E., Saifullah, A., Han, S., Jennehag, U., & Gidlund, M. (2018). Industrial internet of things: Challenges, opportunities, and directions. *IEEE Transactions on Industrial Informatics*, *14*(11), 4724–4734. doi:10.1109/TII.2018.2852491

Stock, T., & Seliger, G. (2016). Opportunities of Sustainable Manufacturing in Industry 4.0. *Procedia CIRP*, *40*, 536–541. doi:10.1016/j.procir.2016.01.129

Tortorella, G. L., & Fettermann, D. (2018). Implementation of industry 4.0 and lean production in brazilian manufacturing companies. *International Journal of Production Research*, *56*(8), 2975–2987. doi:10.1080/00207543.2017.1391420

Tronchoni, D., & Brennan, G. (2017). Industry 4.0: A Key Enabler of the Circular Economy. *Climate Innovation Insights*, *2*(7), 2–5. http://edinn.com/dwn/Insights_2-7_Industry_4 0_EU.pdf

Ulz, T., Pieber, T., Steger, C., Haas, S., Bock, H., & Matischek, R. (2017). Bring your own key for the industrial Internet of Things. *2017 IEEE International Conference on Industrial Technology, ICIT 2017*, 1430–1435. 10.1109/ICIT.2017.7915575

Verma, P., & Sharma, R. R. K. (2016). Structure, Evaluation System, and Employee Resistance: Impediments to Horizontal Strategy Implementation in Diversified Corporations. *Sixth International Conference on Industrial Engineering and Operations Management (IEOM).*

Verma, P., Sharma, R. R. K., & Kumar, V. (2017). The sustainability issues of diversified firms in emerging economies context: A theoretical model and propositions. *International Journal of Process Management and Benchmarking*, *7*(2), 224. doi:10.1504/IJPMB.2017.083107

Vyatkin, V., Salcic, Z., Roop, P., & Fitzgerald, J. (2007). Now That's Smart! *IEEE Industrial Electronics Magazine*, *1*(4), 17–29. doi:10.1109/MIE.2007.909540

Wang, S., Wan, J., Zhang, D., Li, D., & Zhang, C. (2016). Towards smart factory for industry 4.0: A self-organized multi-agent system with big data based feedback and coordination. *Computer Networks*, *101*(June), 158–168. doi:10.1016/j.comnet.2015.12.017

Webster, G. D., Harris, R. L., Hanif, Z. D., Hembree, B. A., Grossklags, J., & Eckert, C. (2018). Sharing is Caring: Collaborative Analysis and Real-Time Enquiry for Security Analytics. *11th IEEE International Congress on Conferences on Internet of Things, 14th IEEE International Conference on Green Computing and Communications, 11th IEEE International Conference on Cyber, Physical and Social Computing, 4th IEEE International Conference*, 1402–1409. 10.1109/Cybermatics_2018.2018.00240

World Economic Forum. (2017). Harnessing the fourth industrial revolution for sustainable emerging cities. *Fourth Industrial Revolution for the Earth Series*. http://www3.weforum.org/docs/WEF_Harnessing_the_4IR_for_Sustainable_Emerging_Cities.pdf

Zhang, P., White, J., & Schmidt, D. (2018). Architectures and patterns for leveraging high-frequency, low-fidelity data in the healthcare domain. *6th IEEE International Conference on Healthcare Informatics, ICHI 2018*, 463–464. 10.1109/ICHI.2018.00104

Zhou, K., Liu, T., & Zhou, L. (2015). Industry 4.0: Towards future industrial opportunities and challenges. *2015 12th International Conference on Fuzzy Systems and Knowledge Discovery (FSKD)*, 2147–2152. 10.1109/FSKD.2015.7382284

Chapter 3
Evolution of Maintenance Processes in Industry 4.0

Adithya Thaduri
https://orcid.org/0000-0002-1938-0985
Luleå University of Technology, Sweden

Stephen Mayowa Famurewa
Luleå University of Technology, Sweden

ABSTRACT

Several industries are looking for smart methods to increase their production throughput and operational efficiency at the lowest cost, reduced risk, and reduced spending of resources considering demands from stakeholders, governments, and competitors. To achieve this, industries are looking for possible solutions to the above problems by adopting emerging technologies. A foremost concept that is setting the pace and direction for many sectors and services is Industry 4.0. The focus is on augmenting machines and infrastructure with wireless connectivity, sensors, and intelligent systems to monitor, visualize, and communicate incidences between different entities for decision making. An aspect of physical asset management that has been enormously influenced by the new industrial set-up is the maintenance process. This chapter highlights the issues and challenges of Industry 4.0 from maintenance process viewpoint according to EN 60300-3-14. Further, a conceptual model on how maintenance process can be integrated into Industrial 4.0 architecture is proposed to enhance its value.

INTRODUCTION

Several industries are looking for intelligent systems, smart methods and functional processes to increase their production throughput and operational efficiency at the lowest cost. At the same time there is a steady need to reduce operational risk and product or service quality considering demands from stakeholders, governments and competitors. In this process, these industries suffer from operational flaws, human errors, systematic failures and process ineffectiveness leading to unanticipated delays in production and other negative incidences. To reduce these technical and operational deficiencies as well as improve their

DOI: 10.4018/978-1-7998-3904-0.ch003

productivity, industries are looking for possible solutions to the above problems by adopting emerging technologies. A foremost concept that is setting the pace and direction for many sectors and services is industry 4.0. The focus is on augmenting machines, infrastructure and systems with wireless connectivity and sensors to monitor, visualize and communicate incidences between different entities for decision making. This new approach is essential for competitiveness in current industrial set up and for assuring a successful business enterprise.

This new trend entails the use of well - designed technologies process, and models in the form of internet of things (IoT), cyber-physical systems (CPS), cloud computing, Big Data and artificial intelligence (AI) to facilitate data exchange and automation. This disruptive revolution has caused substantial evolution in physical asset management. An aspect of physical asset management that has been enormously influenced by the new industrial set-up is the maintenance process. Maintenance practices, perception and prospect have been influenced in different industrial context, thus there is a need to adopt the concept of maintenance process according to the international standard EN 60300-3-14 into Industry 4.0 framework. This book chapter gives an overview on the evolution of maintenance function and the role of maintenance process in the bigger picture of industry 4.0 revolution. It also highlights the issues and challenges of maintenance process within the context of this new business approach. Some of the drawbacks of developing an industry 4.0 solution without adequate emphasis on maintenance process will be discussed. Further, some essential features and assisting technologies of industry 4.0 will be discussed with interest on how they can be used to connect the various elements of maintenance process. These elements include maintenance management, planning, preparation, execution, assessment and improvement according to maintenance standards. The purpose, characteristics and contents of each of these maintenance process elements differs thus there is a need to adequately investigate how they can be addressed in industry 4.0 context. This book chapter also gives a conceptual model on how maintenance process can be integrated into industrial 4.0 architecture. This conceptual model will support a seamless integration of operation and maintenance processes to facilitate effective and efficient maintenance decision making.

Concept of Maintenance

Maintenance is a function that combines technical, administrative and management actions intended to retain an item in, or restore it to, a state in which it can perform as required (CEN, 2001). It is necessary for all physical engineering asset that are intended to add value to an organization or individual to be maintained in relation to its value creation capability. The maintenance function can be defined as "activities for retaining a system in an operating state or restoring it to a state that is considered necessary for its operation and utilization". Hence, the important step in the effective management of the maintenance process is the precise identification of the need of maintenance, that is further demanded by the present and future state of the machine, and the necessary actions that need to be taken to restore it or retain it in an operating condition (Kumar, 2008). These activities cover the period from the creation of an asset to the end of its life. This is against the common misunderstanding of maintenance, where it's only limited to the operation phase of an asset life cycle (Ben-Daya, Kumar, & Murthy, 2016).

With recent advancements in technology, modern systems have become massive in size, extensive in functionality, complex in configuration and connected for automation. The sustainability and dependability demand on such systems are on the increase thus created need for new processes, technology, models and solutions for effective and efficient maintenance of the complex systems. This need is associated to competitive business environment and extremely increasing cost (direct and consequential cost) of unavailability and poor service quality. Further, breakdowns or poor equipment performance also result in a loss of product or service quality. It could also negatively impair safety, health and environment aspects of physical asset management goals (Bream, 2006). Prevention of these consequential incidences have become mandatory to survive the competitive nature of production, manufacturing and service delivery. In short, these are some of the driving forces for the evolution of maintenance discipline to becoming an indispensable area in engineering and technology.

Evolution of Maintenance

In general, the perception and implementation of maintenance in different industries have evolved from unplanned to planned maintenance, fail and fix to monitor and maintain, reactive to proactive maintenance, corrective to preventive maintenance, time based to condition based maintenance, inspection to predictive maintenance, prognostics to prescriptive maintenance. Figure 1 exemplifies the chronological development of the business of maintenance through the years (Lee & Wang, 2008).

Figure 1. Evolution of maintenance concept with time [Adapted from (Lee & Wang, 2008)]

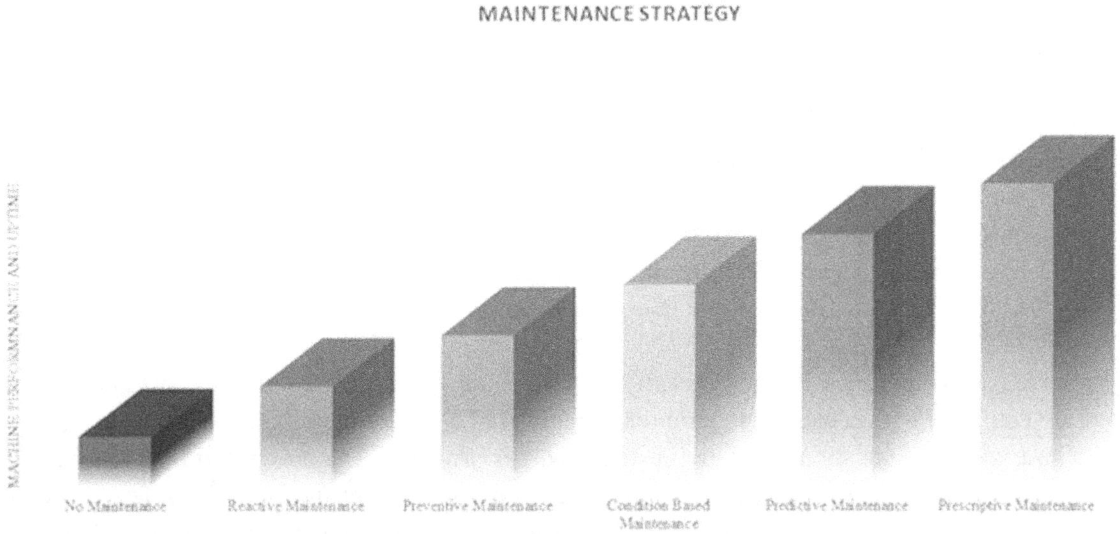

In the past maintenance was perceived as a necessary evil and treated as a dirty, non-value adding and unplanned job. It's neither seen as core function as production/operation nor recognized as a key component of revenue generation and business goal achievement. The paradigm overview in Figure 1 starts with No maintenance scenario where there is neither no way to fix the fault or no finance to fix it (i.e. it is cost effective to discard the failed system or component).

For reactive maintenance, the focus of reactive maintenance is run to failure and then "fix it after it's broken". Basically, knowledge of the equipment degradation behavior is lacking thus, little to no maintenance is performed and the machinery operates up until a failure occurs making maintenance a fire-fighting function. The competitive nature of industrial operation and environmental/safety issues have rendered this approach inappropriate.

Preventive maintenance is a policy which adapts a measure of time such as calendar or machine time, cycles, distance travelled, number of products for triggering replacement, overhauling or adjustment of an item. It is done at fixed or varying intervals, irrespective of its condition at that time. There exists different modifications and adaptation of this policy. This approach is no longer considered efficient especially for large and complex systems and continuous operations where remaining useful life is considered important.

Condition based maintenance is based on the failure limit policy in which maintenance is executed only when a feature describing the health of a unit reaches a predetermined level. This policy requires condition monitoring technology to control certain performance indices periodically or continuously. Maintenance is triggered whenever a feature value crosses predefined threshold to restore the machine to its original state or satisfactory level in comparison to the threshold. It analyzes the current level of measured physical parameters against established engineering limits for the objective of detecting, analyzing and correcting a problem before the occurrence of failure. This method gives information required for failure diagnostics, maintenance planning, and thereby reducing unexpected operating costs and loss. However, the lead time for maintenance is often too short for effective operation and maintenance decision making.

Predictive maintenance is a concept that is rapidly evolving along with the development in data technology. The core is seamless integration of maintenance diagnosis and prognosis of machine health via relevant communication technology. Trend of system behavior and the degradation pattern is very important aspect in this concept. Specifically, there are three main aspects of this concept:

- Sensor technology and intelligent applications to enable the systems monitor, predict, and optimize their performance in an intelligent way.
- Philosophical inclination towards "failure proactive" and not "failure reactive" i.e. capability to find underlying conditions that can lead to machine faults and degradation.
- Effective reporting to complete the maintenance loop by feeding the maintenance information back to re-design, modification and maintenance improvements.

Prescriptive maintenance is a new design and maintenance concept. In this approach, engineering assets are equipped with capability to monitor, diagnose, be aware of their state and propose repair action in case there is indication of failure or degraded performance. The prescribed intervention takes several factors into consideration, such as operating environment, business goals, safety standards, decision scenarios, maintenance context. In some instances, systems are expected to restore themselves to perfect state or give actionable information to maintenance robot or send information directly to the appropriate maintenance service provider. In other instances, prescriptive maintenance implements self-repair by recovering the required function of a degrading system through a tradeoff for other functions. The required capabilities of prescriptive maintenance policy with self-maintenance include the following: monitoring capability, fault judging capability, diagnosing capability, prescriptive capability, repair executing capability, self-learning and improvement (Lee & Wang, 2008). An integral aspect of this approach is automated service trigger function that enables a system to initiate service request after prescription and before a failure occurs, this task can be eventually carried out by a maintenance team.

MAINTENANCE PROCESS

Understanding maintenance function, maintenance support and maintenance process is a pre-requisite for the development of effective operation and maintenance related solutions. Inadequate consideration to maintenance process is one of the major reasons for unsuccessful implementations of several emerging technologies in the field of engineering asset management. Emerging maintenance technologies and methodologies should be connected to necessary actions and support resources required under given conditions. For such technologies to have expected impact especially at the shop floor level, it should give actionable information that can guide the maintenance execution. In addition, it should have clear linkage with necessary resources such as human resources, support equipment, materials and spare parts, maintenance facilities, documentation, information and maintenance information systems.

A well-designed maintenance process can be implemented in the emerging industry 4.0 framework to ensure consistent application of maintenance and maintenance support (Kans, Galar, & Thaduri, 2016). The other recognized applications of maintenance process in the context of industry 4.0 for smart manufacturing (Thoben, Wiesner, & Wuest, 2017), to improve preventive maintenance (Wan et al., 2017), for self-aware machines (Bagheri, Yang, Kao, & Lee, 2015), for process monitoring (Reis & Gins, 2017), for smart grid (Batista, Melício, & Mendes, 2017), fault diagnosis and prognosis (Li, Wang, & Wang, 2017), smart maintenance and logistics (Rakyta, Fusko, Herčko, Závodská, & Zrnić, 2016) and human centric maintenance process (Fantini, Pinzone, & Taisch, 2018) and for linear assets (Seneviratne, Ciani, Catelani, & Galar, 2018).

This will not only make the technology complete in terms of maintenance but will also raise its value-addition to organizational goals. A maintenance process to be integrated into and assured by the promising industry 4.0 technology should include these essential elements: maintenance management, planning, preparation, execution, assessment and improvement. A general description of the constituting elements of this process is shown in Figure 2 and explained thereafter according to the existing European standard (EN 60300-3-14, 2004). It should be noted that the contents and underlying activities of each elements might vary depending on organizational structures, needs, values, processes, strategies and priorities. Therefore, it is necessary to adapt or redesign this generic process to fit the organizational environment in which the technology is to be applied.

Figure 2. Maintenance process (EN 60300-3-14, 2004)

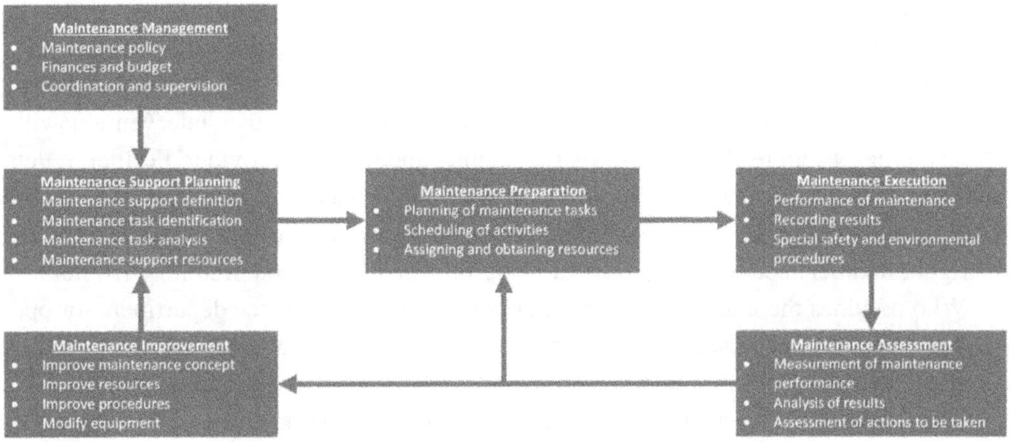

Maintenance Management

There is a need to consider dynamic maintenance management information, concept and activities as an integral aspect of industry 4.0. Some aspects of maintenance management are still left of the current framework and model for industry 4.0. The management activities and its associated information required to assure effective and efficient maintenance in any industry 4.0 deployment should include the following:

- Development of maintenance policy for plant, system and critical components
- Development of finances and budget for maintenance
- Design of maintenance coordination and supervision structures
- Acquisition of adequate information on processes, methods and procedures relevant to the above-mentioned activities.
- Collection of information on operation and maintenance organizational structures.

Maintenance and Maintenance Support Planning

In the development of maintenance centered disruptive technology such as industry 4.0, it is essential to cover the planning aspect of maintenance. This will assist in establishing maintenance concept for critical items at the appropriate life cycle phase. The planning activities and the associated information required in an integrated industry 4.0 deployment should include the following:

- Maintenance task identification using manufacturer's recommendation, conventional approaches (e.g. in-house experience) or structured approaches (e.g. RCM)
- Maintenance task analysis to determine the essential information and required resources for each maintenance significant item. These includes:
    - Description of the maintenance task as required for self-reconfiguration or for intervention by maintenance personnel
    - Task interval in relevant time measure, i.e. elapsed time, operational cycles or distance
    - Task times and skills of personnel required to perform the tasks
    - Maintenance, safety and handling procedures
    - Specials tools, test equipment and support equipment spare parts, materials and consumables required;
    - Observations and recordings to be made;
    - Test and checkout procedures to verify successful completion of the
    - Maintenance task
- Identification of maintenance support resources to determine whether intervention will be carried out by the system itself, robots or by the maintenance service provider. Further, it determines which maintenance organizational level will provide the service required for certain system/items. Items may be maintained on-site, at a local repair shop or by an external repair facility. In case of self-trigger maintenance planning, the following information are required for each item:
    - Who provides the maintenance: internal or external maintenance department, or operators;
    - Who provides the spare parts, materials and consumables, and the information about their stock level;
    - Who provides the special tools, transportation, lifting, testing and support equipment;

   - Who provides the condition monitoring equipment and the associated information management system;

Maintenance Preparation

This is an aspect of maintenance that should be given due attention when developing operation and maintenance enablers. The following activities are important in order to create enough lead time to plan and supply the necessary resources in an I4.0 implementation:

- Identifying and assigning personnel or robots
- Acquiring materials and spare parts
- Ensuring tools, transportation, lifting and support equipment are available
- Preparing workplans and callout procedures for all intervention
- Preparing required operating, maintenance, safety and environmental procedures and
- Providing necessary personnel training and robot calibration
- Scheduling of activities, based on a priority system to ensure the most urgent works are carried out first by either available crew or robots

Maintenance execution

This perspective is not fully addressed in the current frameworks and models for industry 4.0. The execution of maintenance tasks either by robots, operating machine, internal or external maintenance personnel should be performed with due attention to the technical procedure for isolation, disassembling, cleaning, repairing, refurbishing, replacing and testing. For effective integration of this element of maintenance process into an industry 4.0 framework, it is essential to address the following activities and information:

- Recording of observations, readings, measurements, tasks carried out and resources used.
- Gathering of technical data and task description
- Obtaining spare parts, tools and support equipment;
- Recording task times including travel time to the worksite and active maintenance time
- Preparation of the worksite such as equipment shutdown, isolating and lockout
- Gathering of information related to testing, checkout and clearing of worksite;

Maintenance Assessment

The assessment of maintenance interventions should be performed either at the point of intervention or periodically to review overall maintenance performance. A standard method should be established for analyzing the performance of maintenance in terms of logistic, economic, dependability and quality performance of all maintenance efforts. The results could be used to identify and justify improvements. The following activities are considered relevant for this element of maintenance process:

- Measurement of maintenance performance and analysis of result
- Assessment of actions to be taken
- Assessment of maintenance effectiveness

- Assessment of technical aspects of the maintenance task- resource adequacy, effectiveness of operating, safety and environmental procedures.
- Overall review of maintenance procedure in terms of reworks, trends related to operating conditions, design and manufacturing quality problems

Maintenance Improvement

Maintenance improvement is another aspect of maintenance that should be implemented in the emerging industry 4.0 framework to ensure completion of the value-addition process. Information and activities related to maintenance improvement that should be assured in the emerging I4.0 framework include changes in:

- Procedure and process for establishing maintenance concept, strategy and policy
- Specification for skills and training of maintenance and operations personnel
- Specification for robot upgrade, calibrations and update
- Policies for spare parts and materials management
- Quality and type of tools and support equipment
- Operating, safety and environmental procedures
- Equipment and system design

ENABLING TECHNOLOGIES OF INDUSTRY4.0

Conventionally, industries functions in fragmented silos with almost of the departments inside an organization. Due to this dis-integration of silos, it takes lot of time to take decision and information unavailability in the event of a failure, hazard, accident, loss of profit, etc. These lagging decisions will significantly affect the performance of the organization in the coming age of growing demands and dynamic behavior of the market. These lagging indicators also effect the human behavior that lead to lot of stress and have societal implications. Hence, there is a necessary to integrate several systems in the organization to improve the performability to reach the competition in the market.

One of the technologies that industries are focusing is Industry 4.0 (Pascual, Daponte, & Kumar, 2019). Though it is originated from the manufacturing industry to improve the productivity of the production process with high automation and efficiency, this technology is also getting attention from the other industries where they modify according to their own requirements and demands, mostly to improve the performance, efficiency and effectivity. It encapsulates different enabling technologies such as IoTs for data acquisition and connectivity, Artificial Intelligence for data analytics and decision modelling, Big data for handling large amount and variety of data, service-oriented architecture for defining stakeholder's requirements context aware systems for decision support systems and cyber-physical systems for architecture. The unification of these technologies will enable the organizations to improve overall life cycle management of their assets thus also improve the maintenance process, in general to increase the usage life of the assets. Several other technologies are mentioned in (Chen et al., 2017; Lu, 2017; Zhou, Liu, & Zhou, 2015).

Internet of Things (IoTs)

IoT concept facilitates access to internet for any type of "device/thing". An interesting review on IoT is written by (Holler et al., 2015). Each "thing" will have an Internet Protocol (IP) address and will be referred using standard internet technology like Domain Name System (DNS). Applications of the Internet of Things in automation mainly include acquisition of relevant data from the sensors, actuators for actuating a task, Programmable logic controllers (PLCs) for controlling the variables, and Control loops. Hence, the IoT can be perceived as a physical device or as a functionality that is processed in a software that will be executed on any type of thing/device that is having enough computational resources. Presently, there is no coherent or standard technology that can be recognized to an IoT (Trappey, Trappey, Govindarajan, Chuang, & Sun, 2017). Hence, mainly, in the context of Industry 4.0, IoTs can be used to obtain the necessary data from the environment to facilitate the maintenance decision making. These IoTs can be connected through other IoTs by using a standard networking technology such as Bluetooth/ WiFi or other technologies to transfer all acquired data to the cloud. An application of IoT for the application in assessing the condition of roads in winter domain is demonstrated by (Odelius, Famurewa, Forslöf, Casselgren, & Konttaniemi, 2017).

Artificial Intelligence (AI)

Interesting applications of data mining and artificial intelligence (A.I.) in industrial production process are in maintenance (Bastos, Lopes, & Pires, 2014), in predictive maintenance reading Internet of Thing (IoT) sensor data and generally in predictive analytics (Winters, Adae, & Silipo, 2014). External data mining tool (KNIME): this external tool can improve advance analytics related to predictive maintenance of the production lines (Massaro, Maritati, Galiano, Birardi, & Pellicani, 2018). There are case studies that provides AI methodologies for sustainable, predictive maintenance of production equipment (Tretten & Karim, 2016), context-driven maintenance services (Thaduri, Galar, Kumar, & Verma, 2016) and infrastructure maintenance decision support (Famurewa, Zhang, & Asplund, 2017). (Diez-Olivan, Del Ser, Galar, & Sierra, 2019; Seneviratne et al., 2018) provided several artificial intelligence tools for diagnosis and prognosis of general and linear assets. Applicable AI algorithms must be developed and trained to distinguish which specified engineering limits of the monitored parameter requests an immediate maintenance intervention. The most relevant benefits of maintenance strategies based on the continuous condition monitoring of an operating parameter are listed below (Alsina, Chica, Trawiński, & Regattieri, 2018):

- Improved plant system availability
- Reduced total cost of ownership
- Enhancement in the design of complex systems
- RAMS (Reliability, Availability, Maintainability and Safety) analysis, maintenance optimization, and forecasting of part consumption.

Big Data (BD) and Data Mining (DM)

Gartner research defined big data in terms of three Vs i.e., Volume (growing rate of data), Velocity (speediness of data) and Variety (Gartner,). Now, other two more Vs have been attached to it i.e., Veracity and Value. Big Data Analytics can be used to detect and predict the fault/failure in the component/system by analyzing the vast amount of data gathered from the sensors in coherent and real-time and lessen unplanned or unexpected service delays to enhance the efficiency or productivity. There are several potential applications of Big Data in Railway sector (Thaduri, Galar, & Kumar, 2015). Its impact of maintenance in the era of big data within asset management is stipulated by (Galar & Kans, 2017). Some of the big data tools and technologies for transportation systems used to store, clean, integrate, manage, analyze and visualize big data are mentioned in (Kour, Thaduri, Singh, & Martinetti, 2019).

DM can be described as the process of discovering appealing and understandable patterns and to discover the knowledge from large amounts of data (Galar, Gustafson, Tormos Martínez, & Berges, 2012). The evolution in the data mining is an vital process, where several existing intelligent methods are applied on the big data to extract relevant data patterns and obtain knowledge from data to perform maintenance decision support (Galar, Kans, & Schmidt, 2016). These data sources can comprises of several databases, data warehouses, internet, other information sources, or data from the human or data that are flowed into a system dynamically (Thaduri et al., 2015). In additional words, DM is the extracting the answers to the questions by searching through database for with specific rules, relationships, and patterns among different parameters not obtained by conventional query tools using extrapolatory analysis. This can also be helpful for conducting process mining for maintenance decision support system for railways (Thaduri, Famurewa, Verma, & Kumar, 2019).

Cloud Computing (CC)

Cloud computing is a technological product with integration of the networking and traditional computing, such as parallel computing, distributed computing, grid computing, edge computing, utility computing, virtualization, network storage, load balancing, etc. (Zou, Deng, & Qiu, 2013). Cloud computing is a convenient model for enabling on-demand network access to a shared pool of data with configurable computing resources. It can be swiftly provided with minimum effort or service to provider interaction. It consists of mainly three services, (Mell & Grance, 2011) namely

Software as a Service (SaaS): This service has a capability to facilitate the customer's applications running on a cloud infrastructure. These applications can be accessible from any devices through either a client interface. The consumer doesn't need to control the cloud infrastructure such as installed servers, disk operation systems, data storage, etc., but able to do limited application specific configuration settings.

Platform as a Service (PaaS): This service has capability to facilitate consumer to mount onto the cloud infrastructure with customized application using programming languages, libraries, services, and tools. Like previous service, customer doesn't need to control the cloud infrastructure, but has control over the number of applications deployed and application specific configuration settings.

Infrastructure as a Service (IaaS): This service has capability to facilitate consumer for processing, storage, networks, and other essential computing resources where he/she can deploy and run software. The consumer has control over operating systems, storage, and installed applications with full settings.

Service Oriented Architecture (SOA)

Presently, the industries and governments demand for sustainability, flexibility, efficiency and competitiveness because of dynamic nature of societal and market trends. It is important to handle the requirements of multi-stakeholder environment and need for efficient co-operation and collaboration which requires a system wide architecture. This process also needs to be followed through the value chain and the life cycle of the product and its process (Karim, et al, 2016). This also required a digitization procedure to be able to connect through different stakeholders in real-time to acquire data, process the information and disseminate the knowledge among the partners from Industry 4.0. From the perspective of organization, cooperation and managing the operation assets, the three domains required are:

- Product life cycle management (design to support)
- Supply chain management (suppliers to customers)
- Stakeholder integration management (shop floor to business)

These three management domains must work together so that the requirements of multi-stakeholders must be achieved. The dynamic collaboration among each of these domains has the potential possibility of integrated learning among them and transfer of data and information being the key. To provision these developments in these domains, there are still several gaps in administrative, managerial and technological gaps that needs to be address where the present state of the art is not enough. The combination of digitization, digitalization and automation could improve the competitiveness, flexibility and sustainability.

Context-Aware Systems

A context-aware system (CAS) is a system that acclimates actively and autonomously adapts according to the required function that enables for more relevant information to the users based on information gathered from machines/people's contextual information. The concept of context-aware computing was as ''the ability of a mobile user's applications to discover and react to changes in the environment they are situation'' (Schilit and Theimer, 1994, Schilit, et al, 1994). Context-aware systems are complex, and they can do different tasks such as data representation, data modelling, data management, reasoning from data, and analysis of contextual information (Thaduri, et al, 2014). There also exists different context-aware systems that are difficult to provide a generic process though it consists of four main steps are (Thaduri, Kumar, & Verma, 2017):

- **Context Acquisition:** The first step is to select and acquire necessary data from the sensors. The sensors can be physical sensors such as any wired or wireless sensors depending on the availability. The information can also be gathered from so called virtual to get secondary information.
- **Storing Information:** The data will be stored in repository. Before modelling the data, it is necessary to organize the data in taxonomical order so that the modelling because easier. There are several closed and open standards to store the data from bottom to top level characterizing with different entities. These entities can be defined with failure modes, failure effects, etc.
- **Context Abstraction:** The context-aware system requires the abstraction level by interpreting or aggregating them which can be useful for data analytics.
- **Context Utilization:** At the last step, this data is analyzed with data analytics to provide decision support system.

Cyber-Physical Systems

Cyber-physical systems (CPS) is the upcoming jargon resembles the integration of physical systems with cyber capabilities such as cloud, networking, computation, etc. It is being in testing phase in lot of applications mainly, medical, automation, manufacturing and aviation to take advantages of this integration. It is very important to the consider that most of the existing technologies in the capabilities are not fully developed and the expectations from the business is to properly integrate these technologies to meet their demands. Hence, it requires amalgamation of comprehensive variety of scientific areas, substantial quantity of efforts and research are essential to design, develop and implement CPS methodologies.

CPS systems are implemented by the assimilation of physical methods with software and communication with detailed generalizations and design, modelling, and analysis techniques for a sub-system, system or systems of systems. The prediction and control of the dynamic behavior of the systems consisting of multidisciplinary areas of computers, networking, and physical systems and their interaction in ways that necessitate essentially novel innovative design technologies. A starting point for developing such an architecture is for example the five-level architecture of CPS is outlined in Figure 3.

Figure 3. Cyber-physical systems (CPS) architecture of Industry 4.0

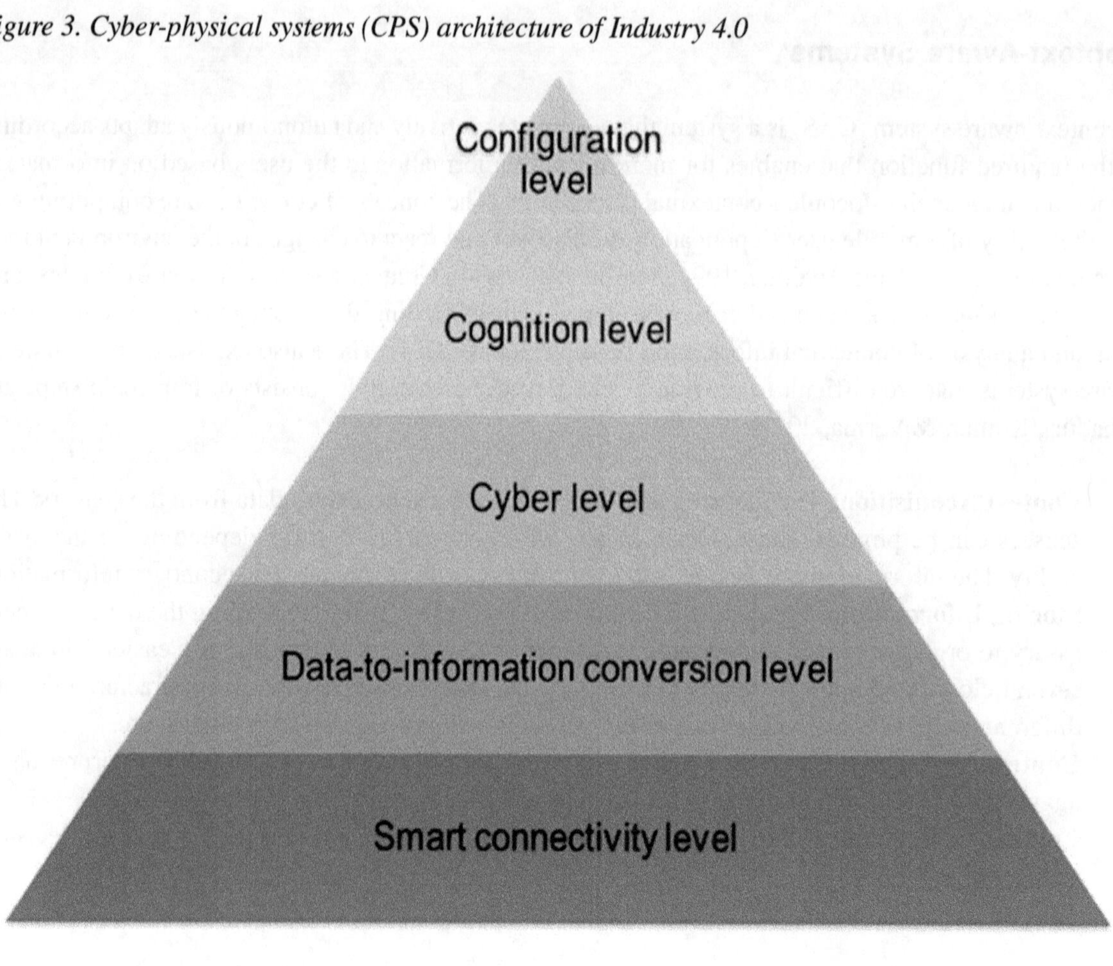

This architecture includes the following basic architectural levels:

- **Connectivity Level:** Data acquisition using condition monitoring techniques using existing sensors or Internet of Things (IoTs) or unmanned vehicles. The information must be acquired from item level to the system level depending on the objective of the function expected from CPS architecture. Due to increase in demands from the real-time communication, the data acquired from the systems must be stored in cloud through different communication protocols using service-oriented architecture.
- **Conversion Level:** The next level in the CPS is to extract the required information from the data obtained in Smart connectivity level. Sometimes, it will be difficult to extract information due to complexity in nature, accuracy of the data and huge storage. Hence, there is a need to perform advanced data manipulation or pre-processing methods such as data quality, data cleaning, feature extraction to implement the efficient data for condition assessment and the prediction. There are also other tools such as data reduction, data normalization and soft sensors to acquire secondary information from the existing data (called as precursors or covariates). When wireless sensor systems with limited power harvesting resources are involved this process is particularly challenging and requires use of energy-efficient computing concepts.
- **Cyber Level:** Once information is obtained from every connected component and system is available, modelling and simulation methods are performed to evaluate different operation and maintenance scenarios. Online maintenance analytics tools are implemented to extract useful information to support the above scenarios (Karim, Westerberg, Galar, & Kumar, 2016). Information about different machines of the same type can be compared and used for prediction.
- **Cognition Level:** Decision-making is performed by analyzing different maintenance scenarios with optimizing the parameters such as required availability, total cost, total risk and maintainability aspects using efficient genetic algorithms. The main intention of this level is to capitalize the artificial cognitive systems that can relate with specific objectives to the strategies and actions. At this level the capabilities to perform detailed simulations at the cyber level is combined with synthesis capabilities enabled by cognitive computation and human collaborative diagnostics and decision making.
- **Configuration Level:** The virtual architecture is organized by required entities and operational requirements/indicators defined by business organizations such as Key Performance Indicators (KPIs). The main purpose of the configuration level is to adapt and adopt the strategies according to the dynamic nature of the environment, contextual requirements and business demands by implementing the self-configured and self-optimizing systems. To implement these systems, the lower systems need to be supported/changed according to the needs of the organization.

All the above enabling technologies mentioned in this section can be integrated into the 5C framework for CPS-based I4.0 as illustrated in Figure 4. The interconnections between the enabling technologies, their applications and techniques in a 5C-based framework might vary based on industrial applications business objectives and other limitations.

Figure 4. Integration of the enabling technologies of industry 4.0 in a CPS-based framework

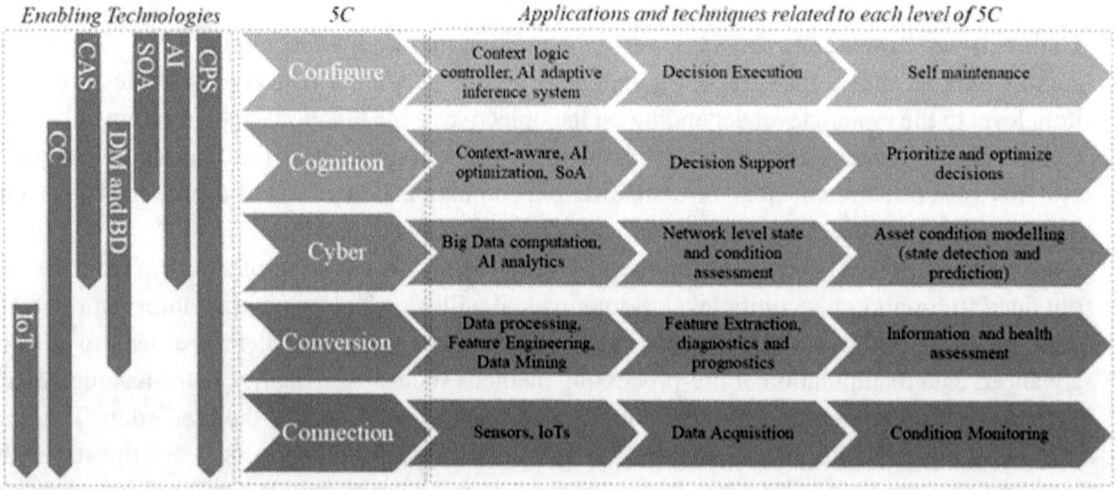

ISSUES AND CHALLENGES IN MAINTENANCE PROCESS WITHIN INDUSTRY 4.0 CONTEXT

Industry 4.0 is rapidly growing in deployment and has been contributing to the operational performance in different sectors. However, the ineffective implementation of well-designed maintenance process is a major setback that should be investigated for improvement. Some of the issues and challenges of industry 4.0 as raised by (Kumar & Galar, 2018) including related to maintenance process are highlighted below:

Domain Knowledge

In order to apply Industry 4.0 and leverage these technologies to specific industrial problems, it is essential to consider domain maintenance knowledge and experience. It is needful to map the industrial process with relevant maintenance process. Adequate knowledge about the maintenance and maintenance support of industrial operation is often an issue that is not well addressed in the development of industry 4.0 solutions. Detail understanding of the elements of maintenance process are often lacking in several industry 4.0 implementations. It is crucial to understand and improve the challenges and difficulties prevailing within the present maintenance functions when designing the enabling technologies of industry 4.0. Hence, there is a need to include specific requirements for effective maintenance process when developing industry 4.0 framework.

Organizational Constraints

The organizational structure of many industries is not supportive enough for the implementation of maintenance process in industry 4.0. Traditionally, decisions are made at different asset life cycle phases and organizational hierarchical levels such as strategic, tactical and operational levels but these levels exist as distinct layers without due co-ordination and diligence. These constraints often lead to multi-level decision making, sub-optimization and multiple information silos. Hence, there is a need of integrated organization framework which enables the integration of maintenance process in industry 4.0 framework.

Self-Configurable

The end focus of many industry 4.0 solutions is high-level configuration in terms of self-adjustment, organization and maintenance. Thus, other possible maturity levels of configuration are often neglected. In some organizations, the achievement of very high level of self-configuration is almost impossible because of requirement of huge investment and involvement of huge risk. Hence, there is a need to consider other self-capabilities in combination with conventional maintenance process i.e. self-maintenance triggering, remote adjustment, self-alignment with remote supervision, set call for maintenance service or intervention and self-resource management and preparation.

Data Management

There are lot of issues and challenges related to data handling, processing, storage and analysis due to the veracity, volume and velocity of data. These issues are due to

- Disorganization of data: At each stage of the maintenance process, data is either generated or leveraged from the previous stages. There are issues related to streamlining the data within whole process due to lack of proper structure and framework interconnectivity-workflow within the maintenance process itself
- Human Machine Interface (HMI): The use of maintenance information, recordings and measurements gathered by the Human is still a challenge within the scope of data management model. These challenges arise due to lack of seamless interface between human and machine, lack of common interaction data platform and format, and control logic for data integration between human and machine, etc.
- Data quality and data quantity: Though the sensors can give lot of information on the condition of the assets, there are still lot of issues related to data quality such as reliability of the sensor data, accuracy and precision of the data, applicability of the data, technical feasibility of data handling, incorrect and non-standard data transformation

Standardization of Framework and Solutions

The effective implement of Industry 4.0 can be achieved by properly integrating the enabling technologies such as IoT, AI, Big data, and CPS, etc. But these technologies itself suffers from standardization issues related to development protocols, methods and processes. Furthermore, the high-level integration of these technologies further complicates the process of standardization of different devices/modules incorporating with the Industry 4.0. These standardization issues will result in problems such as failures within device activations, data accuracy, delay in processing the maintenance process, incorrect mutual understanding of technical jargon, incorrect condition assessment of assets that lead to inaccurate decisions which often doesn't solve main problems.

Static Maintenance Process Information

The assumption of static maintenance process information is an issue that remains unaddressed in industry 4.0 development and implementation. This is a very relevant issue especially in scenarios where external maintenance service is needed. There is a need to handle and model maintenance data such as resources, documentation, standards, equipment, budget, crews in a dynamic way because their availability, content or condition changes with time. This makes the integration of maintenance process into industry 4.0 framework very crucial.

Operation Based Framework

Most of the Industry 4.0 solutions are developed using frameworks based on the operation of the asset with good insight on the health condition but without holistic view of the required maintenance process. This makes the entire loop incomplete in several instances as the decision making is not connected to the maintenance actions and support resources namely machines, consumables, manuals and standards, and maintenance crew, etc. This issue can be solved by adapting standard maintenance process and its element, defining its data and information requirements from relevant sources and incorporating it into the data management protocol. This can be achieved by integrating an effective maintenance process into the Industry 4.0 framework during the design and development stage.

Integration of Maintenance process into I4.0

The effective implementation of maintenance process within Industry 4.0 can address some of the above issues and challenges of industry 4.0. This can be achieved by mapping the individual task to be performed at each stage of the maintenance process to the 5Cs of the cyber-physical systems. A conceptual method for the integration of the different elements of the maintenance process according to EN standard with the five levels (5Cs) of the CPS-based Industry 4.0 framework is presented in Table 1 and described below.

For each stage of the maintenance process, there is a list of activities to be assured at the different levels of the CPS-based Industry 4.0 framework. Incorporating these tasks in the design and development of industry 4.0 framework will make the solution of equal benefit to operation process and maintenance function. The mapping in Table 1 entails identifying, relating and integrating the elements of a standard maintenance process to the industry 4.0 framework in a holistic way. Further, the input and output data for the constituent elements of the maintenance process is specified and linked to the appropriate component of the I4.0 architecture. This linkage also includes the definition, format, structure, type, acquisition frequency and source of the data.

For example, for the "Maintenance Support Planning" elements of maintenance process, the data and information regarding defined maintenance tasks, required resources, task costs, respective operational and environmental conditions are collected at the Connection Level. The data collected for maintenance support planning are thereafter processed and manipulated at the conversion level to extract actionable information and relevant indicators for each system and its critical component. The information extracted can be an index such as item criticality, risk levels, remaining useful life (RUL), inventory levels etc.

At Cyber Level, the indicators from conversion level are computed at higher system level for example plant or network level. Index such as criticality and risk level of the plant can be estimated, and systems criticality comparison can be done at the same time. This will provide the required input for the cognition level or the decision optimization module of the I4.0 architecture. At the cognition level, the output of cyber level process and an overview of the system architecture is required for the optimization and planning of respective maintenance tasks. The cognition level handles the following activities as relevant for maintenance planning: scenario generation, simulation, optimization, analysis and assessment, cost-benefit analysis and collaborative decision modelling. It is important to mention that activities at this level is considered common for all the maintenance process elements but the objectives, input data and final decision may differ.

At the Configure Level, the asset is expected to respond according to the condition and maintenance planning information gathered and processed. In some instances, no response is required as this information will be transferred to subsequent maintenance process stage (i.e. maintenance scheduling) while in other instances specific action is triggered by the asset itself in the form of:

- Self-awareness of the possible consequences of different planning scenarios
- Self-adjustment of loading and operating conditions based on processed planning information
- Self-operational and functional trade-offs to meet support planning limitations

Table 1 presents the constituent activities of all the elements of a standard maintenance process in connection to the CPS framework within an industry 4.0. However, there is a need to adapt it to prevailing implementation conditions and environment. This might require combination of some of the elements of maintenance process since there might be is no distinct difference between these elements in some organizational set-up.

CONCLUSION

A foremost concept that is setting the pace and direction for many sectors and services is industry 4.0. The focus is on augmenting machines, infrastructure and systems with wireless connectivity and sensors to monitor, visualize and communicate incidences between different entities for decision making. This new trend entails the use of enabling technologies process, and models in the form of internet of things (IoT), cyber-physical systems (CPS), big data and data mining, cloud computing, context-aware systems and artificial intelligence (AI) to facilitate data exchange and automation. This disruptive revolution has caused substantial evolution in physical asset management. An aspect of physical asset management that has been enormously influenced by the new industrial set-up is the maintenance process. Hence, there is a need to adopt the concept of maintenance process according to the international standard EN 60300-3-14 into Industry 4.0 framework. This book chapter addressed several issues and challenges in the Industry 4.0 framework with respect to maintenance process that need attention to improve its value adding capability and productivity. Furthermore, a conceptual integration of the different elements of the maintenance process according to EN standard with five levels (5Cs) of CPS based I.40 architecture is proposed to address some of the issues raised. This procedure will help to assure a functioning maintenance process to support effective and efficient decision process within the scope of I4.0.

Table 1. Integration of maintenance process into I4.0

<table>
<tr><th>Process/Level</th><th>Connection</th><th>Conversion</th><th>Cyber</th><th>Cognition</th><th>Configure</th></tr>
<tr><td>Maintenance management</td><td>Acquisition of information regarding
• Maintenance policy
• Organization structure
• Financing and Budgeting
• Insourcing/ outsourcing
• Maintenance supervision structure</td><td>Extraction of
• Maintenance performance index
• CAPEX/OPEX
• Human performance index</td><td>• Assessment of maintenance goal
• Business and finance analytics for assessment of budget effectiveness
• Business risk analysis</td><td rowspan="6">• Scenario generation, simulation, optimization, analysis and assessment
• Visualization to all stakeholders
• Cost-benefit analysis
• Multi-objective optimization
• Collaborative decision modelling</td><td>• Self-optimization for disturbances within organizational limitations
• Self-adjustment of configuration with specific budget</td></tr>
<tr><td>Maintenance Support Planning</td><td>Collection of information about
• Maintenance resource and equipment capacity, reliability and availability
• Relevant maintenance tasks
• Task and component costs
• Operational profile
• Environmental conditions
• Documentation (standards, manuals, thresholds, etc.)
• Inventory</td><td>Estimation of
• Item criticality
• System risk levels
• Remaining useful life (RUL)
• Maintenance task/ support requirements
• Inventory levels</td><td>Assessment of
• Plant or network criticality
• Risk comparison at network level
• Remaining useful life at network level
• Maintenance task/ support requirements at network level
• Inventory levels at network</td><td>• Self-awareness of possible consequences of different scenarios
• Self-adjustment of loading and operating conditions
• Self-operational and functional trade-offs to meet support planning limitations</td></tr>
<tr><td>Maintenance Preparation</td><td>Acquisition of
• Inventory data
• Availability of the asset (maintenance window)
• Communication channels about resources
• Information about tools, transportation, lifting and support equipment
• Safety, Health and Environment procedures and work plan
• Training resources</td><td>Estimation of
• Inventory levels
• Resource availability performance
• Personnel availability performance
• Skill level</td><td>• Inventory levels at network
• Resource availability performance for network
• Personnel availability performance for network
• Skill level</td><td>• Self-preparation for resource training and skill
• Self-allocation of inventory
• Self-optimizing of loading and operating conditions
• Self-operational and functional trade-offs to meet preparation limits</td></tr>
<tr><td>Maintenance Execution</td><td>Collection of
• Realtime maintenance report
• Observations and recordings
• Maintenance process times
• Information about troubleshooting, testing and checkout</td><td>Assessment of
• Maintenance report quality for each system
• Maintenance task effectiveness for each system</td><td>Assessment of
• Overall quality of maintenance reports pat network level
• Maintenance effectiveness at network level</td><td>• Self-trigger for service and intervention</td></tr>
<tr><td>Maintenance Assessment</td><td>Gathering of
• Information about relevant KPIs
• Data required for KPIs assessment
• Benchmarks and standards
• Operational, safety and environment procedures</td><td colspan="2">Assessment of
• Indicators (KPIs)*
• Maintenance efficiency *
• Maintenance effectiveness *
• Safety and environmental performance*
* These activities are carried out at system and network level for conversion and cyber levels respectively.</td><td>• Self-adjustment, correction and alignment</td></tr>
<tr><td>Maintenance Improvement</td><td>Gathering of
• OEM manuals
• Regulations
• Standards improvement models
• Maintenance, safety and environment procedures
• Design specifications</td><td>Assessment of
• Indicator about need for system improvement
• Indicator about required level of system improvement</td><td>Assessment of
• Indicator about need for network improvement (comparison and aggregation)
• Indicator about required level of network improvement</td><td>• Modifications to improve functionality by redesigning and resource management</td></tr>
</table>

REFERENCES

Alsina, E. F., Chica, M., Trawiński, K., & Regattieri, A. (2018). On the use of machine learning methods to predict component reliability from data-driven industrial case studies. *International Journal of Advanced Manufacturing Technology*, *94*(5-8), 2419–2433. doi:10.100700170-017-1039-x

Bagheri, B., Yang, S., Kao, H., & Lee, J. (2015). Cyber-physical systems architecture for self-aware machines in industry 4.0 environment. *IFAC-PapersOnLine*, *48*(3), 1622–1627. doi:10.1016/j.ifacol.2015.06.318

Bastos, P., Lopes, I., & Pires, L. (2014). Application of data mining in a maintenance system for failure prediction. *Safety, Reliability and Risk Analysis: Beyond the Horizon: 22nd European Safety and Reliability, 1*, 933-940.

Batista, N., Melício, R., & Mendes, V. (2017). Services enabler architecture for smart grid and smart living services providers under industry 4.0. *Energy and Building*, *141*, 16–27. doi:10.1016/j.enbuild.2017.02.039

Ben-Daya, M., Kumar, U., & Murthy, D. P. (2016). *Introduction to maintenance engineering: Modelling, optimization and management*. John Wiley & Sons. doi:10.1002/9781118926581

Bream, R. (2006). Lawsuit poses further risk to BP's image. *Financial Times.*

Cen, E. (2001). *13306: 2001-maintenance terminology. European Standard.* European Committee for Standardization.

Chen, B., Wan, J., Shu, L., Li, P., Mukherjee, M., & Yin, B. (2017). Smart factory of industry 4.0: Key technologies, application case, and challenges. *IEEE Access: Practical Innovations, Open Solutions*, *6*, 6505–6519. doi:10.1109/ACCESS.2017.2783682

Diez-Olivan, A., Del Ser, J., Galar, D., & Sierra, B. (2019). Data fusion and machine learning for industrial prognosis: Trends and perspectives towards industry 4.0. *Information Fusion*, *50*, 92–111. doi:10.1016/j.inffus.2018.10.005

EN 60300-3-14. (2004). 60300 (3-14): Dependability Management–Part 3-14: Application Guide–Maintenance and maintenance support.

Famurewa, S. M., Zhang, L., & Asplund, M. (2017). Maintenance analytics for railway infrastructure decision support. *Journal of Quality in Maintenance Engineering*, *23*(3), 310–325. doi:10.1108/JQME-11-2016-0059

Fantini, P., Pinzone, M., & Taisch, M. (2018). Placing the operator at the centre of industry 4.0 design: Modelling and assessing human activities within cyber-physical systems. *Computers & Industrial Engineering*.

Galar, D., Gustafson, A., Tormos Martínez, B. V., & Berges, L. (2012). Maintenance decision making based on different types of data fusion. *Eksploatacja i Niezawodnosc-Maintenance and Reliability*, *14*(2), 135–144.

Galar, D., & Kans, M. (2017). *The impact of maintenance 4.0 and big data analytics within strategic asset management.* Paper presented at the Maintenance Performance and Measurement and Management 2016 (MPMM 2016), Luleå, Sweden.

Galar, D., Kans, M., & Schmidt, B. (2016). Big data in asset management: Knowledge discovery in asset data by the means of data mining. *Proceedings of the 10th World Congress on Engineering Asset Management (WCEAM 2015)*, 161-171. 10.1007/978-3-319-27064-7_16

Holler, J., Tsiatsis, V., Mulligan, C., Avesand, S., Karnouskos, S., & Boyle, D. (2015). *From machine-to-machine to the internet of things*. Academic Press.

Kans, M., Galar, D., & Thaduri, A. (2016). Maintenance 4.0 in railway transportation industry. *Proceedings of the 10th World Congress on Engineering Asset Management (WCEAM 2015)*, 317-331. 10.1007/978-3-319-27064-7_30

Karim, R., Westerberg, J., Galar, D., & Kumar, U. (2016). Maintenance analytics–the new know in maintenance. *IFAC-PapersOnLine, 49*(28), 214–219. doi:10.1016/j.ifacol.2016.11.037

Kour, R., Thaduri, A., Singh, S., & Martinetti, A. (2019). *Big data analytics for maintaining transportation systems. In Transportation systems* (pp. 73–91). Springer.

Kumar, U. (2008). *System maintenance: Trends in management and technology. In Handbook of performability engineering* (pp. 773–787). Springer.

Kumar, U., & Galar, D. (2018). *Maintenance in the era of industry 4.0: Issues and challenges. In Quality, IT and business operations* (pp. 231–250). Springer. doi:10.1007/978-981-10-5577-5_19

Lee, J., & Wang, H. (2008). *New technologies for maintenance. In Complex system maintenance handbook* (pp. 49–78). Springer. doi:10.1007/978-1-84800-011-7_3

Li, Z., Wang, Y., & Wang, K. (2017). Intelligent predictive maintenance for fault diagnosis and prognosis in machine centers: Industry 4.0 scenario. *Advances in Manufacturing, 5*(4), 377–387. doi:10.100740436-017-0203-8

Lu, Y. (2017). Industry 4.0: A survey on technologies, applications and open research issues. *Journal of Industrial Information Integration, 6*, 1–10. doi:10.1016/j.jii.2017.04.005

Massaro, A., Maritati, V., Galiano, A., Birardi, V., & Pellicani, L. (2018). ESB platform integrating KNIME data mining tool oriented on industry 4.0 based on artificial neural network predictive maintenance. *Int.J.Artif.Intell Appl, 9*, 1–17.

Mell, P., & Grance, T. (2011). *The NIST definition of cloud computing*. Academic Press.

Odelius, J., Famurewa, S. M., Forslöf, L., Casselgren, J., & Konttaniemi, H. (2017). Industrial internet applications for efficient road winter maintenance. *Journal of Quality in Maintenance Engineering, 23*(3), 355–367. doi:10.1108/JQME-11-2016-0071

Rakyta, M., Fusko, M., Herčko, J., Závodská, Ľ., & Zrnić, N. (2016). Proactive approach to smart maintenance and logistics as a auxiliary and service processes in a company. *Journal of Applied Engineering Science, 14*(4), 433–442. doi:10.5937/jaes14-11664

Reis, M., & Gins, G. (2017). Industrial process monitoring in the big data/industry 4.0 era: From detection, to diagnosis, to prognosis. *Processes, 5*(3), 35. doi:10.3390/pr5030035

Seneviratne, D., Ciani, L., Catelani, M., & Galar, D. (2018). Smart maintenance and inspection of linear assets: An industry 4.0 approach. Acta Imeko.

Thaduri, A., Famurewa, S. M., Verma, A. K., & Kumar, U. (2019). Process mining for maintenance decision support. In *System performance and management analytics* (pp. 279–293). Springer.

Thaduri, A., Galar, D., & Kumar, U. (2015). Railway assets: A potential domain for big data analytics. *Procedia Computer Science*, *53*, 457–467. doi:10.1016/j.procs.2015.07.323

Thaduri, A., Galar, D., Kumar, U., & Verma, A. K. (2016). *Context-based maintenance and repair shop suggestion for a moving vehicle. In Current trends in reliability, availability, maintainability and safety* (pp. 67–81). Springer. doi:10.1007/978-3-319-23597-4_6

Thaduri, A., Kumar, U., & Verma, A. K. (2017). Computational intelligence framework for context-aware decision making. *International Journal of System Assurance Engineering and Management*, *8*(4), 2146–2157. doi:10.100713198-014-0320-8

Thoben, K., Wiesner, S., & Wuest, T. (2017). "Industrie 4.0" and smart manufacturing-a review of research issues and application examples. *International Journal of Automotive Technology*, *11*(1), 4–16.

Trappey, A. J., Trappey, C. V., Govindarajan, U. H., Chuang, A. C., & Sun, J. J. (2017). A review of essential standards and patent landscapes for the internet of things: A key enabler for industry 4.0. *Advanced Engineering Informatics*, *33*, 208–229. doi:10.1016/j.aei.2016.11.007

Tretten, P., & Karim, R. (2016). *Project: iMain 'Methodologies and tools for the sustainable, predictive maintenance of production equipment'*. Academic Press.

Wan, J., Tang, S., Li, D., Wang, S., Liu, C., Abbas, H., & Vasilakos, A. V. (2017). A manufacturing big data solution for active preventive maintenance. *IEEE Transactions on Industrial Informatics*, *13*(4), 2039–2047. doi:10.1109/TII.2017.2670505

Winters, R., Adae, I., & Silipo, R. (2014). *Anomaly detection in predictive maintenance anomaly detection with time series analysis*. Paper presented at the Knime.

Zhou, K., Liu, T., & Zhou, L. (2015). Industry 4.0: Towards future industrial opportunities and challenges. *2015 12th International Conference on Fuzzy Systems and Knowledge Discovery (FSKD)*, 2147-2152.

Zou, C., Deng, H., & Qiu, Q. (2013). Design and implementation of hybrid cloud computing architecture based on cloud bus. *2013 IEEE 9th International Conference on Mobile Ad-Hoc and Sensor Networks*, 289-293. 10.1109/MSN.2013.72

Chapter 4
Tacit Knowledge Sharing for System Integration:
A Case of Netherlands Railways in Industry 4.0

Yawar Abbas
https://orcid.org/0000-0001-9965-2778
University of Twente, The Netherlands

Mohammad Rajabalinejad
https://orcid.org/0000-0002-1550-2762
University of Twente, The Netherlands

Alberto Martinetti
https://orcid.org/0000-0002-9633-1431
University of Twente, The Netherlands

Lex Frunt
The Netherlands Railways, The Netherlands

Leo van Dongen
University of Twente, The Netherlands

ABSTRACT

Sharing of tacit knowledge is a key topic of research within the knowledge management community. Considering its embodied nature, organizations have always struggled with embedding it into their processes. Proper execution of complex processes such as system integration asks for an adequate sharing of tacit knowledge. Acknowledging the importance of lessons learned for system integration and their presence in tacit and explicit form, a case study was conducted within the Netherlands Railways. It was determined that non-sensitivity to the tacit dimension of lessons learned has resulted in their lack of utilization. Consequently, LEAF framework was developed, where LEAF stands for learnability, embraceability, applicability, and findability. The framework suggests that addressing these four features collectively can eventually lead to an adequate knowledge-sharing strategy for lessons learned. Lastly, the chapter presents an example from the Netherlands Railways to emphasize the key role technological solutions of Industry 4.0 can play in facilitating tacit knowledge sharing.

DOI: 10.4018/978-1-7998-3904-0.ch004

INTRODUCTION

The value of organizational knowledge has grown significantly over the past seven decades. Technological developments after the second world war and research conducted within the management sciences have brought the topic of knowledge management to the center stage. More recently, knowledge is being viewed as an organizational asset and as a source of competitive advantage (Kakabadse et al., 2001). Technology has played a key role in this transition and revolutionized the way in which organizations manage their knowledge. Meanwhile, these developments have also pointed out the limitations of human capability in articulating one's knowledge to explicit form. A deeper understanding of human capabilities and a closer look at human interactions with the technical systems are required to address these limitations adequately. Currently, the knowledge management community, by and large, acknowledges the conceptual distinction of knowledge into two main types namely tacit and explicit knowledge (Nonaka & von Krogh, 2009). The idea of the former knowledge type is mainly attributed to Polyani when he famously stated "We know more than we can tell" (Polanyi, 1966), while the latter can be readily articulated, codified, stored and accessed (Hélie & Sun, 2010). Research within the management sciences has shown that not only is explicit knowledge an important resource for firms (Conner & Prahalad, 1996) but also that tacit knowledge is a source of competitive advantage for firms (Winter, 1987). This chapter builds upon the concept of tacit knowledge sharing and presents ways in which it can be enhanced within an organizational setting.

Knowledge sharing is a major field of research within the knowledge management community, with challenges on multiple fronts. It requires the transfer of knowledge from one entity to another (Argote & Ingram, 2000). Naturally, the transfer of explicit knowledge is easier and more straightforward. Moreover, technology has greatly assisted in optimizing explicit knowledge transfer. A common example, in this regard, is the use of various Information Technology (IT) based knowledge management systems by the organizations. On the other hand, tacit knowledge whose primary source is experience (Bratianu & Orzea, 2010), is rooted in a mix of "action, procedures, routines, commitment, ideals, values and emotions" (Nonaka, 1994), and generally difficult to share. Research has shown that the sharing of tacit knowledge is stimulated by intrinsic motivators (Chena et al., 2011) and facilitated by engaging environments (Muniz et al., 2013). Tacit knowledge management requires a shift towards practice-based approach and more sensitivity to workforce abilities and skills (Ribeiro, 2013). Furthermore, it also requires awareness of the nature of the system under consideration, as different knowledge management approaches are required for complex systems as well for complicated systems (Snowden, 2002). The difference between the complex and complicated system here is the intertwining and separation of the cause and effect relationships of the system respectively (Snowden, 2002). This chapter focuses on tacit knowledge sharing and looks primarily into the railway transportation system of the Netherlands.

Tacit knowledge sharing for a complex, safety-critical system such as railway is a strenuous task. The goal of the railway system as defined by CEN, (2017) is to "achieve a defined level of rail traffic at a given time, safely and within certain cost limits". For the system under consideration, system integration is a prominent process within the system life cycle phases (CEN, 2017). It is aimed at synthesizing the system elements into a realized system (ISO/IEC/IEEE 15288) and a key topic of research within the systems engineering community. Moreover, proper execution of system integration processes is of fundamental importance to key railway system stakeholders such as railway operators and infrastructure managers. Within this context, adequate sharing of relevant tacit knowledge can facilitate in the improvement of system integration processes both academically, managerially, and practically. The research topic

"Tacit knowledge sharing for system integration projects" addresses the research domains of knowledge management, system integration for the railway system, and project and quality management as shown in Figure 1. Moreover, the figure is adapted by the four dimensions of a core capability by Leonard-barton, (1992), where he defines core capability as "an interrelated, interdependent knowledge system". More specifically, system integration for railway system compromises of knowledge related to technical system; knowledge management of skills and knowledge base; project and quality management of managerial systems; and tacit knowledge sharing of values and norms.

Figure 1. Domains of interest for the research topic adapted by Leonard-barton, (1992)

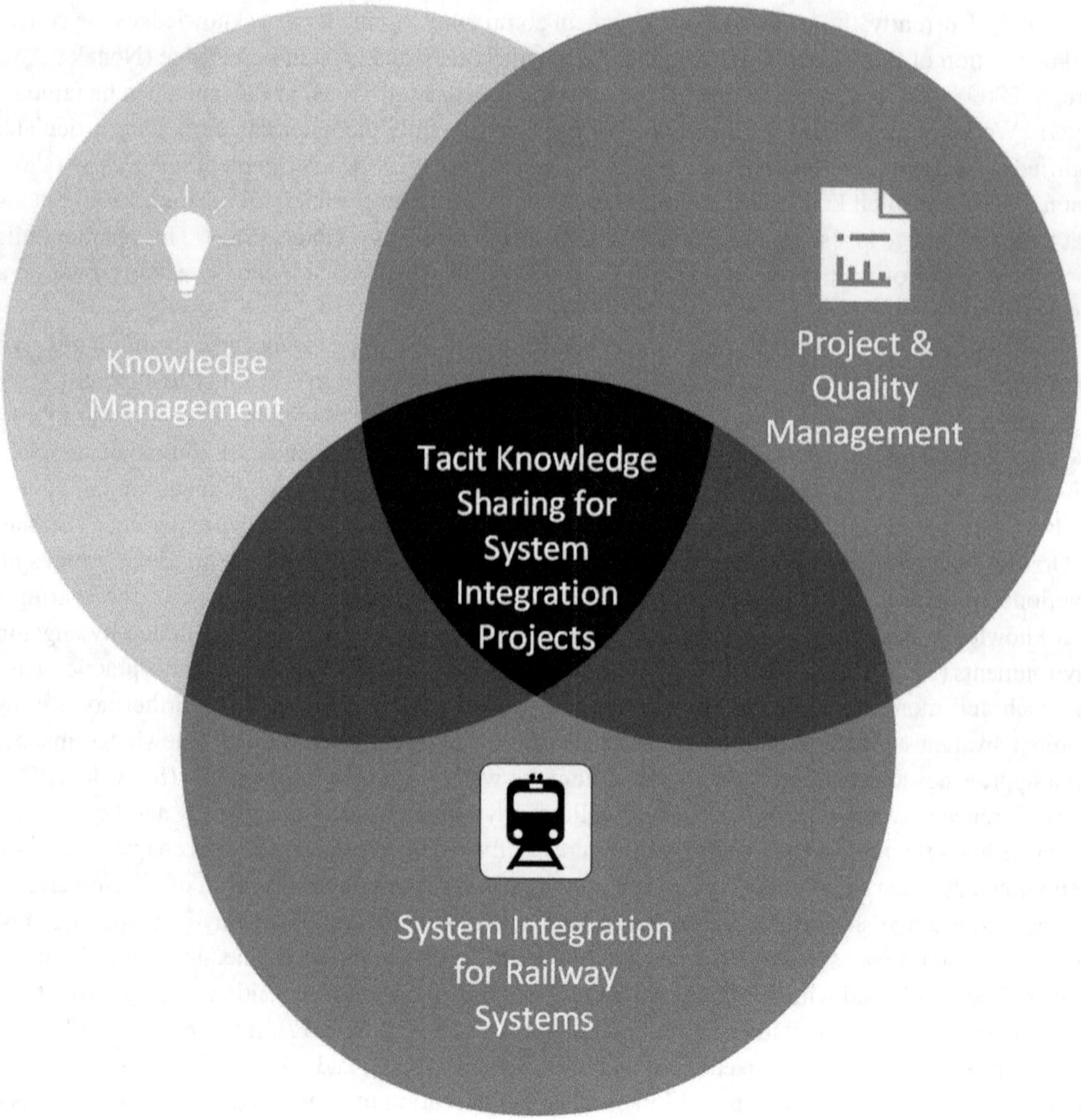

The fundamental goal of this chapter is to demonstrate the importance of tacit knowledge sharing for system integration projects and recommend ways in which IT solutions of Industry 4.0. can facilitate its transfer. The chapter follows a three-step approach to address the presented goal. Firstly, it exhibits the relevance of tacit knowledge sharing for system integration processes. Secondly, it presents a case of Netherlands Railways (main railway operator) to demonstrate the significance of the tacit dimension of lessons learned for the integration of New Rolling Stock (NRS). A knowledge-sharing framework is presented in this regard for the sharing of lessons learned. The framework identifies four key features that require attention to optimizing the sharing of lessons learned within the NRS teams of the Netherlands Railways. Lastly, the chapter presents an example of the Netherlands Railways to demonstrate the central role of Industry 4.0, which is largely IT-driven, in facilitating tacit knowledge sharing, for the railway system.

TACIT KNOWLEDGE SHARING FOR SYSTEMS INTEGRATION

Tacit knowledge sharing in the context of systems integration is a challenging activity. Moreover, systems integration is an evolving and critical topic of research within the systems engineering community (Madni & Sievers, 2013; Rajabalinejad, 2018). In addition to this, Blanchard et al., (2011) accept the lack of a commonly accepted definition for systems engineering and suggest that systems engineering features as "top-down approach; life-cycle orientation; early concentration on defining systems requirements; and an inter-disciplinary or team-based approach in the development process" (Wilson, 2014). Consequently, the scope of system integration in a given context is dependent on the system under consideration. Considering the focus of systems engineering on the development processes enables the use of ongoing technological innovation for developing engineering solutions more accustomed to customer needs and desires. Moreover, life cycle orientation of system engineering provides an opportunity to reflect on the entire lifecycle stages of the system, and design their implementation plan in due time or adapt it in operations due to the changing demands of stakeholders over the whole lifecycle. Besides this, systems engineering inspires to learn from past experiences and embed acquired experienced knowledge in the development of new systems. Within this context, system integration is a key process where embedding of experienced knowledge can play a major role. It can facilitate in learning from past experiences, foreseeing integration challenges and determining the future directions for complex system integration processes. Exposure to the challenges of complex system integration challenges results in the acquisition and enhancement of possessed tacit knowledge. The ability of successful system integrators to understand the complete picture of the safety-critical system and act responsibly, is a result of not just the already determined explicit knowledge, but also the possessed tacit knowledge from experience. Naturally, transition of this knowledge in the development of newer systems is paramount to the future of the system engineering domain. However, sharing of this experienced knowledge is no easy task.

Experienced knowledge from system integration exists in the tacit and explicit form within the system integration project teams of Netherlands rail-sector. ISO/IEC 15288, a systems engineering standard for processes and lifecycle stages (including integration), recommends sharing knowledge assets as a key activity of the knowledge management process. Lessons learned among others is the key knowledge asset identified by ISO/IEC/15288:2015E, and broadly stating is the knowledge acquired through experience on a certain project. Clearly, sharing of acquired lessons learned whether in tacit or explicit form is critical to the proper execution of system integration processes. The challenge, however, comes from the very nature of lessons learned, as contrary to the common perception they exist not just in explicit but also in tacit form. Adequate sharing of lessons learned from requires sensitivity to this attribute and development of a knowledge sharing strategy that addresses not just the explicit but also the tacit dimension of lessons learned. Preliminary investigation within the Netherlands Railways has shown that not much attention has been paid to the tacit dimension of lessons learned. This has resulted in inadequate utilization of determined lessons learned and lack in the enhancement of organizational learning. Paying attention to the tacit dimension of lessons learned is critical in this regard due to the changing roles of individuals within the project teams and the complex nature of the systems integration processes. Moreover, as mentioned by Nonaka & von Krogh, (2009) conversion of knowledge (from tacit to explicit in this case) helps in product and process innovations, enhancement of resilience, and in the evolution of new social practices. This chapter makes explicit key features influencing tacit knowledge sharing of lessons learned within the context of system integration. An example of the introduction of New Rolling Stock (NRS) projects is presented in this regard. NRS projects hold critical importance from an integration standpoint. This is mainly because of the recent shift of the railway undertakings towards buying trains from the shelf mentality. A recent example in this regard is the acquisition of new sprinter series purchased by Netherlands Railways from CAF, which is a Spanish rolling stock manufacturer. This shift in approach towards NRS requires adjusting the organization for the train rather than adjusting the train for the organization. Next section presents the case of tacit knowledge sharing of lessons learned within the NRS teams of the Netherlands Railways.

KNOWLEDGE SHARING OF LESSONS LEARNED: AN INVESTIGATION WITHIN THE NRS TEAMS OF THE NETHERLANDS RAILWAYS

Knowledge sharing of lessons learned plays a central role in the integration process of NRS. The acquisition process of NRS has undergone a major change in the past few years within the Netherlands rail-sector. The so-called "buying a train from the shelf" analogy is quite common within the rolling stock teams. The analogy refers to the shift in the approach of preparing the organization for the train rather than the prior focus of getting the right train for the organization also known as "fit for the processes". The Netherlands Railways as an organization has undergone through a rough period in this transition. Starting from the painful experience of aborted FYRA trainsets introduction and moving towards a rather successful introduction of recent new sprinter series. This ongoing shift is not easy and among other factors owes its success to the continuous embedding of experienced knowledge in the NRS projects. A step towards such embedding was the development of a lessons learned database. The database inspires to share acquired lessons learned from prior projects with the current NRS teams and enhance organizational learning within the Netherlands Railways. It was developed through a process compiling lessons learned from a number of NRS introduction teams. Although the development of such a database is appreciated and a step in the right direction, its development alone does not lead to the adoption of the compiled lessons learned. Preliminary investigation, conducted within the Sprinter New Generation (SNG) NRS team, has manifested that the sole focus on explicating lessons learned and not paying attention to the tacit dimension of lessons learned has resulted in lack of utilization of the acquired lessons learned. Clearly, this lack of utilization is undesired and requires investigation into its root causes. Consequently, a case study was conducted within the SNG team with the sole purpose of identifying the key root causes of this lack of utilization. In this regard, 14 qualitative interviews were conducted.

The interviewees include eight members of the SNG project team, four members from the Programma Materieel Implementatie (PMI) team (the team that developed the lessons learned database), and two new rolling stock test drivers. The interviews were conducted in an open format and together with each interviewee key issues to the sharing of acquired lessons learned were discussed. Consequently, four key issues were identified after reflecting on the gathered notes from these interviews which are presented next. Moreover, to properly address the identified issues a knowledge sharing framework for sharing of lessons learned was also developed. The framework advocates for four key features that require attention while sharing of lessons learned and is presented after the identified issues.

Identified Issues Related to the Sharing of Lessons Learned

This section presents four key issues related to the sharing of lessons learned that were identified during the conducted interviews. A small description of each issue is presented with an example.

Inaccurate Applicability

This issue points to the lack of sensitivity towards the nature of the system under consideration while determining the lessons learned. It has been identified that where some lessons learned are of repeatable nature and can be used to determine best practice others simply don't repeat themselves and are specific to the context. The authors' advocate for distinguishing the lessons learned based on their applicability and repetitive nature. The recommendation is in line with the findings of Snowden (2002), where he distinguishes between managing knowledge for complicated and complex domains. Snowden (2002) argues that best practices are only legitimate in spaces where all the cause and effects relationships are clear which he calls known space. The distinction between the acquired lessons learned based on their applicable domain is crucial to their effective utilization. The current approach of expecting the determined lessons learned to be applicable in the similar scenarios in the future has its drawbacks, especially given the lack of sensitivity towards their applicability. Consider the example of a lesson learned which states "Give training at the right time to all the train drivers for the NRS". The presented lessons learned are related to a complex process of training train drivers and are dependent on a number of key constraints. These include organizational, production, and intrinsic project constraints. Organizational constraints include the availability of train drivers, instructors, budget, etc. Similarly, production constructs include timely delivery of rolling stock by the suppliers, timely exchange of valuable information, etc. Lastly, intrinsic project constraints include the nature of the project (for example, national vs. international use of the acquired NRS), possessed experience about the project by the workforce, etc. All the stated constraints play a major role in the proper implementation of the stated lessons learned in practice. A standard best practice won't be applicable in every scenario simply because not all the constraints are known in advance or will follow the planned path. Consequently, the development of suitable practices for some real-life scenarios is much more effective and can assist in determining the applicability of the stated lessons learned.

Low Sensitivity to Learnability

Low sensitivity demonstrates the importance of learnability for lessons learned as identified during the conducted interviews. Semantically speaking the term lessons learned implies that someone has learned that lesson. Undoubtedly, one of the key aims of sharing lessons learned is for someone else to also learn the same lesson. The key challenge, however, is that people learn in different ways. An identified lessons learned doesn't imply that everyone can easily and adequately learn that lesson. Due attention to this challenge is necessary before any positive effects of sharing lessons learned can be expected. Addressing the stated challenge requires adjustment of the shared lesson learned in accordance with the receiver's prior knowledge. Knowledge sharing is dependent on the cost of disembodiment and levels of acceptable abstraction (Snowden, 2002). It has been noticed that although having a learning environment is one of the key goals of PMI, the currently shared lessons learned require more adaption to the receiver's prior knowledge and mode of optimal learning. For example, by adjusting them based on receivers prior job role and experience. Enhanced learnability from these lessons learned will require among other things development of different acceptable levels of abstraction for the determined lessons learned. Moreover, adapting the shared message based on the receiver's preferred mode of learning can have a significant impact on the utilization of shared lesson learned. For example, where applicable development and conduction of a lessons learned game can enhance learnability and create the curiosity in NRS to learn more about them.

Lack of Findability

Lack of findability can have serious consequences on the overall impact of shared lessons learned. The identified issue points to the two main types of findability issues. The first being the technological issues which imply not being able to find desired lessons learned on a certain subject in an easy manner. Moreover, technological findability addresses both the ability of a certain lesson learned to be findable from a list of stored information and its ability to be easily accessible to the desired audience. The second findability issue is the so-called source findability. Source findability refers to the ability to identify the right person or place who holds the knowledge of certain lessons learned. A frequent example of this issue is not being able to find the source of certain lessons learned when the source has changed the position within the organization. Attention to source findability is very important because it recognizes that the source is the holder of not just explicit knowledge but also possess immense tacit knowledge on the subject matter. Moreover, easy access to the source and platform for communication between the knowledge seeker and source is fundamental for especially sharing of tacit knowledge and adaption of the shared lessons learned.

Lack of Embraceability

The fourth key identified issue is the so-called lack of embraceability. The issue points to the non-adopting of the lessons learned by the relevant individuals and teams. Lack of embraceability hinders the organization from learning from its past mistakes and results in the repetition of the same mistakes in different parts of the organization. It also influences the embedding experienced knowledge in relevant processes and procedures. A number of psychological and organizational factors can be the cause of this issue. Among the most prominent psychological factors are the lack of trust, lack of motivation, lack of recognition, personal grudges and ambitions, etc. Similarly, key examples of organizational factors include lack of awareness, lack of management support, organizational culture, lack of incentives and reward system, etc. Consequently, it is crucial to identify such factors in a given context and address them in an adequate manner. Moreover, considering the embodied nature of tacit knowledge, adequate sharing of the tacit dimension of lessons learned is only possible when the identified psychological and organizations factors in a given context are adequately addressed. Next section presents the developed knowledge sharing framework for enhanced sharing of lessons learned.

LEAF: A Commended Framework for Sharing of Lessons Learned

As a consequence of determining the stated issues, LEAF framework is presented for the sharing of lessons learned from system integration. LEAF stands for Learnability Embraceability Applicability and Findability and represents these four essential features that collectively determine the adequacy of the knowledge sharing approach for the lessons learned. The framework inspires to facilitate sharing of acquired tacit and explicit lessons learned from system integration projects. By focusing on the stated four features the framework provides direction into the required technological and organizational solutions for the embedding of the experienced knowledge in the integration processes. Furthermore, enhanced sharing of acquired lessons learned also facilitates the development of engineering solutions that incorporate better tacit customer expectations. LEAF acknowledges that the identified features are not mutually exclusive and suggests a zone of optimal knowledge sharing strategy, as shown in Figure 2. The framework proposes the overlap of the stated features as the novel zone of optimal knowledge sharing strategy. Moreover, the framework recognizes that the contribution and significance of each of the four features are context-dependent and acknowledges it by the dotted boundary of each feature in Figure 2.

Although each specified feature requires a detailed description of their role in the sharing of lessons learned, this chapter focuses only on the zone of optimal knowledge sharing strategy. This zone can be achieved when a knowledge sharing strategy is developed that collectively addresses the stated four features.

Devising such a strategy requires close collaboration between the knowledge management experts and integration project stakeholders. Adequate knowledge sharing practices need to be put into practice that adheres to both the tacit and explicit nature of experienced knowledge. For example, by developing technological and social platforms where people, with similar job descriptions and experience, working in different teams can communicate easily. Moreover, regular qualitative deep dives on the subject matter and storytelling sessions regarding the acquired lessons learned must be organized and made digitally available to facilitate the tacit knowledge transfer. Moreover, careful attention must be paid to the level of acceptable abstraction when sharing a lesson learned with the target audience. In addition to this, proper characterization of lessons learned based on their repetitive nature is the key to their actual utilization. Besides this, for the system under consideration acquired lessons learned can be classified into lessons learned related to the technical system, or its interaction with its surroundings. Such classification can enhance the applicability and findability of the acquired lessons learned. The strategy must also determine and address adequately all the psychological and organizational factors based on the on-ground contextual conditions. Formulating a strategy in accordance with the stated recommendations can facilitate entry into the proposed zone of optimal knowledge sharing strategy. A zone where lessons learned are actively shared and with full recognition to both the tacit and explicit nature of lessons learned. Lastly, for future work the authors would like to explore the utility of identified features in the LEAN framework for the technology acceptance model by Davis et al. (1989), which models user's acceptance and use of technology. Specifically, the connection of stated features with the perceived ease of use and perceived usefulness of the technology acceptance model. Next section presents some ways in which the technological solutions of Industry 4.0. can facilitate the discussed active knowledge sharing approach.

Figure 2. LEAF framework

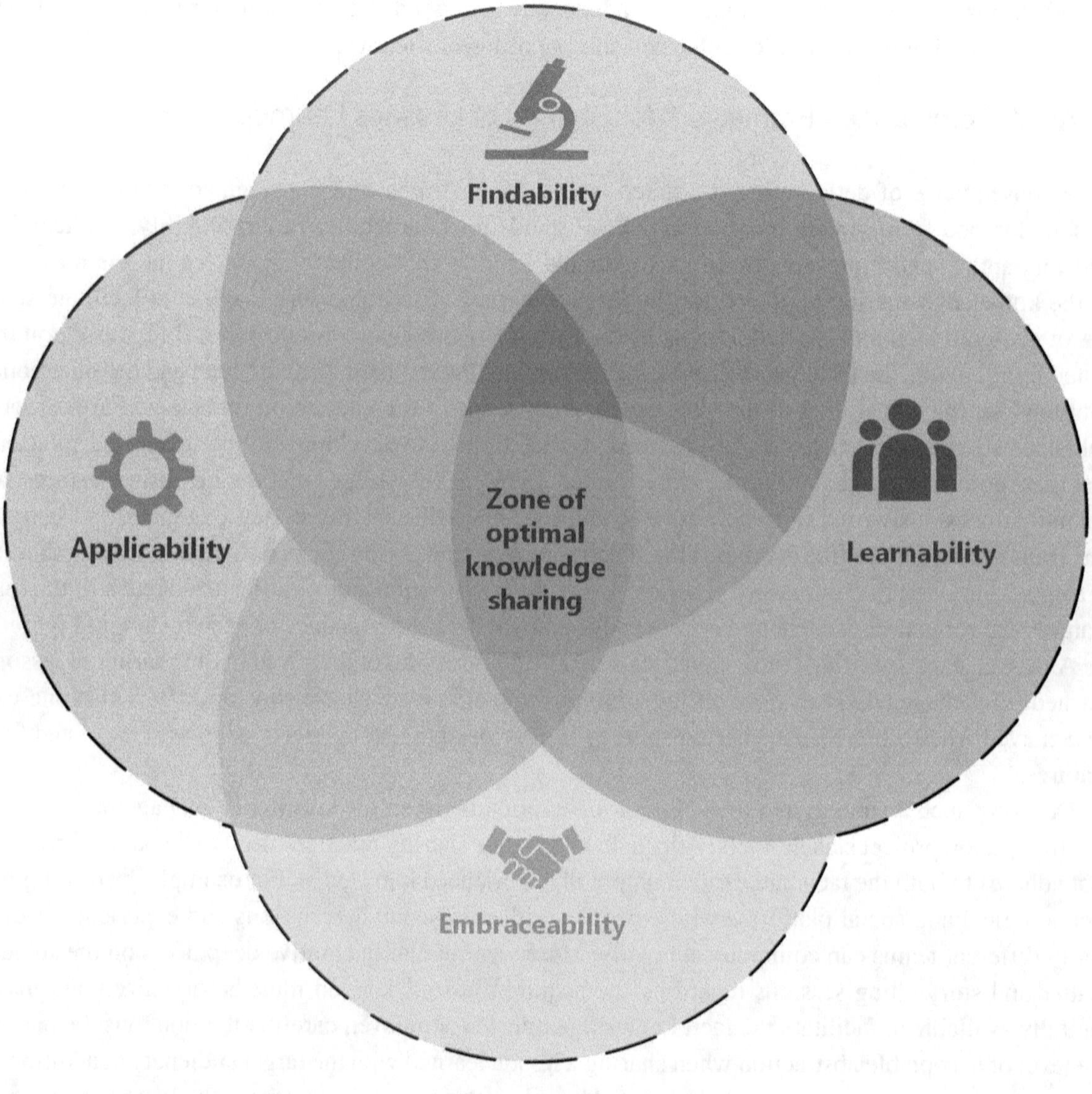

INDUSTRY 4.0 AND EMBEDDING TACIT KNOWLEDGE SHARING IN NETHERLANDS RAILWAYS IT SYSTEMS

The term Industry 4.0, as summarized by Lasi et al. (2014), describes "different – primarily IT-driven – changes in manufacturing systems". The ongoing maturity of technological innovations and growing customer expectations have resulted in a shift of many industries towards data-driven approaches for systems engineering and asset management. Whether one's focus is on the design process, integration process, or on operations and maintenance, industries are increasingly relying more on data-driven policies. Concepts such as condition-based maintenance, digital twins, adaptive learning are becoming more mainstream and are being researched to enhance system performances. Kagermann et al. (2013) foresee huge potential in meeting individual customer requirements through Industry 4.0 initiatives. Embedding of experienced knowledge is fundamental to addressing the ever-growing customer demands and developing desired functionalities in the technical systems. The relevance of tacit knowledge for adequately adapting to these innovations and developing the right engineering solutions cannot be emphasized enough. Currently, many industries struggle with the identification of right data that must be collected for optimizing their system processes. Moreover, the generation of metadata has made distinguishing valuable data from large datasets a challenging task. Consequently, experience based predictions still hold significant position in key system lifecycle processes such as maintenance within the era of Industry 4.0 (Diaz, 2019).

Embedding of experienced knowledge in the ongoing IT-driven changes in the manufacturing systems is fundamental to the proper execution of system integration processes. However, coupling experienced knowledge with digital transformations is no easy task. From the knowledge management perspective, Peinl et al. (2017) have investigated the impact of digital transformations on knowledge management approaches and have acknowledged the distinction between data, information, and knowledge. Moreover, Peinl et al. (2017) recognized the influence increased volume of knowledge and pace of changes have on organizational learning. The significance of tacit and explicit knowledge distinction is well understood within the domain of information sciences. Buckland (1991) has categorized information into "Information as thing", "Information as process", and "Information as knowledge". His work advocates that stored information (Information as a thing) facilitates learning (Information as a process) resulting in the appropriate action (Information as knowledge). Considering the fast pace of changes in the digital world and growing volumes of gathered information, the authors call for a deeper analysis of human interaction with these digital transformations. Conceptually, the authors believe that for complex safety-critical systems such as railway transportation system digital transformations of Industry 4.0 should go hand in hand with acquired tacit knowledge. In this regard, the authors build upon the concepts of "Information as process" and "Information as knowledge" by Buckland (1991). An example from Netherlands Railways customer services is presented in this regard to demonstrate the ways in which IT solutions of Industry 4.0 are facilitating this viewpoint.

Quality of service is of key importance to the rail-sector, and special attention is paid to it by the Reliability, Availability, Maintainability, and Safety (RAMS)- Railway applications standard (CEN, 2017). Moreover, among the key social objectives of the Netherlands Railways passengers are regarded as the "first, second and third priority" (NS, 2016). Consequently, the Netherlands Railways is always looking for opportunities to enhance the quality of their services. Technological transformations of Industry 4.0. provide a great opportunity in this regard. By developing IT solutions which can optimize system processes and address tacit expectations of the passengers, Netherlands Railways can improve the quality of its services. A recent example of such effort is the introduction so-called "Seat location finder" within the "NS Travel Planner" application as shown in Figure 3b. The purpose of the "Seat location finder" is to inform passengers about the status of congestion in different parts of the train, and to direct them towards less crowded sections for seat search. Such information assists passengers in looking for a seat in less crowded sections of the train and resulting in the use of information as knowledge. Figure 3 shows an example of a train journey from Utrecht Central Station to Rotterdam Central Station. As can be seen in Figure 3a the overall status of congestions is reported as quiet for this train journey and less congested sections are displayed as green color in Figure 3b.

Figure 3. (a and b)Travel information available to the passenger on the NS Travel Planner application

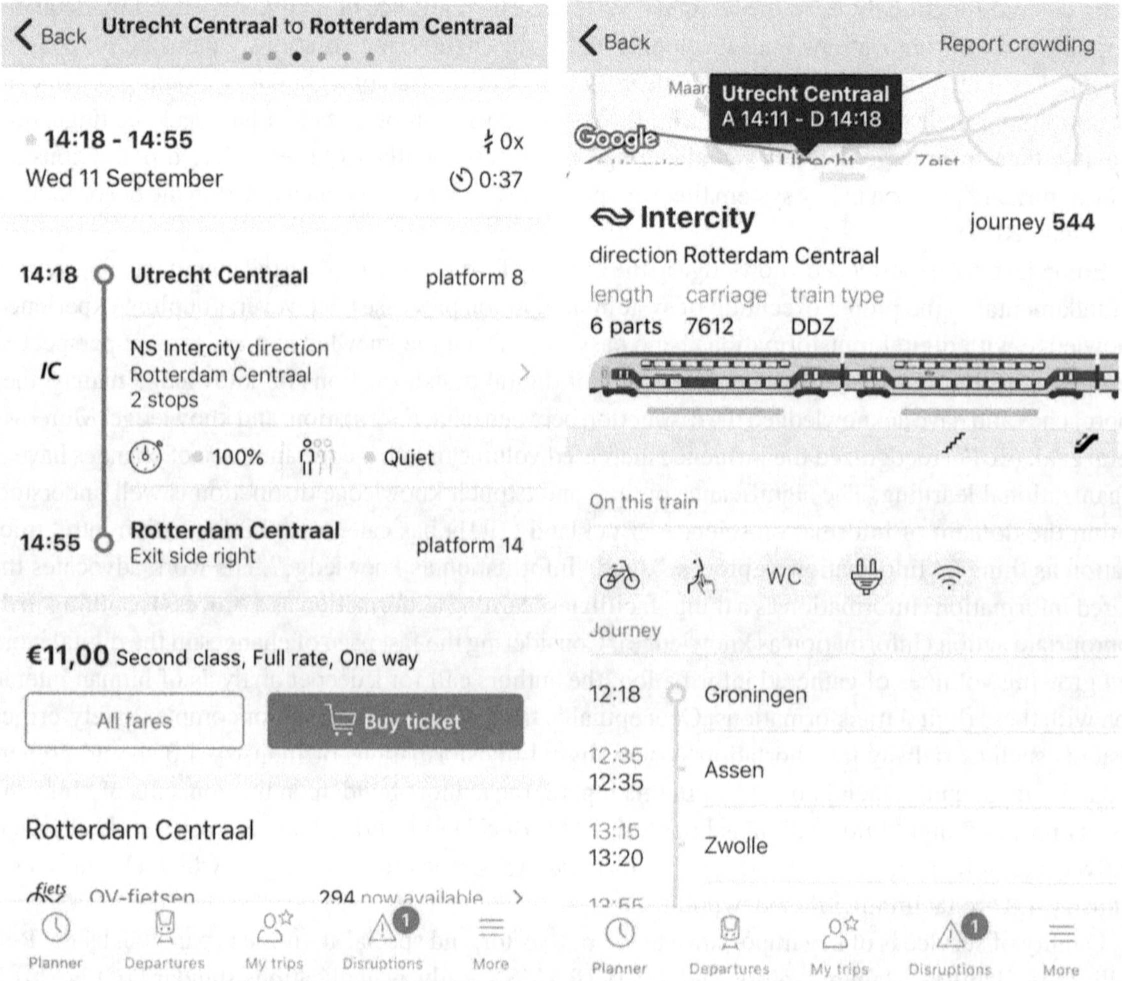

(a) **(b)**

Such information is made available by processing the stored information, from axle load data and check-in/out data, through the developed explicit model and sharing it with the passengers via "NS Travel Planner" application. The example is a good use of using stored sensorial data and processing it through explicit models to enhance overall knowledge of the desired audience. Especially, considering the fact that the axle load sensors in the track were originally installed for measuring flat spots on the wheels and axle load distribution in the bogie for the safety analysis. The example shows how technological transformations facilitate using the same information for multiple purposes. Another important aspect of these technological transformations is the possibility to grasp the tacit expectations of the customers. For example, as shown in Figure 3b the passengers can also report crowding which provides an opportunity to analyze the accuracy of the developed explicit model and understand better the tacit expectations of the passengers. Summing up, the presented example manifests two important aspects. Firstly, it shows how IT solutions can facilitate in utilizing gathered sensorial data for multiple purposes and using stored information as knowledge. Secondly, it shows how tacit expectations of the passengers can be better inferred through the IT solutions and assist in checking the validity of developed explicit models.

CONCLUSION AND FUTURE AREAS OF RESEARCH

The chapter presented the case for tacit knowledge sharing in the era of Industry 4.0. It reflected on the relevance of tacit knowledge for system integration projects and demonstrated the value of embedding tacit knowledge in such processes. Moreover, the chapter recognizes the significance of lessons learned for the proper execution of system integration processes and propose to pay attention to not just the explicit but also the tacit dimension of lessons learned. To back the suggestion, the chapter presents the LEAF framework for the sharing of lessons learned from system integration. The framework consists of four key features Learnability, Embraceability, Applicability and Findability which require addressing when developing a knowledge sharing strategy for lessons learned. Lastly, the chapter presents an example to demonstrate ways in which technological solutions of Industry 4.0 can facilitate in effective knowledge sharing and developing a knowledge sharing strategy which can bridge the gap between tacit expectation and developed explicit models. Future work stemming from this research includes testing the LEAF framework in various case studies and developing technological solutions that can facilitate tacit knowledge sharing for system integration projects.

ACKNOWLEDGMENT

The presented chapter was supported and sponsored by the Netherlands Railways and TKI under the project System Integration for Railway Advancement (SIRA). Furthermore, it was made achievable with the precious suggestions by the member colleagues of the SIRA project from the Netherlands Railways, ProRail and the University of Twente.

REFERENCES

Blanchard, B. S., Fabrycky, W. J., & Fabrycky, W. J. (2011). *Systems engineering and analysis* (Vol. 5). Prentice Hall.

CEN. (2017). *NEN-EN 50126-1 Railway Applications - The specification and Demonstration of Reliability, Availability, Maintainability and Safety (RAMS)- Part 1: Generic RAMS process* (Vol. 1). Brussels: CEN-CENELEC Management Center.

Chena, G. L., Wu, W. C., Ling, W. Y., Yang, S. C., & Tang, S. M. (2011). Explicit knowledge and tacit knowledge sharing. *International Conference on Management and Service Science, MASS 2011*, 1–4. 10.1109/ICMSS.2011.5998951

Conner, K. R., & Prahalad, C. K. (1996). A Resource-Based Theory of the Firm: Knowledge Versus Opportunism. *Organization Science*, *7*(5), 477–501. doi:10.1287/orsc.7.5.477

Davis, F. D., Bagozzi, R. P., & Warshaw, P. R. (1989). User Acceptance of Computer Technology: A Comparison of Two Theoretical Models. *Management Science*, *35*(8), 982–1003. doi:10.1287/mnsc.35.8.982

Diaz, V. G. (2019). Handbook of Research on Industrial Advancement in Scientific Knowledge. doi:10.4018/978-1-5225-7152-0

Hélie, S., & Sun, R. (2010). Incubation, insight, and creative problem solving: a unified theory and a connectionist model. *Psychological review* (Vol. 117). doi:10.1037/a0019532

Kagermann, H., Wahlster, W., & Helbig, J. (2013). Recommendations for Implementing the Strategic Initiative INDUSTRIE 4.0 - Final Report of the Industrie 4.0 Working Group. *Industrie*, 82. doi:10.13140/RG.2.1.1205.8966

Kakabadse, N. K., Kouzmin, A., & Kakabadse, A. (2001). Fromtacit knowledge to knowledgemanagement: Leveraging invisible assets. *Knowledge and Process Management*, *8*(3), 137–154. doi:10.1002/kpm.120

Lasi, H., Fettke, P., Kemper, H.-G., Feld, T., & Hoffmann, M. (2014). Industry 4.0. *Business & Information Systems Engineering*, *6*(4), 239–242. doi:10.100712599-014-0334-4

Leonard-barton, D. (1992). Core Capabilities and Core Rigidities : A Paradox in Managing New Product Development. *Strategic Management Journal*, *13*(S1), 111–125. doi:10.1002mj.4250131009

Madni, A. M., & Sievers, M. (2013). System of Systems Integration: Key Considerations and Challenges. *Systems Engineering*, *17*(3), 330–347. doi:10.1002ys.21272

Michael, K. (1991). Information as Thing. *Journal of the American Society for Information Science*, *42*(5), 351–360. doi:10.1002/(SICI)1097-4571(199106)42:5<351::AID-ASI5>3.0.CO;2-3

Muniz, J., Dias Batista, E., & Nakano, D. (2013). Engaging environments: Tacit knowledge sharing on the shop floor. *Journal of Knowledge Management*, *17*(2), 290–306. doi:10.1108/13673271311315222

Nonaka, I., & von Krogh, G. (2009). Perspective—Tacit Knowledge and Knowledge Conversion: Controversy and Advancement in Organizational Knowledge Creation Theory. *Organization Science*, *20*(3), 635–652. doi:10.1287/orsc.1080.0412

NS. (2016). *Back on Track The passenger as our first, second and third priority NS Strategic Document*. Academic Press.

Peinl, R. (2017). Knowledge management 4.0 - Lessons learned from IT trends. *CEUR Workshop Proceedings*, *1821*(July), 112–117.

Polanyi, M. (1966). *The tacit dimension*. Chicago: University of Chicago Press.

Rajabalinejad, M. (2018). System Integration : Challenges and Opportunities. In *13th System of Systems Engineering Conference* (pp. 6–11). IEEE.

Snowden, D. (2002). Complex acts of knowing: Paradox and descriptive self-awareness. *Journal of Knowledge Management*, *6*(2), 100–111. doi:10.1108/13673270210424639

Wilson, J. R. (2014). Fundamentals of systems ergonomics/human factors. *Applied Ergonomics*, *45*(1), 5–13. doi:10.1016/j.apergo.2013.03.021 PMID:23684119

Winter, S. (1987). Knowledge and Competence as Strategic Assets. The Strategic Management of Intellectual Capital. doi:10.1016/b978-0-7506-9850-4.50013-0

KEY TERMS AND DEFINITIONS

Core Capability: It is an "interrelated, interdependent knowledge system" (Leonard-Barton, 1992).

Explicit Knowledge: A knowledge that can be readily articulated, codified, stored and accessed (Hélie & Sun, 2010).

Integration Process: The process of synthesizing a set of system elements into a realized system (product or service) that satisfies system requirements, architecture, and design (ISO 15288:2015).

Knowledge Assets for System Engineering: Assets such as system elements or their representations (e.g., reusable code libraries, reference architectures) architecture or design elements (e.g., architecture or design patterns), processes, criteria, or other technical information (e.g., training materials) related to domain knowledge, and lessons learned (ISO 15288:2015).

Knowledge Management: The process of creating, sharing, using and managing the knowledge and information of an organization (Girard, 2015).

Lessons Learned: Tacit and explicit knowledge acquired through experience from a system integration project.

System Integration Project: Projects aimed at the implementing the integration process of the system lifecycle for the railway system as specified in RAMS standard (CEN, 2017).

Tacit Knowledge: Knowledge which is difficult to articulate in explicit form and is fundamentally acquired through experience.

Chapter 5
Cybersecurity Issues and Challenges in Industry 4.0

Ravdeep Kour
https://orcid.org/0000-0003-0734-0959
Luleå University of Technology, Sweden

ABSTRACT

The convergence of information technology (IT) and operational technology (OT) and the associated paradigm shift toward fourth industrial revolution (aka Industry 4.0) in companies has brought tremendous changes in technology vision with innovative technologies such as robotics, big data, cloud computing, online monitoring, internet of things (IoT), cyber-physical systems (CPS), cognitive computing, and artificial intelligence (AI). However, this transition towards the fourth industrial revolution has many benefits in productivity, efficiency, revenues, customer experience, and profitability, but also imposes many challenges. One of the challenges is to manage and secure large amount of data generated from internet of things (IoT) devices that provide many entry points for hackers in the form of a threat to exploit new and existing vulnerabilities within the network. This chapter investigates various cybersecurity issues and challenges in Industry 4.0 with more focus on three industrial case studies.

INTRODUCTION

The evolution from Industry 1.0 as steam-powered machines towards Industry 4.0 as cyber physical systems (CPS) has brought many benefits in productivity, efficiency, revenues, customer experience, and profitability, but also imposes many challenges as managing human factors, often a critical element in several domains (Fontaine et al, 2016). One of the challenges is to manage and secure large amount of data generated from Internet-of-Things (IoT) devices that provide many entry points for an intruder (a person who attempts to gain unauthorized access to a system in order to compromise system availability, data Integrity or data Confidentiality) in the form of a threat to exploit new and existing vulnerabilities within the IoT network. Today, more and more organizations and businesses understand that an efficient flow of secured information creates major benefits, both economically and with greater customer satisfaction. To remain proficient and responsive, business processes must permanently transform themselves in this technological world of Industry 4.0 (Figure 1).

DOI: 10.4018/978-1-7998-3904-0.ch005

Figure 1. Industry 4.0

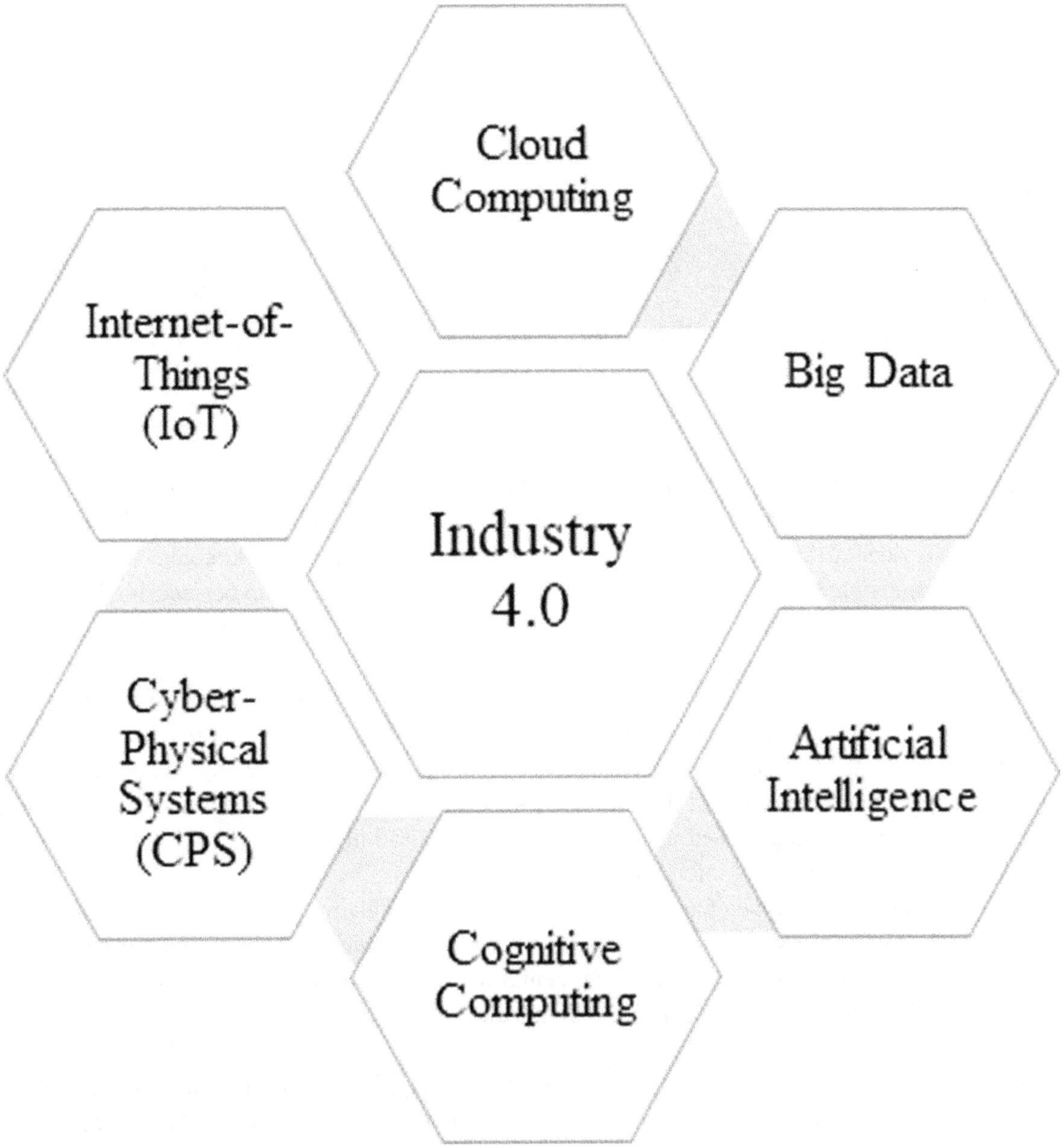

Industry 4.0 is a national strategic initiative from the German government where numerous elements comprising industrial systems are being interfaced with internet communication technologies to form the smart factories and manufacturing organizations of the future (Thames and Schaefer, 2017). The IoT connected devices itself is a superb innovation, but it also presents numerous points of entry for malicious activities. Figure 2 shows the number of connected IoT devices from year 2012 to 2025.

The IoT and internet communication technologies are plagued by cybersecurity issues that will present major challenges and barricades for adopters of Industry 4.0 technologies. If these challenges are not addressed, the true potential of Industry 4.0 may never be attained. Cybersecurity is defined as

Figure 2. Number of connected IoT devices from year 2012 to 2025 (Columbus, 2016)

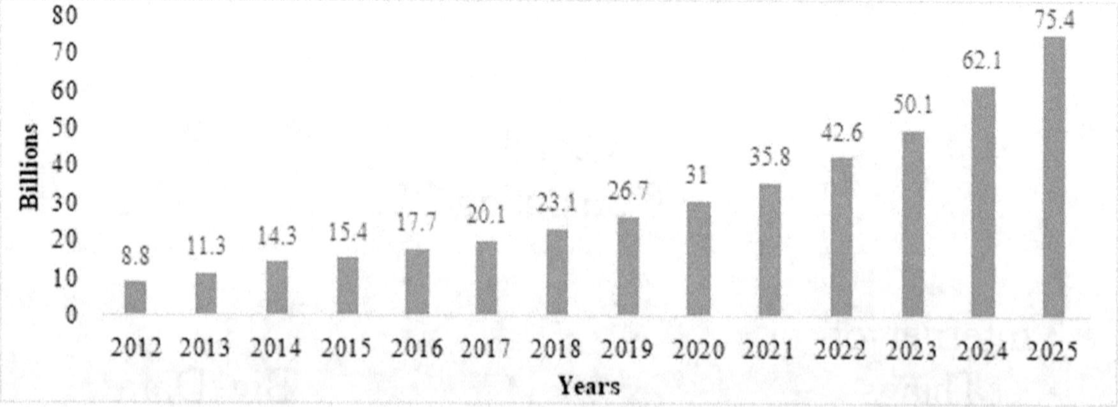

"preservation of confidentiality, integrity and availability of information in the Cyberspace" (ISO/IEC 27032) and Cyberspace is defined as "the complex environment resulting from the interaction of people, software and services on the Internet by means of technology devices and networks connected to it, which does not exist in any physical form" (ISO/IEC 27032). The main focus of the cybersecurity discourse is cyber-attacks (both passive and active), which are possibly destructive events. These attacks are:

- **Passive Attack**: In this type of attack, attacker's goal is to obtain information only. He does not modify data or harm the system. This type of attack is difficult to detect until sender or receiver finds out about the leaking of confidential information. It can be prevented by one of the methods like encipherment. Examples of this attack are tapping, snooping, traffic analysis, eavesdropping, port scanning Espionage based attacks that steal data and information etc. This type of attack harms the confidentiality of information.
- **Active Attack**: In this type of attack, attacker's goal is not only to obtain information but he will modify it or harm the system. This attack is easier to detect than to prevent. Examples of this attack are modification, replay, repudiation, denial of service (DOS / DDOS), Man-in-the-middle attack, SQL Injection, virus, worm, logic bomb, etc.

Some more examples of cyberattacks are Malware, Phishing, Cross-Site scripting, Botnets, Social Botnets, Espionage based attacks that steal data and information, Drive-by-downloads, Last Mile Interceptions, Transmission Bugs / Intercepts, Critical Infrastructure, Cyber Kidnapping, Cyber Extortion, Hacktivisim, etc. The impact of these cyber-attacks on industries are:

- Threat to the safety of the employees or the public in general.
- Loss of confidential or sensitive information.
- Embarrassment and reputational damage.
- Monetary loss
- Loss of public confidence
- Legal action against the industry
- Data inaccuracy

- Fraud
- Erroneous decisions
- Reduction in assurance of an IT system
- Loss of production time, Reliability, safety and Continuity

INTERNET HITS FOR CYBERSECURITY TERMS IN INDUSTRY 4.0

To obtain initial estimates of the current role played by cybersecurity within Industries 4.0, a web-based search has been conducted. The popular databases used were Google scholar, Ieee Xplore, Scopus, and Web of Science (Table 1).

The criteria for searching are based on the following terms like; "Cybersecurity OR Cyber security" AND "Industry 4.0"; "Cyber-attack" AND "" Industry 4.0"; and so on.

Numbers in the table 1 shows search results for the specific terms like "cyber security" AND "Industry 4.0" in the full text of literature and number in the braces () shows literature containing specific terms like "cyber security" AND "Industry 4.0" in the title of the literature.

Table 1. Internet hits for cybersecurity terms in Industry 4.0

Databases	Keywords used for the search								
	Cyber security	**cyber-attack**	**cyber breach**	**Hack**	**Cyber crime**	**Cyber Threats**	**Computer security**	**Network security**	**Information security**
Google scholar	6500(20)	797(0)	18(0)	3290(0)	229(0)	611(0)	754(0)	1910(1)	3350(4)
Ieee Xplore	520(5)	1174(4)	915(0)	664(0)	33(0)	44(2)	336(0)	927(0)	878(0)
Scopus	13(13)	0	0	0	0	0	0	0	37(2)
Web of Science	43(5)	24(4)	11(0)	5(1)	0	12(1)	1(0)	25(0)	23(2)

Table 1 results show that cybersecurity research in Industry 4.0 has been started and active. The chapter has identified major cybersecurity challenges from these literatures.

CYBERSECURITY KEY PRINCIPLES/ELEMENTS

ENISA (2018) has provided list of challenges and recommendations related to cybersecurity in Industry 4.0, which are associated with People, Processes, and Technologies, components of an Information System (IS). An Information System (IS) is a socio-technical system which delivers information and communication services required by an industry in order to attain business objectives and goals. IS play major role to coordinate activities within industries and they connect manufacturing industry, their customers, suppliers and service providers. An IS contains six components: (1) information (data), (2) people, (3) business processes (procedures), and ICT, which includes (4) hardware (5) software and (6) networks (Alter, 2008; Whitman and Mattord, 2012). These systems must be integrated and secured in order to operate better in rapidly changing and competitive market. These components of IS can be connected with three cybersecurity elements i.e. Confidentiality, Integrity and Availability (CIA) as shown in figure 3 to assure security guarantee in the entire IoT system.

Figure 3. Cybersecurity key principles (Willett, 2008)

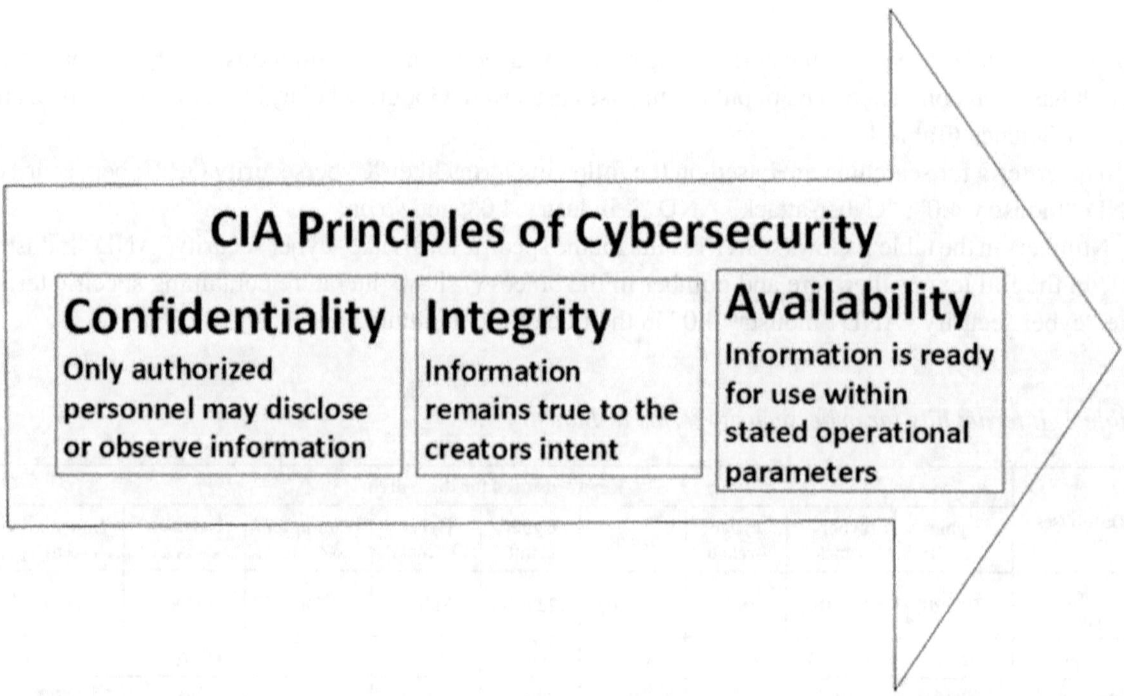

Data Confidentiality: According to Willett (2008), "confidentiality ensures the disclosure of information only to those persons with authority to see it". Data confidentiality is not only a security issue, but also of juristic concerns in some practical application systems (Yu et al., 2010). For example, according to Act (1996) in healthcare, use and disclosure of protected health information should meet the requirements of this act and it must keep the user data confidential on the servers. Thus, Confidentiality is an important security element in IoT. Encryption is one of the methods to ensure confidentiality of information in transit and access controls ensure confidentiality of data at rest (Willett, 2008). The risk associated with this security element is disclosure of confidential information and its direct impact is loss of public confidence, embarrassment, and legal action against the industry. Further, countermeasures to mitigate this risk can be cryptography, PKI, access control, identity management or privilege management, etc.

Data Integrity: It ensures the original form of the information. Integrity means that data can be modified only by authorized person(s) or the data owner to prevent misuse. With the support of IoT and cloud technology, users are provided with opportunity to store and manage their data in cloud data centers. Therefore, applications must ensure data integrity. Moreover, one of the main challenges that must be addressed is to ensure the correctness of user data in the cloud. The risk associated with this security element is corruption of information and its direct impact on the system is data inaccuracy, fraud, erroneous decisions, and reduction in assurance of an IT system. Further, countermeasures to mitigate this risk can be backups, integrity checks (e.g., cyclical redundancy check), hashing, PKI, etc.

Availability: Availability refers to the resources of the system accessible on demand by an authorized individual (Zissis, 2012). Thus, the IoT resources must be timely available to avoid significant losses. One of the main issues in the cloud environment concerning cloud service providers (CSP) is the availability of the data stored in the cloud. Therefore, services must remain available and operational even in the case of a security breach (Zissis, 2012). In addition, with the increasing number of cloud users, CSP must address the issue of providing the requested data available to users to deliver high-quality services. The risk associated with this security element is denial of service and its direct impact on the system is loss of production time, reliability, safety and continuity. Further, countermeasures to mitigate this risk can be OS Security, host configuration, IDS, anti-malware, etc.

Table 2 shows significance of three security elements for the components of IS along with risks, impacts and countermeasures.

CYBERSECURITY ISSUES AND CHALLENGES IN INDUSTRY 4.0

Following are the identified cybersecurity issues and challenges in Industry 4.0.

1. **Malware:** Malware means malicious software. In a broad term, it refers to a variety of malicious programs. These programs can steal confidential data, bypass access controls or cause harm to the host computer. This challenge can be minimized by regular follow-up on reported threats and vulnerabilities and installation of security patches or upgrades for closing the security gaps left open by system vulnerabilities. Malware like Downad, Gamarueraises, and WannaCry poses significant challenge to manufacturing environments. Figure 4 shows top malware types in the manufacturing industry.
2. **Increased use of IoT Devices:** The escalation in the implementation of IoT devices and objects in industries has established itself as both excellent invention and a security problem. This move towards sensor-based technology can be complex, from reforming secure business processes to reeducating employees and investing in technologies. For adding value to business, smart sensors are collecting data for use in machine learning algorithms, and hence, the volume of this data generated from IoT devices is enormous and, therefore, provides a significant number of entry points for hackers to steal, corrupt, delete or even modify it. Thus, it is advisable for the industries to setup strong cybersecurity strategy and employ cybersecurity professionals to secure IoT networks and devices against unwanted infiltration. According to SonicWall report (2019), that IoT malware attacks jumped 215.7% to 32.7 million in 2018 (up from 10.3 million in 2017) (Figure 5).
3. **Insiders Attack:** Today, malicious insider attack is one of the biggest cybersecurity challenges associated with IoT devices (Khan et al., 2019). According to CERT Guide to Insider Threats, “A malicious insider threat is a current or former employee, contractor, or business partner who has or had authorized access to an organization’s network, system, or data and intentionally exceeded or misused that access in a manner that negatively affected the confidentiality, integrity, or availability of the organization’s information or information systems” (Cappelli et al, 2012*)*. According to Privileged Access Threat Report (2019), insider threats remain top-of-mind and 64% of respondents believe they have suffered a breach due to misused or abused employee access. Thus, an

Table 2. Significance of security elements for components of an information system along with risks, impacts and countermeasures

Security Elements	Risk	Impact	Components of an information system					Attack example and its type	Countermeasures
			Information	People	H/W	S/W	Network		
Confidentiality	Disclosure	Loss of public confidence, embarrassment, legal action against the industry	Safeguard access, transfer, use, modification and deletion of information.	Insider attack or malicious outsider	Camera is compromised	Safeguard against making unauthorized software copy	Attack against communication channels for traffic analysis	Eavesdropping, Escalation of privilege snooping, (Passive attack)	Cryptography, PKI, access control, identity management, privilege management
Integrity	Corruption	Data inaccuracy, fraud, erroneous decisions, Reduction in assurance of an IT system	Ensure the integrity of data at rest, in motion and in use for the system	Insider attack or malicious outsider	IoT connector device is compromised to tamper data	Ensure working program should not be modified, either to do some unintended work or to cause it to fail. For example, 3D design file may be modified during the process of communication	Attack against communication channels and access to the cloud	Modification, Masquerading, Replay, Repudiation (Active attack)	Backups, integrity checks (e.g., cyclical redundancy check), hashing, PKI
Availability	Denial of service	Loss of production time, Reliability, safety and Continuity	Ensure that secured information must be available to the end users	Insider attack or malicious outsider	Equipment should not be stolen or disabled to deny service. For example, DoS attack make 3D printer unavailable	Ensure that programs are available so that users can access them	Ensure that communication lines and networks are available	Denial of services (Active attack)	Operating System Security, host configuration, Intrusion Detection System, anti-malware

Figure 4. Top malware types in the manufacturing industry

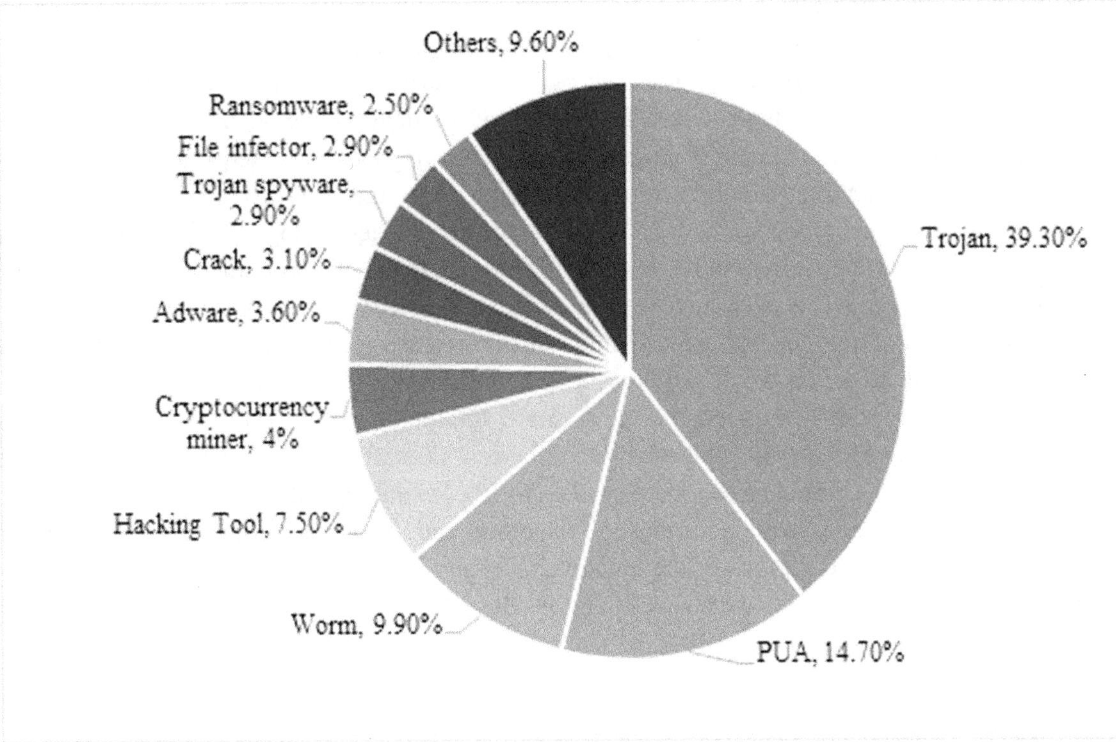

Figure 5. Global IoT malware attacks (SonicWall, 2019)

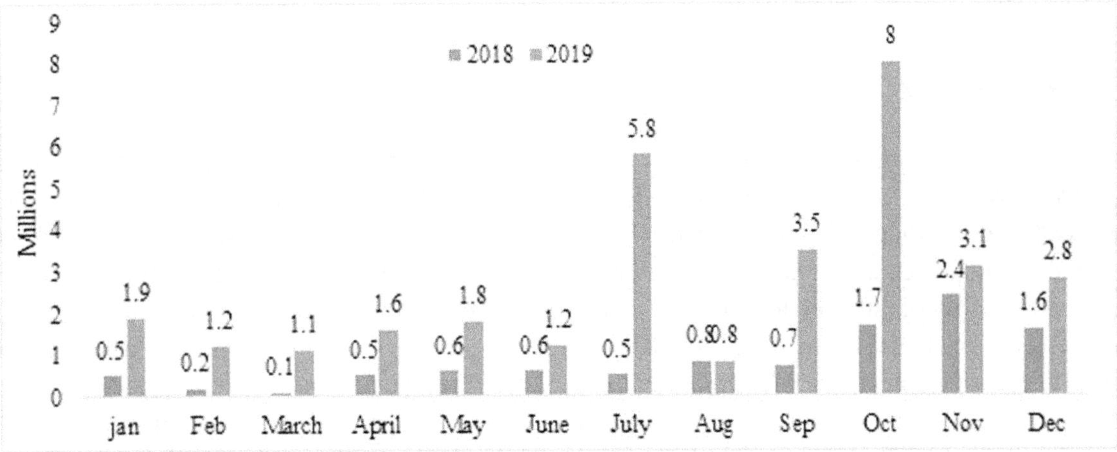

employee can be a threat to the industry, if he leaks or steals sensitive or confidential data to harm the targeted IoT networks.

4. **Cybersecurity Workforce:** Cybersecurity education and training plays an important role towards the contribution of cybersecurity. The lack of cybersecurity education among the workforce is creating a more severe challenge with the adoption of IoT devices and other smart devices which expose industries and individuals to new threats where the impact and consequences are incredible.
5. **Cloudification:** Cloudification may introduce several advantages, but cybersecurity is still one of the biggest challenges (Monshizadeh et al, 2015). Due to cloud technology applications in Industry 4.0, the traditional applications of the system will migrate to the cloud platform and uses Big Data analytics to analyse and visualise huge volume of data in the cloud. According to CSA report (2018), the top barrier to faster cloud adoption is cloud security concerns which include protection against data loss, threats to data privacy, and breaches of confidentiality. Further, this report says lack of qualified security staff is the second biggest barrier to cloud adoption, and more than half of organizations are looking to train and certify their current IT staff to address the shortage, followed by partnering with a managed service provider (MSP), leveraging software solutions, and hiring dedicated staff. As more workloads move to the cloud, industries are recognizing that traditional security tools are not designed for the unique challenges cloud adoption presents and, therefore, strong security management and control solutions designed specifically for the cloud are required for the data protection.

INDUSTRIAL CASE STUDIES

This section presents three industrial case studies of cybersecurity for condition monitoring of safety critical systems.

Railway Industry

As pinpointed by Martinetti et al. (2017), the railway is a complex system which consists of railway infrastructure, rolling stock and operators. Railway infrastructure includes signalling system, track, electrical system, bridges & tunnels, and telecommunication system. Rolling stock consists of both powered and unpowered vehicles that move on the rail track. Maintenance of rail and trackside assets are of high priority of rail operator. Moreover, unplanned trackside maintenance affect train service and inconvenience to thousands of people by delays. Thus, it is always required to predict when maintenance is required to ensure the best service and greatest safety to the people. For adding value to railway business, smart sensors are collecting condition monitoring and predictive maintenance data for use in machine learning algorithms, and the volume of this data generated from IoT devices is enormous and, therefore, provides a significant number of entry points for hackers to steal, corrupt, delete or even modify the data. Thus, cyber-attacks on railway maintenance systems may affect the intensity of the underlying data, which in turn would influence the data driven models and thus affect the maintenance decision-making process. Ultimately, these cyber-attacks may have an impact on railway stakeholders, e.g., threat to the safety of the employees, passengers or the public in general, loss of sensitive railway information, reputational damage, monetary loss, erroneous decisions, loss of dependability, etc. Kour et al. (2019) has provided statistical review of cybersecurity incidents in the transportation sector with

Table 3 Activities related to cybersecurity in railway

<table>
<tr><th>Activities/Frameworks/Standards/Guidelines</th><th>Literature</th><th>Countermeasures (Hutchins, 2011; Malone, 2016; Shift2rail report, 2017; Tarnowski, 2017)</th></tr>
<tr><td>EN 50159:2010, which addresses a particular subject of cybersecurity communications and identifies threats against transmission systems used in the railway sector.</td><td>EN 50159, 2010</td><td rowspan="13">• Virtual private networks (VPN)
• Wavelength-division multiplexing (WDM)
• Cryptography (PE26)
• Firewall
• Demilitarized Zone (DMZ)
• Network segmentation
• Network intrusion detection system/ host intrusion detection system that checks the signalling traffic
• Contingency plans for cyber attack
• Adoption of security standards
• Real-time functional monitoring system
• Double check of received commands by onboard units
• Collaboration with national Community Emergency Response
• Forensics
• Breach insurance
• Data diode
• Behavioral analysis of successful login events
• Decoy servers</td></tr>
<tr><td>A network design for securing data Communication system for automatic train control.</td><td>Bantin and Siu, 2011</td></tr>
<tr><td>APTA SS-CCS-004-16 standard, which covers recommended practices for securing control and communications security systems in rail transit environments for the North America.</td><td>SS-CC, A. P. T. A, 2015</td></tr>
<tr><td>European Union has put into place the network and information security (NIS) directive which aims at safeguarding the key critical infrastructures</td><td>Van, 2015</td></tr>
<tr><td>Rail Cyber Security Guidance to Industry, which supports the rail industry in reducing its vulnerability to cyber-attacks prepared by department for transport, UK.</td><td>Department for transport, 2016</td></tr>
<tr><td>Cybersecurity in the RAILway (CYRAIL) project.</td><td>Shift2Rail, 2016</td></tr>
<tr><td>A high-level cybersecurity risk assessment of a national ERTMS (The European Rail Traffic Management System) implementation of the GB rail industry.</td><td>Bloomfield et al., 2016</td></tr>
<tr><td>Rail Cyber Security Strategy, a cybersecurity vision for the rail industry, provided by the Rail delivery group in UK.</td><td>Rail Delivery Group, 2017</td></tr>
<tr><td>A framework for risk assessment and High-Level Security Assessment (HLSA) based on the IEC 62443 standard.</td><td>Braband, 2017; Masson and Gransart, 2017</td></tr>
<tr><td>AS 7770 Rail Cyber Security, an Australian Standard, which was prepared by a Rail Industry Safety and Standards Board (RISSB).</td><td>AS 7770,2018</td></tr>
<tr><td>Nokia IOT & Analytics for railways.</td><td>Nokia (2019)</td></tr>
<tr><td>Cylus is also providing cybersecurity solution for railways, which aims to be a step ahead of the cyber threat.</td><td>Cylus, 2020</td></tr>
<tr><td>Thales' supported railway sector in its fight against cyber-attacks by participating in the development of CERTs (Computer Emergency Response Teams) as part of the Shift2Rail program of the European Commission.</td><td>Thales, 2020</td></tr>
</table>

a focus on railways and identified various cybersecurity issues and challenges in railway maintenance. Henceforth, organizations operating railway system must establish procedures and plans to safeguard against cyber-attacks, and the research community is active in this area. Table 3 provides list of activities related to cybersecurity in railway.

Power Grid

The power grid enables the transformation of electricity from various sources through power lines and transformers to the consumers. Power transmission is supervised, monitored and performed at control and supervisory substations which are critical to national power grid infrastructure (Jarmakiewicz et al.,

2017). An imminent future power network called as "Smart Grid" is incorporating information processing, fault identification, and self-healing capabilities. According to United States Department of Energy (DOE) (2003) Smart Grid is:

... a fully automated power delivery network that monitors and controls every customer and node, ensuring a two-way flow of electricity and information between the power plant and the appliance, and all points in between. Its distributed intelligence, coupled with broadband communications and automated control systems, enables real-time market transactions and seamless interfaces among people, buildings, industrial plants, generation facilities, and the electric network.

Therefore, due to the distributed and remote control features of power grid, it is vulnerable to cyber-attacks. In December 2015, a first-of-its-kind cyber-attack against Ukraine's power grid cut the lights to 225,000 people to lose power across various areas (Case, 2016). In addition to this, during the maintenance operation, an insider can deliver malicious software via USB device which can control the equipment and affect the national power grid. For example, exploitation of vulnerabilities in supervisory control and data acquisition (SCADA) systems and human-machine interfaces (HMIs) (Alert, 2013). In addition to this, control and supervisory substations consist of various equipment designed by different manufacturers and, therefore, allowing external entities to supervise and monitor these devices. Thus, it is required that cybersecurity system should detect unauthorized actions by authorized people and also report changes to the configurations of smart sensors. Table 4 provides list of activities related to cybersecurity in smart grids.

Table 4 Activities related to cybersecurity in smart grids

<table>
<tr><th>Title of the literature</th><th>Literature</th><th>Countermeasures
(Li et al, 2012; Johansson, 2019; Gunduz & Das, 2020)</th></tr>
<tr><td>The smart grid and cybersecurity: Regulatory policy and issues</td><td>Campbell, 2011</td><td rowspan="7">• Role-based access control (RBAC)
• Authentication scheme that should involve minimal message exchange between grid devices
• IDS (intrusion detection system)
• System Vulnerability Analysis
• Network Obfuscation
• Baiting the Attacker
• Cybersecurity Information Sharing
• Early warning systems
• Dynamic reconfiguration systems
• Demilitarized zones</td></tr>
<tr><td>Preventive maintenance for advanced metering infrastructure against malware propagation</td><td>Guo et al., 2015</td></tr>
<tr><td>Securing the smart grid network: A review</td><td>Kumar et al., 2016</td></tr>
<tr><td>Cybersecurity protection for power grid control infrastructures</td><td>Jarmakiewicz et al., 2017</td></tr>
<tr><td>Challenges of the Existing Security Measures Deployed in the Smart Grid Framework</td><td>Ganguly et al., 2019</td></tr>
<tr><td>Survey of Smart Grid Concepts and Technological Demonstrations Worldwide Emphasizing on the Oman Perspective</td><td>Al-Badi et al., 2020</td></tr>
<tr><td>Cyber-security on smart grid: Threats and potential solutions.</td><td>Gunduz et al., 2020</td></tr>
</table>

Wind Turbine

Wind turbine is one of the most promising renewable energy resources but because of the harsh operating environment its failure rate is more compared to other renewable energy resources. As a result, there is increase in operation cost due to unplanned maintenance and downtime of the wing turbines. Currently, many condition based systems for wind turbines have been developed. These systems use AI, smart sensors, and IoT-based monitoring systems (Lee et al., 2015; Leahy et al., 2016; Zhao et al., 2019). This move towards IoT technology in wind turbine adds value to business, where smart sensors are collecting data for use in machine learning algorithms for the discovery of opportunities to lower maintenanec and operating costs (Figure 6). But, the volume of this data generated from IoT devices is enormous and, therefore, provides a significant number of entry points for intruders to steal, delete or even modify it. However, researchers are active in this area.

Figure 6. IoT technology in wind turbine adapted from Froese, 2016

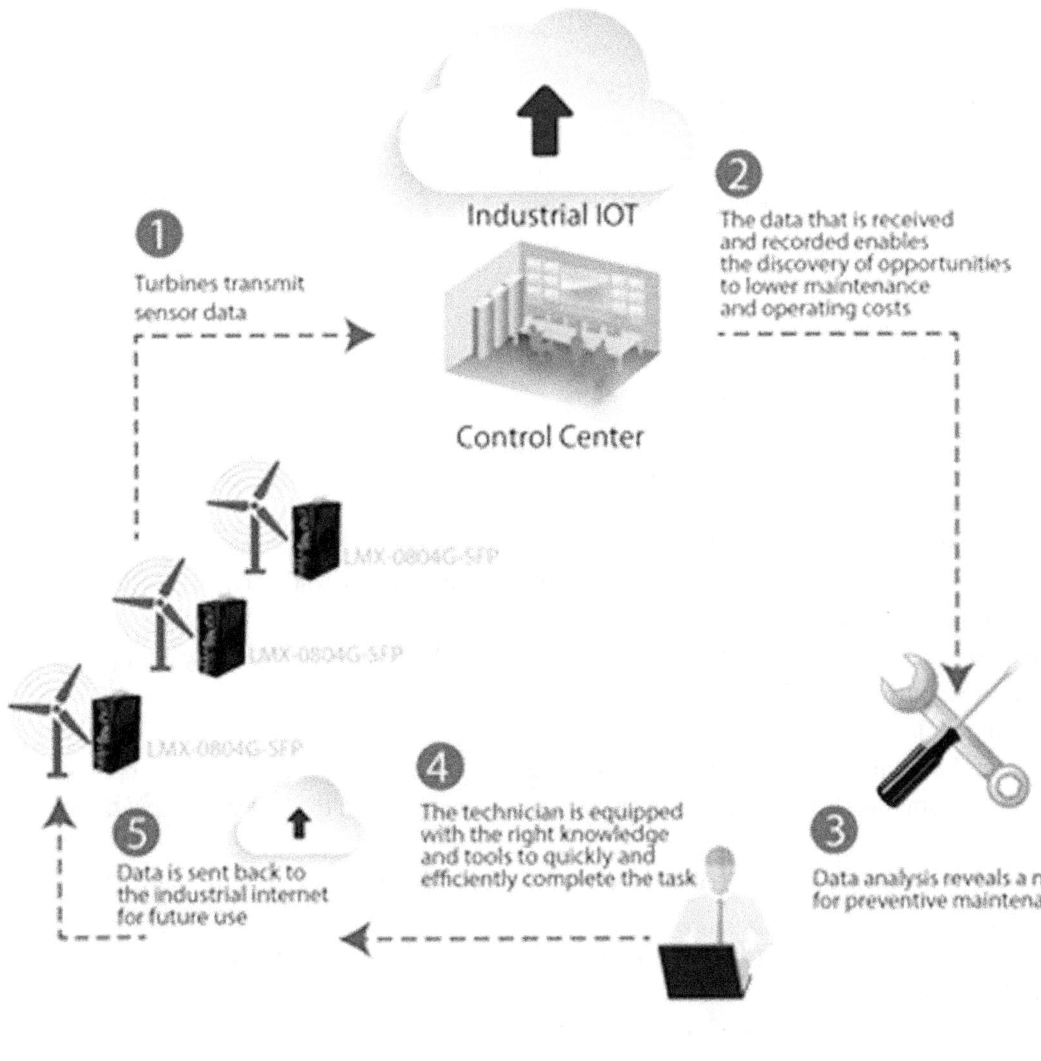

Table 5. Activities related to cybersecurity in wind turbines

Title of the literature	Literature	Countermeasures (Staggs et al., 2017)
Wind farm security: attack surface, targets, scenarios and mitigation.	Staggs et al., 2017	• Multi-factor authentication • Motion sensors • Security cameras • Remote alarm notification systems • Network segmentation • System hardening • System assurance • Strong policies and procedures
Identifying Security Challenges in Renewable Energy Systems: A Wind Turbine Case Study.	Jindal et al., 2019	
Non-resilient behavior of offshore wind farms due to cyber-physical attacks.	Kulev et al., 2019	
Abbaszadeh, M., General Electric Co, 2019. *System and method for anomaly and cyber-threat detection in a wind turbine.* (U.S. Patent Application No. 15/988,515).	Abbaszadeh, 2019	
Interdependent strategic cyber defense and robust switching control design for wind energy systems.	Chen and Zhu, 2017	

Table 5 lists various available literature related to cybersecurity activities in wind turbines along with security countermeasures.

OTHER MORE COUNTERMEASURES

The current literature suggested following countermeasures for cybersecurity in Industry 4.0 (Lezzi et al., 2018). In addtion to these countermeasures, more specific in railway domain were provided by kour et al. (2020).

- Encryption
- Fuzzing to detect software safety errors
- Obfuscation to obscure the intended meaning of communication by making the message difficult to understand
- Patching to solve security vulnerabilities or bugs in the systems
- Vulnerability scan
- Firewall/zoning, gateway and proxy
- Access control (i.e. user authentication, multiple-terminal authorization)
- Quarantine to isolate infected files on a computer's hard disk
- Isolation of data, language, sandbox, Virtual Machine (VM) and Operating System (OS)
- Physical resource-based isolation, also known as air gap
- Software updates
- Machine Learning (ML) method on physical data, covering k-Nearest Neighbours (kNN) algorithm, random forest algorithm and anomaly detection algorithm to real-time detect the malicious attacks
- Physical protection through shielded wires for physical links and utilizing separated racks or spaces
- Tamper proof hardware to prevent attackers from altering system operations and conducting falsification of data
- Keep data distributed instead of centralising them into one, more vulnerable central storage point

- Local storage and analytics so that the raw data do not leave the hardware, and any analytics can be run locally
- Intrusion Detection System (IDS) and Intrusion Prevention System
- Secure Communication (Virtual Private Network, Secure Sockets Layer, IP Security)
- Antivirus and antimalware
- Vulnerability assessment (process of defining, identifying, classifying and prioritizing vulnerabilities in computer systems, applications and network infrastructures)
- Known vulnerability testing
- Static source code analysis to look for software weaknesses
- Penetration testing to evaluate and exploit vulnerabilities

CONCLUSION

Cybersecurity becomes a serious issue due to paradigm shift toward Industry 4.0. With the adoption of IoT technologies, it increases the vulnerability of industries and plants towards cyber threats. It becomes hard to find cybersecurity attacks in the industries because they have become more sophisticated and more adaptable. This chapter focuses on various cybersecurity issues and challenges in Industry 4.0 along with three industrial case studies. This chapter also suggested significance of three security elements i.e. Confidentiality, Integrity, and Availability for the components of information system along with risks, impacts and countermeasures.

REFERENCES

Abbaszadeh, M. (2019). *U.S. Patent Application No. 15/988,515*. Washington, DC: US Patent Office.

Act, A. (1996). Health insurance portability and accountability act of 1996. *Public Law*, *104*, 191.

Al-Badi, A. H., Ahshan, R., Hosseinzadeh, N., Ghorbani, R., & Hossain, E. (2020). Survey of Smart Grid Concepts and Technological Demonstrations Worldwide Emphasizing on the Oman Perspective. *Applied System Innovation*, *3*(1), 5. doi:10.3390/asi3010005

Alert, I. C. S. (2013). *Industrial Control Systems Cyber Emergency Response Team*. Alert 13-164-01. Cited 19.01. 2015, https://ics-cert. uscert. gov/alerts/ICS-ALERT-13-164-01

Alter, S. (2008). Defining information systems as work systems: Implications for the IS field. *European Journal of Information Systems*, *17*(5), 448–469. doi:10.1057/ejis.2008.37

AS 7770:2018. (2018). *Rail cyber security*. Rail Industry Safety and Standards Board.

Bakuei, M., Flores, R., Kropotov, V., & Yarochkin, F. (2017). *Securing Smart Factories*. Academic Press.

Bantin, C. C., & Siu, J. (2011). Designing a secure data communications system for automatic train control. *Proceedings of the Institution of Mechanical Engineers. Part F, Journal of Rail and Rapid Transit*, *225*(4), 395–402. doi:10.1243/09544097JRRT390

Bloomfield, R., Bendele, M., Bishop, P., Stroud, R., & Tonks, S. (2016, June). The risk assessment of ERTMS-based railway systems from a cyber security perspective: Methodology and lessons learned. In *International Conference on Reliability, Safety, and Security of Railway Systems* (pp. 3-19). Springer. 10.1007/978-3-319-33951-1_1

Braband, J. (2017, November). Cyber security in railways: Quo vadis? In *International conference on reliability, safety and security of railway systems* (pp. 3-14). Springer.

Campbell, R. J. (2011, June). *The smart grid and cybersecurity: Regulatory policy and issues*. Congressional Research Service, Library of Congress.

Cappelli, D. M., Moore, A. P., & Trzeciak, R. F. (2012). *The CERT guide to insider threats: how to prevent, detect, and respond to information technology crimes (Theft, Sabotage, Fraud)*. Addison-Wesley.

Case, D. U. (2016). Analysis of the cyber attack on the Ukrainian power grid. *Electricity Information Sharing and Analysis Center, 388*.

Chen, J., & Zhu, Q. (2017, July). Interdependent strategic cyber defense and robust switching control design for wind energy systems. In *2017 IEEE Power & Energy Society General Meeting* (pp. 1-5). IEEE.

Columbus, L. (2016). Roundup of internet of things forecasts and market estimates, 2016. Forbes Magazine.

Cylus. (2020). Retrieved from https://cylus.com/

Department for transport. (2016). *Rail Cyber Security Guidance to Industry*. Retrieved from https://www.rssb.co.uk/Library/improving-industry-performance/2016-02-cyber-security-rail-cyber-security-guidance-to-industry.pdf

EN 50159:2010 (or IEC 62280). (2010). *Railway applications – signalling, telecommunication and processing systems – safety communication in transmission systems*. Brussels: European Committee for Electrotechnical Standardization.

Enisa. (2018). *Industry 4.0 Cybersecurity: Challenges & Recommendations*. Retrieved from https://www.enisa.europa.eu/publications/industry-4-0-cybersecurity-challenges-and-recommendations

Fischer, E.A. (2016). *Cybersecurity Issues and challenges: in brief*. Academic Press.

Fontaine, O., Martinetti, A., & Michaelides-Mateou, S. (2016). Remote pilot aircraft system (RPAS): Just culture, human factors and learnt lessons. In *Chemical engineering transactions, AIDIC* (pp. 205–210). The Italian Association of Chemical Engineering. doi:10.3303/CET1653035

Froese, M. (2016). *Talking with turbines through the Internet of Things*. Retrieved from https://www.windpowerengineering.com/talking-turbines-internet-things/

Ganguly, P., Nasipuri, M., & Dutta, S. (2019, August). Challenges of the Existing Security Measures Deployed in the Smart Grid Framework. In *2019 IEEE 7th International Conference on Smart Energy Grid Engineering (SEGE)* (pp. 1-5). IEEE. 10.1109/SEGE.2019.8859917

Gunduz, M. Z., & Das, R. (2020). Cyber-security on smart grid: Threats and potential solutions. *Computer Networks, 169*, 107094. doi:10.1016/j.comnet.2019.107094

Guo, Y., Ten, C. W., Hu, S., & Weaver, W. W. (2015). Preventive maintenance for advanced metering infrastructure against malware propagation. *IEEE Transactions on Smart Grid*, *7*(3), 1314–1328. doi:10.1109/TSG.2015.2453342

Hutchins, E. M., Cloppert, M. J., & Amin, R. M. (2011). Intelligence-driven computer network defense informed by analysis of adversary campaigns and intrusion kill chains. *Leading Issues in Information Warfare & Security Research*, *1*(1), 80.

ISO/IEC 27032. (2012). Information technology–Security techniques–Guidelines for cybersecurity.

Jarmakiewicz, J., Parobczak, K., & Maślanka, K. (2017). Cybersecurity protection for power grid control infrastructures. *International Journal of Critical Infrastructure Protection*, *18*, 20–33. doi:10.1016/j.ijcip.2017.07.002

Jindal, A., Marnerides, A. K., Scott, A., & Hutchison, D. (2019, June). Identifying Security Challenges in Renewable Energy Systems: A Wind Turbine Case Study. In *Proceedings of the Tenth ACM International Conference on Future Energy Systems* (pp. 370-372). 10.1145/3307772.3330154

Johansson, J. (2019). *Countermeasures Against Coordinated Cyber-Attacks Towards Power Grid Systems: A systematic literature study*. Academic Press.

Khan, A. Y., Latif, R., Latif, S., Tahir, S., Batool, G., & Saba, T. (2019). Malicious Insider Attack Detection in IoTs Using Data Analytics. *IEEE Access : Practical Innovations, Open Solutions*, *8*, 11743–11753. doi:10.1109/ACCESS.2019.2959047

Kissel, R. (2012). Glossary of key information security terms (draft). National Institute of Standards and Technology. http://dx.doi.org/.IR,7298 doi:10.6028/NIST

Kour, R., Aljumaili, M., Karim, R., & Tretten, P. (2019). eMaintenance in railways: Issues and challenges in cybersecurity. *Proceedings of the Institution of Mechanical Engineers. Part F, Journal of Rail and Rapid Transit*, *233*(10), 1012–1022. doi:10.1177/0954409718822915

Kour, R., Thaduri, A., & Karim, R. (2020). Railway Defender Kill Chain to Predict and Detect Cyber-Attacks. *Journal of Cyber Security and Mobility*, *9*(1), 47–90. doi:10.13052/jcsm2245-1439.912

Kulev, N., Reuter, A., Eichhorn, O., Engler, E., & Wrede, C. (2019). Non-resilient behavior of offshore wind farms due to cyber-physical attacks. *REA Symposium on Resilience Engineering Embracing Resilience.*

Kumar, G. E. P., Baskaran, K., Blessing, R. E., & Lydia, M. (2016, December). Securing the smart grid network: A review. In *2016 IEEE International Conference on Computational Intelligence and Computing Research (ICCIC)* (pp. 1-6). IEEE. 10.1109/ICCIC.2016.7919570

Leahy, K., Hu, R. L., Konstantakopoulos, I. C., Spanos, C. J., & Agogino, A. M. (2016, June). Diagnosing wind turbine faults using machine learning techniques applied to operational data. In *2016 IEEE International Conference on Prognostics and Health Management (ICPHM)* (pp. 1-8). IEEE. 10.1109/ICPHM.2016.7542860

Lee, J., & Bagheri, B. (2015). *Cyber-physical systems in future maintenance. In 9th WCEAM Research Papers* (pp. 299–305). Cham: Springer. doi:10.1007/978-3-319-15536-4_25

Lezzi, M., Lazoi, M., & Corallo, A. (2018). Cybersecurity for Industry 4.0 in the current literature: A reference framework. *Computers in Industry*, *103*, 97–110. doi:10.1016/j.compind.2018.09.004

Li, X., Liang, X., Lu, R., Shen, X., Lin, X., & Zhu, H. (2012). Securing smart grid: Cyber attacks, countermeasures, and challenges. *IEEE Communications Magazine*, *50*(8), 38–45. doi:10.1109/MCOM.2012.6257525

Malone, S. (2016). *Using an expanded cyber kill chain model to increase attack resiliency*. Black Hat, US.

Martinetti, A., Braaksma, A. J. J., & van Dongen, L. A. M. (2017). Beyond RAMS Design: Towards an Integral Asset and Process Approach. In L. Redding, R. Roy, & A. Shaw (Eds.), Advances in Through-life Engineering Services (pp. 417-428). Springer. doi:10.1007/978-3-319-49938-3_25

Masson, É., & Gransart, C. (2017, May). Cyber security for railways–a huge challenge–Shift2Rail perspective. In *International workshop on communication technologies for vehicles* (pp. 97-104). Springer. 10.1007/978-3-319-56880-5_10

Monshizadeh, M., Yan, Z., Hippeläinen, L., & Khatri, V. (2015, September). Cloudification and security implications of TaaS. In *2015 World Symposium on Computer Networks and Information Security (WSCNIS)* (pp. 1-8). IEEE.

Nokia. (2019). *IoT & Analytics for railways*. Retrieved from https://onestore.nokia.com/asset/206522

PrivilegedA. T. R. (2019). Retrieved from https://www.beyondtrust.com/resources/whitepapers/privileged-access-threat-report

Rail Delivery Group. (2017). *Rail Cyber Security Strategy*. Retrieved from https://www.raildeliverygroup.com/component/arkhive/?task=file.download&id=469772253

Shift2Rail. (2016). *Cybersecurity in the railway sector*. Retrieved from https://shift2rail.org/project/cyrail/

Shift2rail report. (2017). *CYbersecurity in the RAILway sector D2.1 – Safety and Security requirements of Rail transport system in multi-stakeholder environments.*

SonicWall Cyber Threat Report. (2019). Retrieved from https://cdw-prod.adobecqms.net/content/dam/cdw/on-domain-cdw/brands/sonicwall/2019-sonicwall-cyber-threat-report.pdf

SS-CC. A. P. T. A. (2015). Securing control and communications systems in rail transit environments. Washington, DC: American Public Transportation Association.

Staggs, J., Ferlemann, D., & Shenoi, S. (2017). Wind farm security: Attack surface, targets, scenarios and mitigation. *International Journal of Critical Infrastructure Protection*, *17*, 3–14. doi:10.1016/j.ijcip.2017.03.001

Tarnowski, I. (2017). How to use cyber kill chain model to build cybersecurity? *European Journal of Higher Education IT*. Available: http://www. eunis. org/download/TNC2017/TNC17-IreneuszTarnowski-cybersecurity. pdf

Thales. (2020). *Railway Digitalization: Cybersecurity*. Retrieved from https://www.thalesgroup.com/en/spain/magazine/railway-digitalization-cybersecurity

Thames, L., & Schaefer, D. (2017). *Cybersecurity for industry 4.0.* New York: Springer. doi:10.1007/978-3-319-50660-9

United States Department of Energy. (2003). *"Grid 2030" A National Vision for Electricity's Second 100 Years*. Office of Electric Transmission and Distribution. Retrieved from https://www.energy.gov/sites/prod/files/oeprod/DocumentsandMedia/Electric_Vision_Document.pdf

van Beek, M. V. (2015). Proposal for a Directive of the European Parliament and of the Council on Improving the Gender Balance among Non-Executive Directors of Companies Listed on Stock Exchanges and Related Measures: Is It Courageous and Visionary Or Is It. *RRDE*, 109.

Whitman, M. E., & Mattord, H. J. (2012). *Hands-on information security lab manual*. Cengage Learning.

Willett, K. D. (2008). *Information assurance architecture*. Auerbach Publications. doi:10.1201/9780849380686

Yu, S., Wang, C., Ren, K., & Lou, W. (2010, March). Achieving secure, scalable, and fine-grained data access control in cloud computing. In 2010 Proceedings IEEE INFOCOM (pp. 1-9). IEEE. doi:10.1109/INFCOM.2010.5462174

Zhao, L., Zhou, Y., Matsuo, I., Korkua, S. K., & Lee, W. J. (2019, May). The Design of a Holistic IoT-Based Monitoring System for a Wind Turbine. In *2019 IEEE/IAS 55th Industrial and Commercial Power Systems Technical Conference (I&CPS)* (pp. 1-7). IEEE. 10.1109/ICPS.2019.8733375

Zissis, D., & Lekkas, D. (2012). Addressing cloud computing security issues. *Future Generation Computer Systems*, *28*(3), 583–592. doi:10.1016/j.future.2010.12.006

Section 2

Chapter 6
Safety 4.0:
Analysing the Impact of Digital Technologies

Gabriele Baldissone
https://orcid.org/0000-0001-7015-8995
Politecnico di Torino, Italy

Eleonora Pilone
Politecnico di Torino, Italy

Lorenzo Comberti
Politecnico di Torino, Italy

Vincenzo Tarsitano
Politecnico di Torino, Italy

ABSTRACT

In recent years augmented reality has begun to be a presence in various industrial sectors. In augmented reality the operator's perception of reality is enriched through virtual information useful to help him in his working activity. Augmented reality can be generated through various technical solutions. A first classification can be made based on how the equipment is used: head mounted displays, handheld displays, and spatial displays. Maintenance can benefit from the introduction of augmented reality as it can help operators in activities characterized by variability and in the risky activities. This is because augmented reality allows to remember the steps of the procedures and highlight the dangers if present. However, the use of augmented reality devices can bring new dangers including ergonomic problems or visual fatigue or information overload. This chapter presents an index methodology for assessing the risks introduced by augmented reality devices.

DOI: 10.4018/978-1-7998-3904-0.ch006

INTRODUCTION

Maintenance is one of the most important activities in the lifetime of an industrial plant; i.e., a proper maintenance can help keeping under control the energy consumption and the safety (Darabnia & Demichela, 2013; Demichela et al., 2018).

Maintenance can be preventive or reactive. As demonstrated by Lee (2005) et al., preventive maintenance is the most convenient type, because an adequate and e careful planning allows the managers to optimize plant shutdown (Gits, 1992) and storage of spare parts. In this way, maintenance results economically convenient and the probability of equipment failure that can cause accidents is reduced.

Sometimes, risk assessment process neglects or treats superficially the maintenance tasks, equating them to normal routine activities, however Maintenance is an activity that implies more variables and greater risks than routine tasks.

The adoption of Augmented Reality systems can help reducing the risks related to maintenance activities; i.e. Neumann and Majoros (1998) proposed its use for Aircraft maintenance, while Henderson and Feiner (2011) suggested its application for the military equipment maintenance. Augmented Reality promotes risk reduction during maintenance procedures in several ways: it highlights the dangers, it reproduces warning signals and it monitors the respect of the procedures (Tatic & Tešic, 2017). However, the adoption of Augmented Reality can also bring new risks (Grabowski, Rowen, & Rancy, 2018), including stress and excessive vigilance, or ergonomic problems. For these reasons, some methodologies for risk assessment related to Augmented Reality have been proposed, such as Luts (2018), but a global balance of the effects of Augmented Reality can be obtained only taking into account its effectiveness in reducing the risks inherent in the activities.

In this chapter, a review of the technologies related to Augmented Reality is provided; then, an index methodology is proposed for the assessment of the risks associated with Augmented Reality. An expeditious approach was adopted to facilitate its use and diffusion; in case the results show unacceptable risks, or are ambiguous, more advanced risk assessment methods can be adopted.

MIXED REALITY

Thanks to the advancement of information technology and the advancement of computing power, it is possible to make the user perceive a different or enriched reality. In addition, to the operator can interact in real time with digitally generated information. These technologies are called Mixed Reality: reality is expanded with virtual information. The Mixed Reality presents a continuous spectra enrichment of reality with virtual information (Milgram et al., 1994). In Figure 1 a graphical representation of Mixed Reality is presented.

Figure 1. Mixed reality spectrum

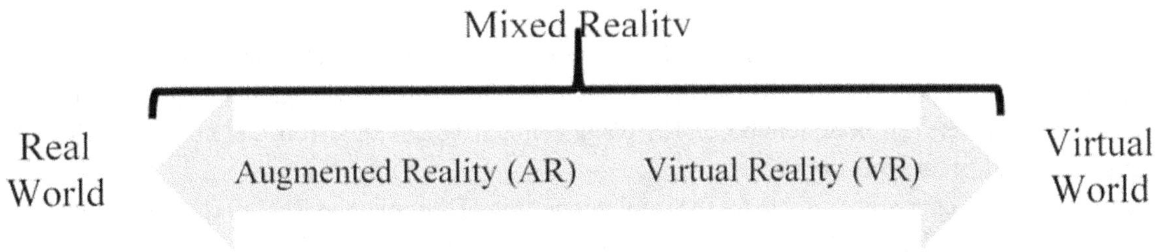

Virtual Reality stands at one end of the Mixed reality spectrum: here, the user perceives a reality generated by IT systems and can interact with it in real time. Virtual Reality has been used in various areas of work and play: in the workplace, most applications concern training, for example Cha et al (2012) proposed V.R. for firefighter training, Lucas and Thabet (2008) for mine safety training and Lewis et al. (2011) for training in surgical area.

Augmented Reality stands at the opposite side of Mixed Reality.

Augmented Reality

Augmented Reality is a branch of Mixed Reality. In Augmented Reality, the operator's perception of reality is enriched with digital information (Feiner, 1994) (Figure 2).

Figure 2. Augmented reality scheme

Augmented Reality increases the users' knowledge by providing him with more information on reality or by helping him to correctly carry out the procedures (Tatic & Tešic, 2017). Augmented Reality can find applications related to leisure or work, i.e. several museums have developed systems for Augmented Reality to facilitate the fruition of the collections (Miyashita, et al., 2008). Kamat and El-Tawil (2007) envisages the use of Augmented Reality in the damage assessment following disasters.

Augmented Reality was also proposed for different productive sectors: by Sakas (2002), for medical field; by Behzadan and Kamat (2013) for operator training activities;. by Shin and Dunston (2008) for supervision and inspection in the construction field; by Behzadan et al. (2015) for excavation activities, to highlight the presence of hidden obstacles.

The use of Augmented Reality has also been proposed in the manufacturing industry (a review of these applications is presented by Bottani and Vignali (2019)), from the design phase (Nee et al., 2012) to the assembly phase (Stork & Schubö, 2010; Wang et al., 2016). (Vanderroost, et al., 2017). Proposed Augmented Reality for food logistics.

Several technologies are available for the implementation of Augmented Reality (Huang et al., 2013); they can be distinguished on the basis of the Tracking and Registration method, that means the process used by the equipment to recognize the information to be proposed and aligned with real objects. The Tracking and Registration system can be Sensor-based or Visual-based. In the Sensor-based system, the equipment identifies the position of digital information on the basis of a series of different electronic sensors: examples of applications are inertial-based (Hallaway et al., 2010) and GPS-based (Reitmayr & Schmalstieg, 2003). The Visual-Based method determines the information and its position thanks to images taken by the equipment video-camera. The Visual-based method is divided into Nature Feature methods (Neumann & You, 1999) and Marker based methods (Kato & Billinghurst, 1999). This short example makes clear the complexity and the variety of the technologies used for the Augmented Reality.

A more macroscopic classification of the technologies used for Augmented Reality is related to the method of use: head-mounted, hand held and spatial displays.

Head Mounted Displays

This category includes various devices, characterized by an interaction with the operator that does not require the use of the hands. This entails many advantages, especially from the point of view of safety and productivity: since the user is free to use both hands and can continue his/her activity without manual interaction with the device, he/she is subjected to minor distraction and fatigue, and the level of attention remains high and focused on the activity. The use of these devices has been proposed in various areas, for example in the medical field (Chen, et al., 2015) or in the military one (Ferrin, 1999).

Head mounted displays usually have a viewer positioned on the head, with a transparent (*Optical see through*) or not transparent (*Video see through*) display. The latter type is more complex, because the visualization of the real world takes place on a display connected to the computer, that requires the presence of two cameras: a first camera that acquires real images and a second monitor that allows the user to view the real scene enriched with virtual information. *Optical see through* devices, on the other hand, give a more natural perception of reality by superimposing virtual information on the real scene, using semi-transparent display. In this case, the digital information is projected onto a semi-transparent and semi-reflexive surface, where digital images (reflected from the surface) are mixed with the vision of the real world (which crosses the surface). An example of *optical see through* display is described by Liu et al. (2010).

One of the hand-free technologies currently most developed and used is Smart Glasses, a device that thanks to recent technological innovations can be worn as normal sunglasses. The Smart Glass can be defined as *wearable Augmented Reality (AR) devices that are worn like regular glasses and merge virtual information with physical information in a user's view field* (Rauschnabel et al., 2015). Smart Glasses are easy to use and have a smaller encumbrance in comparison to other technological solutions; also, the use of voice-activated system is simpler and has a more intuitive interaction with the operator. Sometimes, operators make resistance to the use of Smart Glasses, because of technical problems (limited battery life, misunderstandings of voice recognition systems, smaller display, ...) that limit their use. Some studies evaluate which features of Smart Glasses should be implemented to extend their acceptability (Rauschnabel & Ro, 2016) and diffusion.

Hand-Held Displays

This category of Augmented Reality devices includes all those devices that the user holds in his/her hands: usually, (see examples in Wagner and Schmalstieg (2006). The main classes of hand-held displays are smartphones, personal digital assistant and tablets. Smartphones and personal digital assistant are popular and equipped with powerful processors, but due to their small size and limited data entry capabilities, they are not convenient Augmented Reality platforms, especially in cases of three-dimensional interfaces. Tablets have a greater potential for Augmented Reality applications, even if their size and the difficulty of holding them with one hand greatly limit their use.

These types of devices work as *see through video* systems: the device camera captures the real world, the obtained images are enriched with digital information, and finally the complete images are projected on the device display. This approach presents positive and negative aspects: on one side, it is possible to manage the brightness and control the interaction between real and virtual objects, but on the other side these devices have limited resolution and visual field, and the short distance between the camera and the operator's eyes could cause disorientation.

Spatial Displays

This last category includes devices that use projections, holograms and other elements to represent graphics on a screen. Unlike other types of devices, these displays combine technology with the environment to leave the operator free from devices. Three different types of approach to Spatial displays are identified, depending on the interaction with the environment and the technology used for Augmented Reality (Bimber & Raskar, 2004). In the first case, the display is set and the operator's field of view is limited by the size of the monitor and its distance; despite of these disadvantages, this solution is the simplest and most economical, because it uses standard computer equipment. The second type of approach includes devices that generate images that can align with the real environment through transparent screens or optical holograms. These spatial devices are characterised by a wider field of view, high resolution and better eye positioning; also, they are quite easy to use. The last type the direct augmentation displays are based on the use of projectors, that project the images on the surfaces of physical objects, thus exceeding the limit relative to the possibility of displaying the image exclusively within the user's field of view. The advantages of these systems mainly concern the increase of the viewing area and the improvement of the image quality, while the negative note is related to the darkening of the images in case they are not projected on flat surfaces.

METHODOLOGY

The adoption of new technologies like the Augmented Reality can introduce new risks to the workplace; several methodologies were introduced to assess the risks, like explained by Luts (2018).

This chapter presents an index methodology for assessing the risks associated with the use of the Augmented Reality systems: it can be applied both to head-mounted systems (eg Smart Glasses) and to hand-held systems (eg smartphones and tablets).

The proposed methodology aims at being expeditious and easy to apply; in case its results should be not clear, more detailed analysis methodologies can be adopted to perform the risk analysis.

In order to evaluate the opportunity to adopt Augmented Reality, the risk assessment should not be limited to the intrinsic risks connected to it; the advantages that the adoption of A.R. produces in terms of risk reduction for the activity in which it is implemented should be considered. However, the latter activity is not included in the methodology presented, because the analysis of the positive correlation with each single activities require a major level of detail and in-depth analyses.

Input Parameter

In order to be able to fully analyze the risks associated with Augmented Reality systems, three main components have been taken into account: ergonomics, technology, type of working activity An expert judgement and the available experience on the adoption of A.R. tools demonstrated that these were the most influent components.

Each component is divided into a series of parameters.

1. Ergonomics: Postures to be adopted during work activities (e.g. postures to allow correct operation of Augmented Reality systems), and interferential aspects between the device and the operator (e.g. weight and size of the device, display light exposure) are considered. To evaluate ergonomics, the following three parameters are used:
   a. Posture parameter;
   b. Device weight and size;
   c. Display light parameter.

Table 1 shows the index used to evaluate the ergonomic parameters.

Table 1. Index for ergonomic parameters

Index	a. Posture parameter	b. Device weight and size	c. Display light parameter
1	The devices during the work does not require to adopt uncomfortable positions or the uncomfortable position is limited to a short time.	Weight < 0.1 kg.	The display light is not particularly bright, and the operator often looks away from the display
2	The devices can require to adopt uncomfortable positions for certain periods, but for the majority of the work activity is carried out in comfortable position.	0.1 kg > Weight >0.2 kg.	The display is kept at an adequate distance from the eyes; in addition the light is not very intense and occasionally the operator can look away from the display
3	The device forces the operator to adopt uncomfortable positions for the entire work shift or for long time, causing i.e. muscle pain, back pain and neck pain;	Weight > 0.2 kg or the size of the device is not comfortable for the operator.	The display is placed close to the operator's eyes, the display light is very intense, the operator fixed his/her gaze on the display for a long time.

2. Technology: The technological parameters taken into account are:
   d. Equipment and internet connection availability factor: a good and stable Internet connection is required for a good functioning of A.R. applications. The absence of an adequate connection may cause to the operator performing an activity without adequate support. Also the equipment available are take into account because the equipment fault may cause to the operator performing an activity without adequate support;
   e. Cyber security: ability to resist at cyber intrusions, that can cause loss or theft of sensitive data. In addition, a possible intrusion may represent a potential hazard for the operators, i.e. through the silencing of the warning signs.

Table 2 describes the index used to analyze the technological parameters.

Table 2. Index for technological parameters

Index	d. Equipment and internet connection availability factor	e. Cyber security
1	Availability of between 70% and 100% of a connection speed with communicative protocols that guarantee higher bandwidth and lower latency AND equipment unavailability lower than 10^{-3}	Adequate technical and organisational measures to avoid unauthorised intrusion and to avoid loss, destruction or damage
2	Internet coverage between 50% and 70% to ensure fluidity and continuity of operations AND equipment unavailability between 10^{-3} and 10^{-2}	Measures to ensure the accuracy and reliability of data during their life cycle, standard measure to avoid unauthorised intrusion in the company cyber system
3	Internet coverage in the working environment not exceeding 50% with consequent problems related to the continuity of operations AND equipment unavailability higher than 10^{-2}	Minimum, incomplete and not applicable measures for the data protection and for the IT company technology

3. **Working Activity Aspects:** The following parameters are analyzed:

   f. Information overload: the amount of information to which the operator is exposed. A prolonged exposure to large amounts of information can cause stress or confusion.
   g. Isolation parameter: the use of Augmented Reality systems may monopolize the operator's attention, bringing he/she to neglect the presence of possible hazards in the surroundings.

Table 3 shows the index related to the working activity parameters.

Analysis Structure

The analysis proceeds through binary comparisons according to the scheme reported in Figure 3.

Binary comparisons are developed through comparison matrices. An example comparison matrix is shown in the Table 4, the other comparison matrices are shown in the appendix.

The compilation of comparison matrices was developed through an expert judgment, that considered the weight of each parameter over the other.

Table 3. Index for working activity parameters

Index	f. Information overload	g. Isolation parameter
1	The quantity of information provided by the devices is neither too high nor too low. The information allows the operator to work in a comfortable way.	The operator takes systematic breaks during work and consequently is able to perceive possible hazards in the work environment and manage critical situations in the correct way
2	The information are not always clear and complete. The incorrect information can cause to the operator discomfort and frustration.	The operator spends more time to perform a given task and takes few breaks that do not allow him/her to obtain a complete knowledge of the possible hazards related to the work environment
3	The information provided by the devices are too much or too poor. The operator is exposed to stressful conditions and the work activity is hindered.	The operator spends his/her entire work shift in the vicinity of electronic devices; he/she is concentrated exclusively on the activity and is not able to perceive the other risks present within the work environment.

Table 4. Comparison matrix A

Matrix A	a. Posture parameter			
b. Device weight and size	Index	1	2	3
	1	1	1	2
	2	1	2	3
	3	2	3	3

Methodology Output

The method returns an index of risk associated with the use of Augmented Reality systems.

The risk indices are collected in the following risk levels:

1. Negligible risk (R=1) the Augmented Reality device does not introduce new risks in the work space;
2. Low risk (R=2) the Augmented Reality device can be adopted together with procedural measures to monitor the risk associated;
3. Medium risk ($3 \leq R \leq 4$) before the adoption of the Augmented Reality device, technical and procedural measures to reduce the new risks must be planned;
4. High risk (R=5) the adoption of the Augmented Reality device has to be stopped. New associated risks must be analyzed in detail, and adequate protective and preventive measures must be adopted.

CASE STUDY

An industrial company started to evaluate the use of Augmented Reality systems during maintenance. The introduction of Augmented Reality systems aimed at supporting maintenance for the management of dynamic activities, that sometimes complex and risky.

Figure 3. Analysis methodology structure

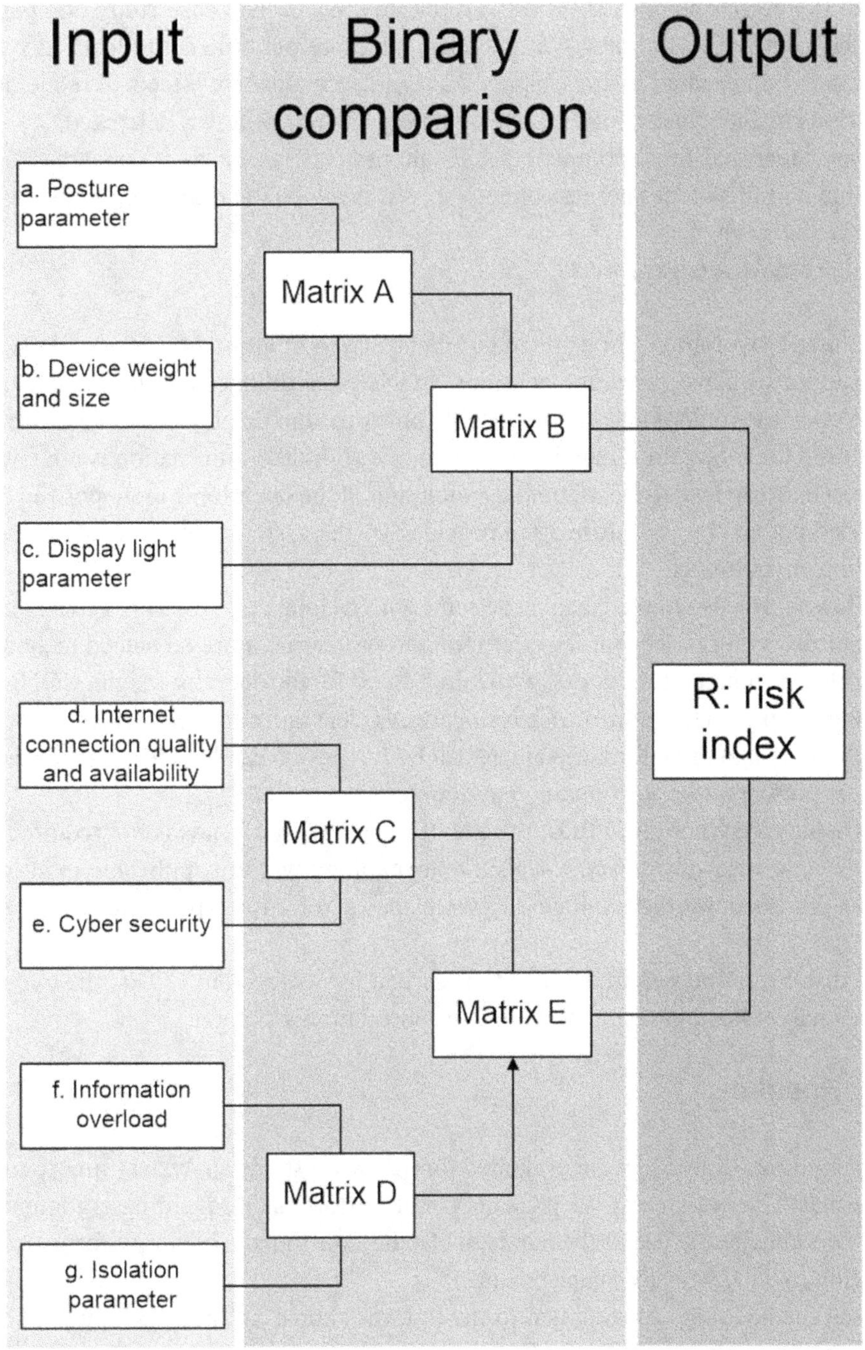

In a preliminary phase, the choice on the adoption of new technology systems management includes parameters like the economic cost, the system reliability, etc.... The case-study company decided to evaluate also the safety effects of new A.R. devices, therefore possible additional risks were assessed with the methodology presented in this chapter. As regards the positive effects of reducing the specific risks of the various maintenance activities, the company HSE system was delegated.

The company identified three competing A.R. alternatives; as far as it concerns Technology parameters, a stable and sufficient internet connection was available, and the corporate IT systems had a discreet Cyber Security level.

The three alternatives analyzed were:

1. **Smartphone:** Use of applications on smartphones (weight around 0.15 kg). PROs: well-known, reliable and inexpensive hardware solutions. CONs: smartphones require the use of at least one hand; the operator could adopt uncomfortable postures; the display can have a strong brightness which, if used for a long time, can tire the operator's sight; the information is not always clear and complete, which can lead to frustration; the attention of the operator is monopolized by the activity to be carried out and by the information provided by the system, with possible isolation from the surrounding environment.
2. **Smart Glasses:** Head-mounted display systems with weight about 0.15 kg. PROs: the use of hands is not required, with consequent fewer ergonomic problems; more advanced technology; display system with a less intense brightness, with minor stress for the operator's sight. CONs: higher costs; the information provided by the device is not always clear and complete, with operator discomfort; the work activity and the information provided by the device can lead the operator not to perceive the dangers present in the surrounding environment.
3. **Spatial Display Technology:** PROs: the use of the operator's hands is not required, the operator does not have to wear any device. CONs: the technology was not applicable in all rooms subject to maintenance, both for the brightness of the places and for the type of surfaces present.

The spatial display option was immediately discarded for its relevant CONs; the two remaining alternatives were analyzed using the methodology proposed in this chapter.

Case Study Results

Firstly, the input indexes (Table 5) were evaluated for the two systems analyzed (Smartphone and Smart glasses). The Smart Glasses alternative presents better performance regarding ergonomic conditions, because it does not require the use of the hands; it also demonstrated a better performance about visual fatigue, according to the Company management.

The proposed methodology was applied to the indexes shown in Table 5: Figure 4 schematize the application for the Smartphone and Smart Glasses.

A risk index equal to 3 (medium risk) was obtained for the Smartphone, while for the Smart Glasses, the risk index was 2 (low risk); this result is due to a better ergonomic approach, and a minor eyes fatigue for the operators.

Table 5. Case studies indexes

Index	Smartphone	Smart Glasses
a. Posture parameter	2	1
b. Device weight and size	2	2
c. Display light parameter	3	2
d. Equipment and Internet connection availability factor	1	1
e. Cyber security	2	2
f. Information overload	2	3
g. Isolation parameter	3	3

However, considering a wider point of view, the Smartphone presents significant advantages over the other case, related to lower costs and the use of more consolidated technologies. For these reasons, management is also taking into account other parameters to assess whether and which Augmented Reality systems to adopt.

CONCLUSION

In the industrial field, applications of Augmented Reality technologies are starting to spread. Maintenance can also take advantage of these new technologies: considering the peculiarities of maintenance activities, like i.e. the variability of activities and the presence of various dangers, Augmented Reality systems should bring great benefits. Augmented Reality could help in checking compliance with procedures and highlighting, via warning signs, potential dangers. But at the same time, Augmented Reality systems could introduce new dangers, such as visual discomfort related to the prolonged use of displays or stress problems related to information overload.

Therefore, the adoption of Augmented Reality systems must take into consideration various aspects such as: economic, technology and safety. As far as it concerns the safety aspects, both the advantages in terms of reduction of the specific risks of the activity, and the emerging risks due to the introduction of the new technology should be analyzed.

This chapter presented an expeditious Index methodology aimed at analyzing new risks introduced by the adoption of Augmented Reality systems. In particular this methodology was intended to be a decision making support for any company Management involved in AR system introduction providing an expeditive quantification of new risk introduced. The proposed methodology takes into account aspects related both to ergonomics, technology, and working activity, through 7 different indexes. The methodology returns a risk index, whose magnitude is evaluated on the basis of four levels (Negligible, Low, Medium, High).

Due to the expeditious nature of this methodology in case it would return ambiguous results, or in case of very complex cases, a more detailed methodology should be developed and applied.

As regards the analysis of safety benefits associated with the use of Augmented Reality systems, the assessment is strongly case-sensitive and it should be done case by case, since the fields of application are wide and it is difficult to develop a more general framework with a general validity.

Figure 4. Case study analysis, (a) smartphone case; (b) smart glass case

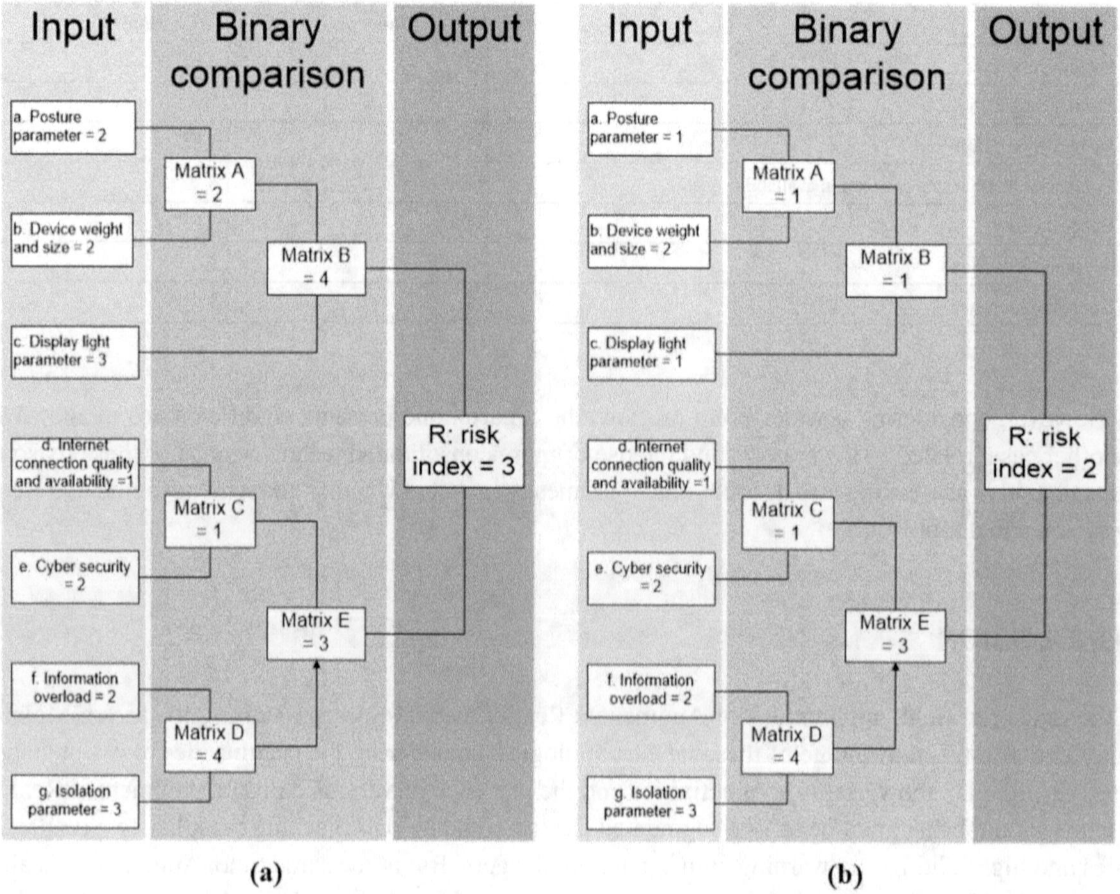

The presented methodology was tested on a case study which compared three Augmented Reality alternatives applied to a maintenance process. The first alternative was related to Smartphone applications, the second consisted of Smart Glasses and the last of Spatial display systems. The method allowed a quick quantified comparison between the tree proposal highlighting immediately the worst one. The following comparison of the two remained showed that the Smart Glass-based alternative introduced less new risks.

In conclusion, the use of Augmented Reality systems applied to maintenance can bring advantages both in terms of production and safety. But before proceeding to their adoption to carefully evaluate both the pros and cons is necessary.

REFERENCES

Behzadan, A., Dong, S., & Kamat, V. (2015). Augmented reality visualization: A review of civil infrastructure system applications. *Advanced Engineering Informatics*, *29*(2), 252–267. doi:10.1016/j.aei.2015.03.005

Behzadan, A., & Kamat, V. (2013). Enabling discovery-based learning in construction using telepresent augmented reality. *Automation in Construction, 33*, 3–10. doi:10.1016/j.autcon.2012.09.003

Bimber, O., & Raskar, R. (2004). *Spatial Augmented Reality. Merging Real and Virtual Worlds*. Wellesley, MA: K Peters Ltd.

Bottani, E., & Vignali, G. (2019). Augmented reality technology in the manufacturing industry: A review of the last decade. *IISE Transactions, 51*(3), 284–310. doi:10.1080/24725854.2018.1493244

Cha, M., Han, S., Lee, J., & Choi, B. (2012). A virtual reality based fire training simulator integrated with fire dynamics data. *Fire Safety Journal, 50*, 12–24. doi:10.1016/j.firesaf.2012.01.004

Chen, X., Xu, L., Wang, Y., Wang, H., Wang, F., Zeng, X., ... Egge, J. (2015). Development of a surgical navigation system based on augmented reality using an optical see-through head-mounted display. *Journal of Biomedical Informatics, 55*, 124–131. doi:10.1016/j.jbi.2015.04.003 PMID:25882923

Darabnia, B., & Demichela, M. (2013). Data field for decision making in maintenance optimization: An opportunity for energy saving. *Chemical Engineering Transactions, 32*, 259–264.

Demichela, M., Baldissone, G., & Darabnia, B. (2018). Using Field Data for Energy Efficiency Based on Maintenance and Operational Optimisation. A Step towards PHM in Process Plants. *Processes (Basel, Switzerland), 6*(3), 1–15. doi:10.3390/pr6030025

Feiner, S. (1994). Redefining the user interface: Augmented reality. *ACM SIGGRAPH. Course Notes, 2*, 1–18.

Ferrin, F. (1999). Update on optical systems for military head-mounted displays. In *Proc. SPIE 3689, Helmet- and Head-Mounted Displays IV* (pp. 178–185). Orlando, FL: SPIE. doi:10.1117/12.352829

Gits, C. (1992). Design of maintenance concepts. *International Journal of Production Economics, 24*(3), 217–226. doi:10.1016/0925-5273(92)90133-R

Grabowski, M., Rowen, A., & Rancy, J. (2018). Evaluation of wearable immersive augmented reality technology in safety critical systems. *Safety Science, 103*, 23–32. doi:10.1016/j.ssci.2017.11.013

Hallaway, D., Feiner, S., & Hollerer, T. (2010). Bridging the gaps: Hybrid tracking for adaptive mobile augmented reality. *Applied Artificial Intelligence, 18*(6), 477–500. doi:10.1080/08839510490462768

Henderson, S., & Feiner, S. (2011). Exploring the Benefits of Augmented Reality Documentation for Maintenance and Repair. *IEEE Transactions on Visualization and Computer Graphics, 17*(10), 1355–1368. doi:10.1109/TVCG.2010.245 PMID:21041888

Huang, Z., Hui, P., Peylo, C., & Chatzopoulos, D. (2013). Mobile augmented reality survey: a bottom-up approach. *The Computing Research Repository (CoRR)*, 1309.4413.

Kamat, V., & El-Tawil, S. (2007). Evaluation of Augmented Reality for Rapid Assessment of Earthquake-Induced Building Damage. *Journal of Computing in Civil Engineering, 21*(5), 303–310. doi:10.1061/(ASCE)0887-3801(2007)21:5(303)

Kato, H., & Billinghurst, M. (1999). Marker tracking and HMD calibration for a video-based augmented reality conferencing system. In *Proceedings 2nd IEEE and ACM International Workshop on Augmented Reality (IWAR'99)* (pp. 85-94). San Francisco, CA: IEEE. 10.1109/IWAR.1999.803809

Lee, H. (2005). A cost/benefit model for investments in inventory and preventive maintenance in an imperfect production system. *Computers & Industrial Engineering*, *48*(1), 55–68. doi:10.1016/j.cie.2004.07.008

Lewis, T., Aggarwal, R., Rajaretnam, N., Grantcharov, T., & Darzi, A. (2011). Training in surgical oncology – The role of VR simulation. *Surgical Oncology*, *20*(3), 134–139. doi:10.1016/j.suronc.2011.04.005 PMID:21605972

Liu, S., Hua, H., & Cheng, D. (2010). A Novel Prototype for an Optical See-Through Head-Mounted Display with Addressable Focus Cues. *IEEE Transactions on Visualization and Computer Graphics*, *16*(3), 381–393. doi:10.1109/TVCG.2009.95 PMID:20224134

Lucas, J., & Thabet, W. (2008). Implementation and evaluation of a VR task-based training tool for conveyor belt safety training. *Journal of Information Technology in Construction*, *13*(40), 637–659.

Lutz, R. (2018). Safe-AR: Reducing Risk While Augmenting Reality. In *IEEE 29th International Symposium on Software Reliability Engineering (ISSRE)* (pp. 70-75). Memphis, TN: IEEE.

Milgram, P., Takemura, H., Utsumi, A., & Kishino, F. (1994). Augmented reality: a class of displays on the reality-virtuality continuum. In *Photonics for industrial applications* (pp. 282–292). Boston, MA: SPIE.

Miyashita, T., Meier, P., Tachikawa, T., Orlic, S., Eble, T., Scholz, V., ... Lieberknecht, S. (2008). An Augmented Reality Museum Guide. In *IEEE International Symposium on Mixed and Augmented Reality 2008* (pp. 103-106). Cambridge, UK: IEEE. 10.1109/ISMAR.2008.4637334

Nee, A., Ong, S., Chryssolouris, G., & Mourtzis, D. (2012). Augmented reality applications in design and manufacturing. *CIRP Annals - Manufacturing Technology, 61*(2), 657-679.

Neumann, U., & Majoros, A. (1998). Cognitive, performance, and systems issues for augmented reality applications in manufacturing and maintenance. In *Proceedings. IEEE 1998 Virtual Reality Annual International Symposium* (pp. 4-11). Atlanta, GA: IEEE. 10.1109/VRAIS.1998.658416

Neumann, U., & You, S. (1999). Natural feature tracking for augmented reality. *IEEE Transactions on Multimedia*, *1*(1), 53–64. doi:10.1109/6046.748171

Rauschnabel, P. A., Brem, A., & Ro, Y. K. (2015). *Augmented Reality Smart Glasses: Definition, Conceptual Insights, and Managerial Importance*. Dearborn, MI: The University of Michigan-Dearborn, College of Business.

Rauschnabel, P. A., & Ro, Y. K. (2016). Augmented reality smart glasses: An investigation of technology acceptance drivers. *International Journal of Technology Marketing*, *11*(2), 123–148. doi:10.1504/IJTMKT.2016.075690

Reitmayr, G., & Schmalstieg, D. (2003). Data Management Strategies for Mobile Augmented Reality. In *Proceedings of the International Workshop on Software Technology for Augmented Reality Systems* (pp. 47-52). Tokyo, Japan: IEEE.

Sakas, G. (2002). Trends in medical imaging: from 2D to 3D. *Computers & Graphics*, *26*(4), 577–587. doi:10.1016/S0097-8493(02)00103-6

Shin, D., & Dunston, P. (2008). Identification of application areas for Augmented Reality in industrial construction based on technology suitability. *Automation in Construction*, *17*(7), 882–894. doi:10.1016/j.autcon.2008.02.012

Stork, S., & Schubö, A. (2010). Human cognition in manual assembly: Theories and applications. *Advanced Engineering Informatics*, *24*(3), 320–328. doi:10.1016/j.aei.2010.05.010

Tatic, D., & Tešic, B. (2017). The application of augmented reality technologies for the improvement of occupational safety in an industrial environment. *Computers in Industry*, *85*, 1–10. doi:10.1016/j.compind.2016.11.004

Vanderroost, M., Ragaert, P., Verwaeren, J., De Meulenaer, B., De Baets, B., & Devlieghere, F. (2017). The digitization of a food package's life cycle: Existing and emerging computer systems in the logistics and post-logistics phase. *Computers in Industry*, *87*, 15–30. doi:10.1016/j.compind.2017.01.004

Wagner, D., & Schmalstieg, D. (2006). Handheld Augmented Reality Displays. In *IEEE Virtual Reality Conference (VR 2006)* (pp. 321-321). Alexandria, VA: IEEE. 10.1109/VR.2006.67

Wang, X., Ong, S., & Nee, A. (2016). A comprehensive survey of augmented reality assembly research. *Advances in Manufacturing*, *4*(1), 1–22. doi:10.100740436-015-0131-4

APPENDIX

In this appendix the comparison matrix for the proposed methodology are presented.

Table 6. Comparison matrix B

Matrix B	c. Display light parameter.			
Result of Matrix A	Index	1	2	3
	1	1	2	3
	2	1	3	4
	3	2	4	5

Table 7. Comparison matrix C

Matrix C	d. Equipment and internet connection availability factor			
e. Cyber security	Index	1	2	3
	1	1	1	2
	2	1	2	3
	3	2	3	3

Table 8. Comparison matrix D

Matrix D	f. Information overload			
g. Isolation parameter	Index	1	2	3
	1	1	1	2
	2	2	3	4
	3	3	4	5

Table 9. Comparison matrix E

Matrix E	Result of matrix D					
Result of matrix C	Index	1	2	3	4	5
	1	1	2	2	3	3
	2	1	2	3	4	5
	3	2	3	4	5	5

Table 10. Comparison matrix R

Matrix D	Result of matrix B					
Result of matrix E	Index	1	2	3	4	5
	1	1	1	2	2	3
	2	1	2	2	3	4
	3	2	2	3	3	4
	4	2	3	3	4	5
	5	3	4	4	5	5

Chapter 7
Operator 4.0 Within the Framework of Industry 4.0

Sarbjeet Singh
https://orcid.org/0000-0001-7229-4050
Luleå University of Technology, Sweden

Phillip Tretten
Luleå University of Technology, Sweden

ABSTRACT

Operator 4.0 is a smart and skilled operator who augments the symbiosis between intelligent machines and operators. Better integration of Operator 4.0 in Industry 4.0 can bring emphasis on human-centric approach, allowing for a paradigm shift towards a human-automation cooperation for inspiring the compulsion of human-in-the-loop. This further enhances the domain knowledge for the improvement of human cyber-physical systems for new generation automated systems. This cooperation of humans and automation makes stability in socio-technical systems with smart automation and human-machine interfacing technologies. This chapter discusses the design principles of Industry 4.0 and Operator 4.0 human-cyber physical systems.

INTRODUCTION

Aa described by Reyes Garcia et al. (2019), the industrial scenario is radically changing due to the technology innovations of the last decades. Industry 4.0, allows collaboration between operators and machines by integrating robotics, automation and data driven technologies into intelligent workspace. This interaction offers significant impact on transforming industrial tasks to accommodate production variability by introducing collaboration between operators and production systems for the development of future workplaces improving safety aspects during the design phase (Martinetti et al., 2017; Martinetti et al. 2019). Integration of operator 4.0 in Industry 4.0 (Figure 1) brought more emphasis on human-centricity, allowing for a paradigm shift towards a human-automation (David Romero et al. 2015) cooperation. This shift emphasis on human cyber-physical systems i.e. more efficient and effective cooperation of system

DOI: 10.4018/978-1-7998-3904-0.ch007

with humans instead of substituting human skills and abilities. Operator 4.0 (David Romero et al. 2016) work will be qualitatively developed and flexible, and will require new qualifications to understand the digital technology in Industry 4.0. It is important that smart factories should motivate the operators in gaining knowledge of new skills. The Operator 4.0, paradigm shift cannot flourish just by presenting new technologies. Work tasks needs to be redesigned and new approaches to training are desirable to support continuous development of skills.

Figure 1. Integration of operator 4.0 in Industry 4.0

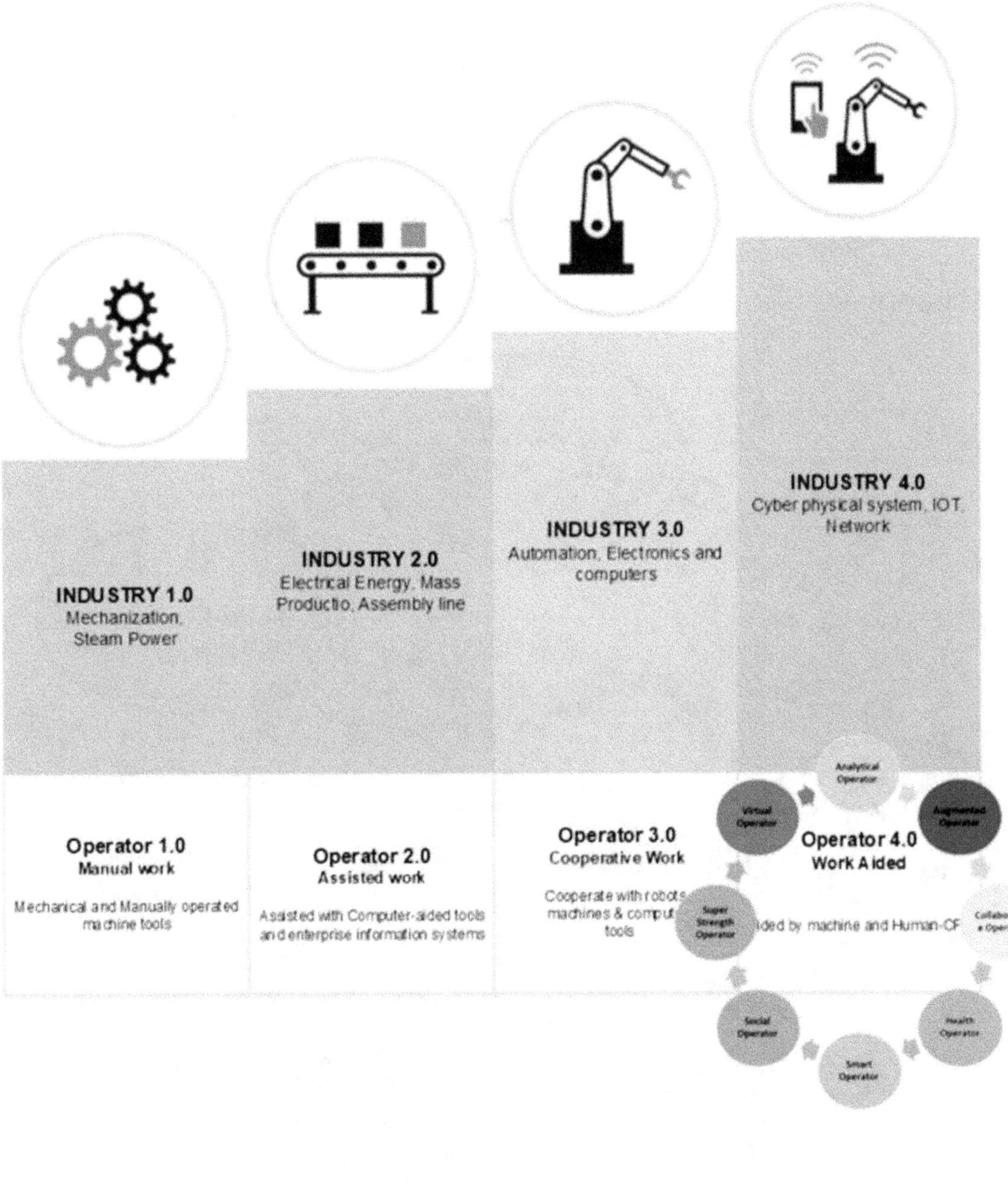

DESIGN PRINCIPLES OF INDUSTRY 4.0: OPERATOR 4.0 PRESPECTIVE

The existing design principles of Industry 4.0 (Figure 2) needs to addressed from user centric perspective for better designing and developing Industry 4.0. As indicated by several works (Borchiellini et al. (2013); Labagnara et al. (2013)), the starting point of a successful project where human operators and technical systems need to interact is the Prevention through Design (PtD) approach in order to minimize the likelihood of unplanned situations that could damage the overall system safety.

Interoperability and System integration

Operator 4.0 will have access to smart products and smart factory and be able to connect, communicate and work together. Interoperability (David & João, 2020) involves accessing real-time data that leads the way to a new approach for manufacturing units to improve their nproduction operations. It allows manufacturing partners (including customers, suppliers, and other departments) and their machines to share information accurately and quickly. It allows operators, resources, smart products and smart factories to connect, communicate and work together. The standardization of data is a critical factor

Figure 2. Design principles of industry 4.0 for better development of operator 4.0

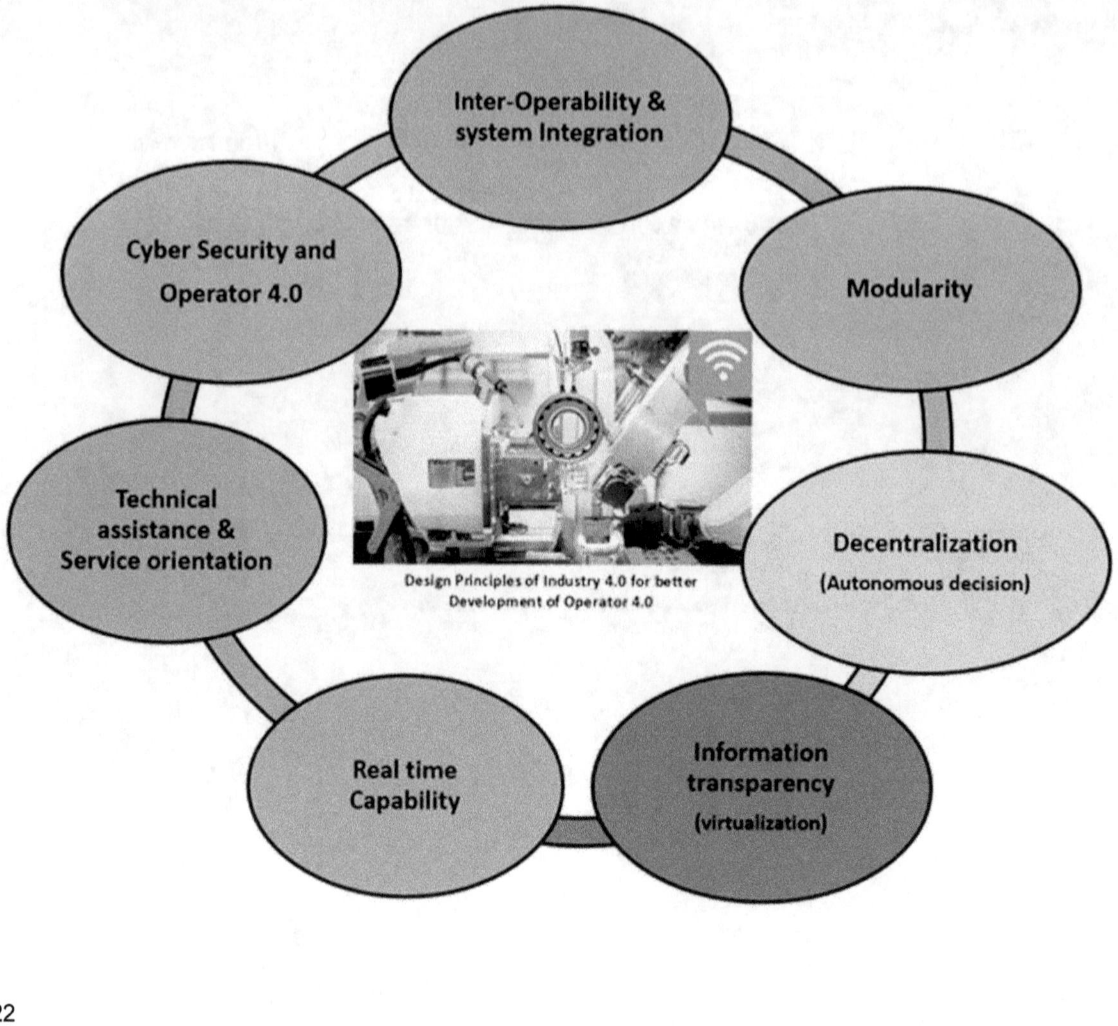

for interoperability because this will boost the user centric approach also helps in the components to understand each other.

System integration: Human System Integration (HSI)

Human system integration is the integration of human needs within system design. Its human-related considerations during system design, development, test, production, use and disposal of systems, subsystems, equipment and facilities (SAE International. 2019 & INCOSE, 2010). In order to have better integration of system and operator, HIS process objectives should include (a) user centric design for best human-machine system performance, (b) the system should comply with limitations and capabilities of the operator, (c) better control on life cycle costs of the system, (d) warrant system safety (Clark & Goulder, 2002). Figure 3 reflects describes human system integration of railway system

Figure 3. Human system integration: railway system

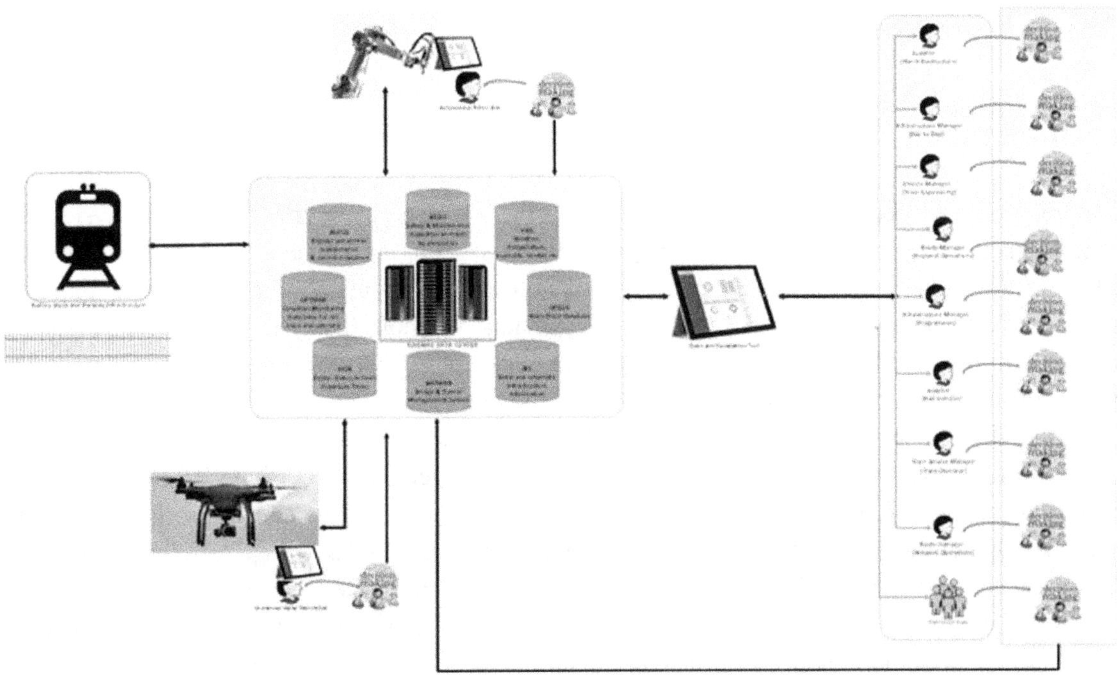

Modularity

Modularity allows the system's components to get separated and recombined (Merriam-Webster Inc, 2018.*)*. Industry 4.0 emphasis on creation of smart factories on modular lines engaging cyber-physical systems to monitor physical processes, creating digital twins of physical processes and hybrid decentralizing decision-making. Modularity assist a system to organize into sub modules, which can scaled up or down as production requirements change from functional or a production capacity standpoint. The fundamental of any system is in its modularity, which enables great flexibility by combining modules in different configurations. Standardization of interfaces helps modules to interchange in smallest possible time. In manufacturing system, modularity helps in ease of configuring different products and manufacturing processes. As shown in Figure 4, it is also possible to integrate several modules of the same type into one production system to optimize the productivity. Modularity helps in minimizing the cost, interoperability, shorter learning time, flexibility in design, non-generationally constrained updating.

Figure 4. Concept of the modular production system (MPS) (Stefan et al., 2014)

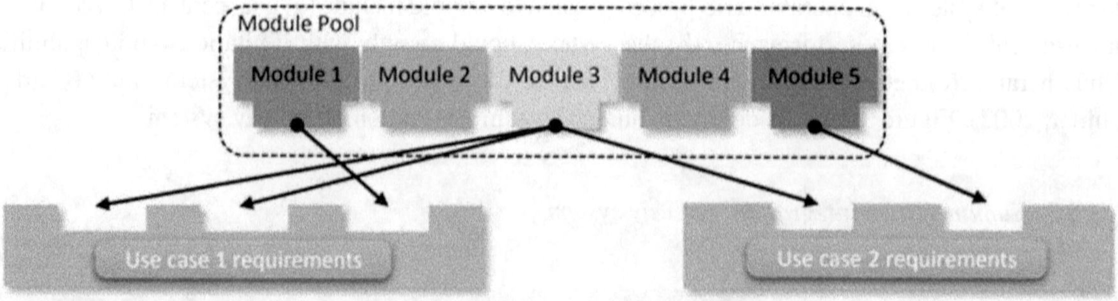

Decentralization, Autonomous Decisions and Autonomy

Decentralization advocates for decentralized decisions. It emphases that cyber-physical subsystem to be able to independently make a decisions and to perform their tasks as autonomously as possible. Decentralized decision is based on distributive approach, where the system consists of autonomous components, which act as an autonomous agent. The decisions spread throughout the system to maximize response time and optimize flexibility while continuing to operate. The cyber physical system of new generation manufacturing systems has integration of AI technology and human with manufacturing domain knowledge. It proposes an excellent example of decentralization in the system. The decision-making is not at centralize level but both at cyber system and human level. The cyber physical system develops self-growing knowledge base, fusion of expert knowledge, System intelligence, Interactive Information and assists Operator 4.0 with the required perception, cognition, analysis, decision-making, and control (David Romero et al. 2016) of Human Cyber Physical System (HCPS) for successful operation of physical systems. Intelligent machines in Industry 4.0, comprises of intelligent sensing, autonomous cognition, intelligent decision-making, and intelligent control. Intelligent decision-making in new generation manufacturing system is based on symbiosis between cyber systems and humans. The task of intelligent decision- making uses intelligent decision-making technologies such as artificial intelligence (Figure 5) for exact calculation of system status, optimization of the decision-making model, and the predictive analysis. It is pertinent to mention that combination of AI and human capabilities, together with an appropriate organization design and decision making, allow humans to be augmented by AI and make smart decisions. However, lot of work is still required to be done on decentralization and autonomy from the human and decision-making perspective.

Figure 5. Process of decision making between AI and humans (adapted from Eric Colson 2019, and Satya Sachdeva)

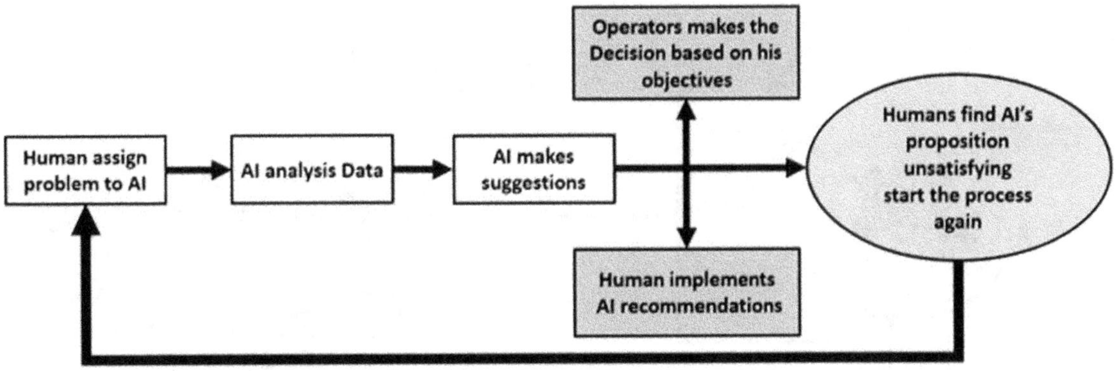

Information transparency

Digital twin, provides a new chance to address the problems of Industry 4.0 by providing more information for understanding cyber-physical integration. This development of digital version will further augments the physical version of the facility with additional information and formulates an interactive system between the physical and digital versions. This design principal of Industry 4.0 focuses on taking human-centered intelligent manufacturing to a new level by giving the operator a major role in adapting and designing and developong own task. In orger to develop smart factories of future, which are suited for operators with different skills, capabilities and preferences, it becomes essential to empower and engage the work community (Figure 6). The development of future smart factory must empower workforce for evaluating adaptive human-automation (David Romero et al. 2015) interaction solutions to increase the flow for supporting the worker and developing competences. The monitoring and measuring of workforce and manufacturing environment helps in developing user model, which provides input for adaption solutions for changing the automation level and other system features. The effective development of smart factory in Industry 4.0 is only possible with active involvement of worker force in the development of smart factory. The workforce must be engaged to share knowledge and to participate in designing the work processes, design of layout, adaption of user centered design approach, task design. For that reason, a virtual model of the factory will be created, demonstrating all functionalities of the real factory. The virtual model of the factory will work as a platform for ngaging the workforce to participate in the design activities. The model supports observing the role of individual and others in the overall framework of the manufacturing process. The virtual factory model also acts as a motivating and easy-to- use contextual platform for knowledge sharing between designer and user.

Figure 6. Empowering the worker and engaging the work community (modified from Kaasinen et al., 2020)

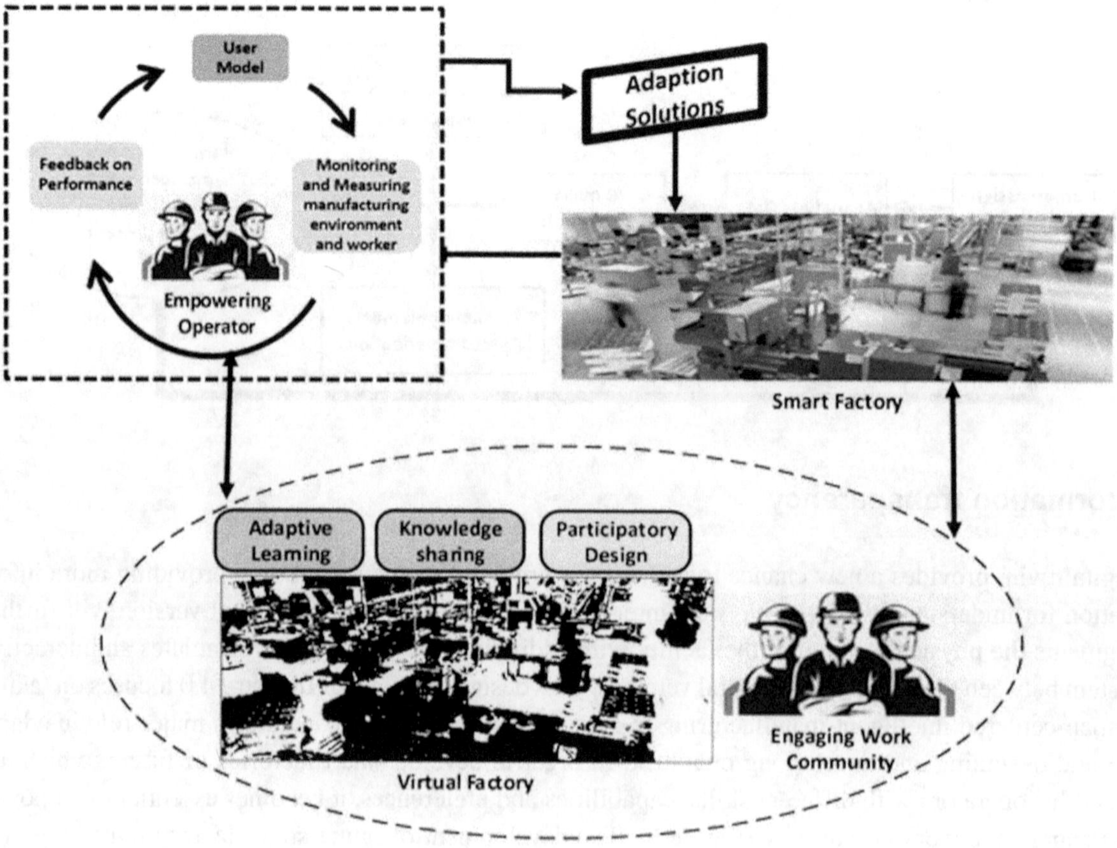

Real-Time Capability

This design principle refers to the collection and analysis of real time information for instant decision making through different communication system. Smart sensors and internet assists many industrial sectors for accessing large amounts of real-time data. This data is further processed and analyzed for continuous optimization of any business unit.

Operator 4.0 and Decision Support Using Real Time Data

The data varies in volume, velocity and variety, which means it, becomes very important to organize and analyze the data for better visualization for better results. The objective is to combine data analytics techniques and Decision Support Tools, in order to enable a high level of automation in the decisional process for different assets in Industry 4.0. The DSS is able to suggest a set of best options to the operators, supporting the decisional process. In this context, the evolution of static HMIs towards dynamic HMIs is an important steps to achieve situation-awareness, reducing human effort in decision making. HMI is the tool of information transperancy for operator 4.0. The formation of virtlualised or digital twin driven by real-time data will assist HMI to generate more flexible interactive mode to improve machine efficiency and operation safety. This will furhter help in meeting the requirements of of intelligent manufacturing along with new emerging technologies (e.g., big data, artificial intelligence, augmented reality, etc.)

Cyber-Security and Operator 4.0

Human factors have significant role in the process of information security (Fahey, 2013). It is essential to be aware that the cyberattack is a human issue, rather than an IT issue. It means that an employee or an individual failing to follow cybersecurity practices is the biggest contributor of cyberattacks. During a possible system failure, workforce can anticipate not only its own course of action but also the system behavior, which makes the operator rely on the fact that they will receive an alert during the event of malfunctioning/failure. In contrast, during cyber-attack, the human operator does not know whether the signs/alerts are trustworthy or clear alert that a system is under a cyber-attack. In case of emergency, the human factors follow the standard operating procedure, but in case of cyber-attack, it is hard to say whether the problem is solvable with the procedures or it can cause another problem right afterwards. Therefore, it is critical to ensure that workforce of an organization or individuals of a civil society are vigilant, fully aware of cyber-threats, and trained to follow cybersecurity practices at all times. Cyber-security education and training plays an important role towards the contribution of cybersecurity. As organizations are adopting new digital technologies and, therefore, it become a challenge to improve the expertise of the existing workforce and appoint workers with the suitable level of cybersecurity education, experience and training. On the other hand, it becomes very important to understand the maliciousness for attackers. The maliciousness for attackers depends on personality traits, mental instability, emotions, self-perception, attitudes, biases, interpersonal behavior, marginality, government structure etc. A human factor framework on cyber security has been proposed (Henshel et al., 2015 & Henshel et al., 2016) for risk assessment.

OPERATOR 4.0 SYMBOSIS WITH CYBER PHYSICAL SYSTEM

Cyber- Physical System (CPS)

New digitalized manufacturing system increases automation, efficiency, quality, stability, and the ability to solve complex tasks. This digitalization in manufacturing system brought a new system called as cyber system, which transformed traditional human–physical system (HPS) (Figure 7) into Cyber- physical system (CPS)

Figure 7. Human Physical System (HPS) for traditional manufacturing system; (Zhou, 2019)

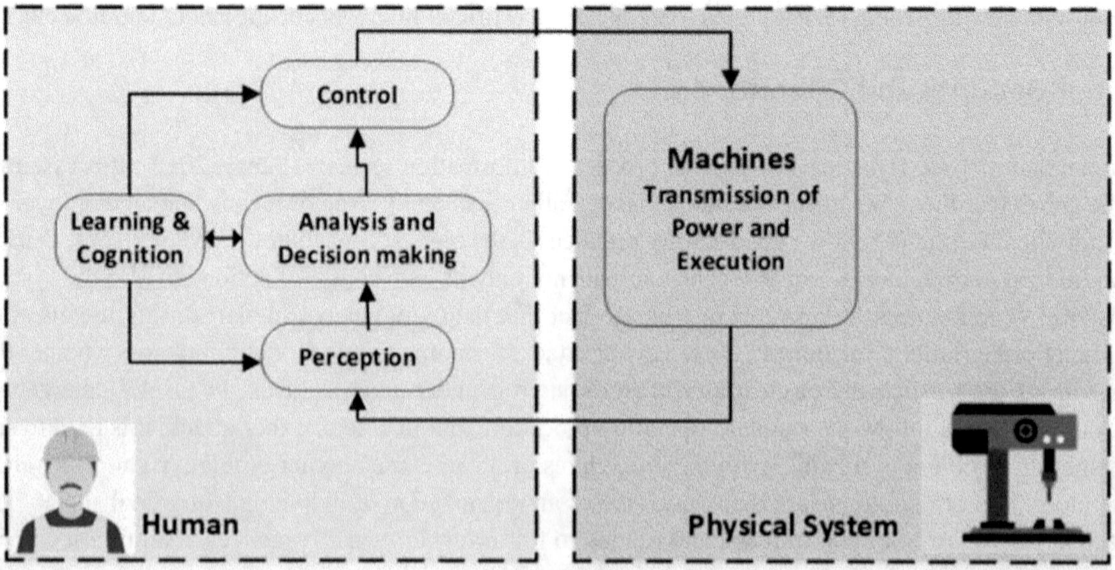

In general, a CPS has five levels (Figure 8) viz. connection, conversion, cyber, cognition and Configuration (Lee et al., 2015).

Smart connection Level

Data can be measured by sensors or acquired from controllers, or enterprise manufacturing systems. Sensors convert physical parameters to electrical signals (transducers), signal conditioning circuitry to convert analog to digital or digital to analog (signal conditioning). In the complex network of modern machining systems, the technology of smart sensors can monitor the integrity of the system and predict network failures. This level of Cyber Physical System (CPS) has to consider to factors i.e. that there should smart sensors for real time monitoring of machine condition and continuous data acquisition and data transfer to main server. In Industry 4.0, data acquisition serves as a pivotal element for binding together a wide variety of products.

Data-to-Information Conversion Level

The real-time data needs to be converted to meaningful information to diagnose the condition and parameters of a process to recognize any deviation from the normal parameters and thus a necessary control action can be taken. There are several tools and methodologies available for the data to information conversion level. Data acquisition systems are very important and efficient systems for gathering and processing of information to analyze and make decisions based on data. This is an important step in developing a Cyber-Physical System application. In Industry 4.0, data acquisition serves as an important element for binding together a wide variety of products and are made more precise, adaptable, and dependable. It is the process of sampling data from sensors and converting them into digital numeric values that can be operated by a computer. There are different types of data acquisition systems, wireless data acquisition system, serial communication data acquisition system, USB data acquisition system, data acquisition plug-in boards (Singh et al., 2014)

Cyber

The Cyber level in Cyber Physical system (Figure 8) acts as crucial information center. Information is being pushed to it from machine and components, connections. Huge information is collected and intelligently analyzed to extract useful information that provide better understanding over the position of system under consideration. This further assist in self-growing knowledge base using system intelligence for better decision making.

Figure 8. Practices associated with various levels of Cyber Physical System (CPS) (modified from J. Lee et al., 2015)

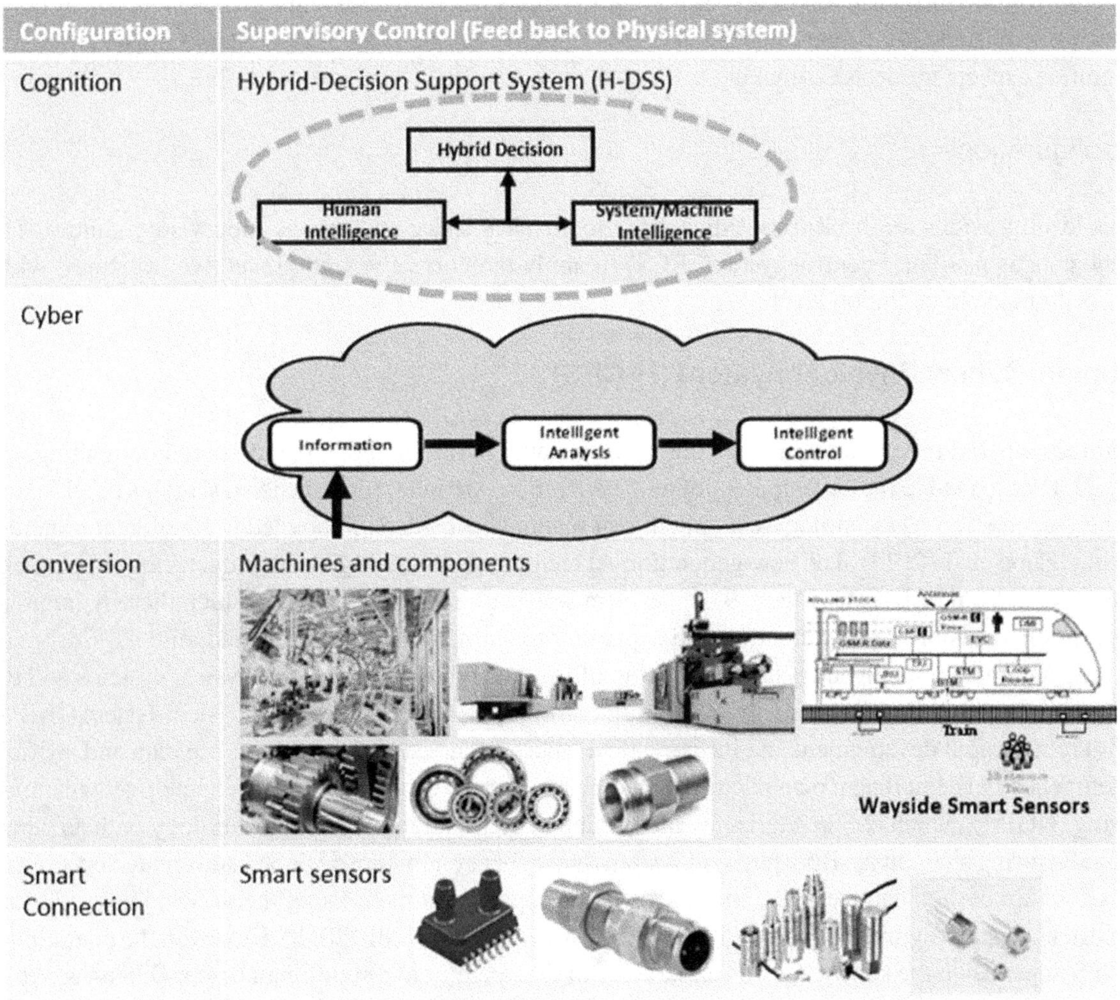

Cognition

Technological advancement with Internet of Things (IoT), smart sensors, GPS devices, radio tags assist in automatically monitoring by computers leading to effective tracking and line management in industry 4.0. No doubt, these advanced sensors and support system helped in automating industrial operations but this also brought challenges of Decision support at human level. Cognitively driven Hybrid-Decision Support System (H-DSS) in Industry 4.0 been recognized as a paradigm in the research and development. H-DSS aims to help system managers in decision making from human cognitive aspects, such as thinking, sensing, understanding and predicting, and fully reuse their experience. The concept of operator 4.0, has to have a smart decision-making approach and will prove to be a new milestone in industrial development that will certainly mark significant changes in the coming years in the digital transformation of all industrial sector. Decision making capabilities have to be enhanced for the implementation of digital operator assistance technology by projecting digital information onto the field of vision such as in Augmented/Virtual Reality, Wearables, Life waist, AR for Maintenance. In order to make a successful decision, maintenance managers' SA about their business environments becomes a critical factor. That is the reason why, Situation awareness (SA) and mental models considered two important human cognitive aspects for decision-making.

Configuration

This level provides feedback from cyber space to physical space and acts as supervisory control. This stage acts as resilience control system (RCS) to apply the corrective and preventive decisions, which has been made in cognition level.

Human-Cyber- Physical System (H-CPS)

Fourth industrial revolution witnessed enormous growth in the field of internet, cloud computing, and big data, which led to the development of new-generation Artificial Intelligence (AI). This combination of new-generation AI technology with intelligent manufacturing technology led to Intelligent Manufacturing (Zhou et al., 2018). The new-generation AI technology in Intelligent Manufacturing supplements the cyber system of H-CPS (Figure 9) to perform self-learning and cognition, which thereby improves perception, decision-making, control, and capability to learn and produce knowledge.

It is a complex intelligent system consisting of people, cyber systems and physical systems and this system composition makes an intelligent manufacturing system a human cyber physical system (HCPS). With recent rapid development and influential advances occurred in the internet, big data and artificial intelligence (AI), Intelligent manufacturing has achieved the level of new-generation intelligent manufacturing (NGIM), which is characterize by integration of artificial intelligence technology with advanced manufacturing technology. Both physical systems and cyber systems are designed and created by humans. It is a matter of fact that, methods, and rules are developed by humans using theoretical knowledge and experience, and programming these into the cyber systems (Zhou et al., 2018). However, the operation of HCPS depends on the knowledge and experience of the operator to a significant extent (Zhang & Wang, 1994). The end of the 20th century viewed rapid emergence of internet technology applied to manufacturing industry, driving a transformation from digital manufacturing to digital-networked manufacturing.

The HCPS in digital-networked manufacturing system has industrial Internet and the cloud platform in the cyber system, which connects relevant cyber systems, physical systems, and humans, thus serving as a tool for system integration.

Figure 9. HCPS for intelligent manufacturing system (modified from Zhou et al., 2018)

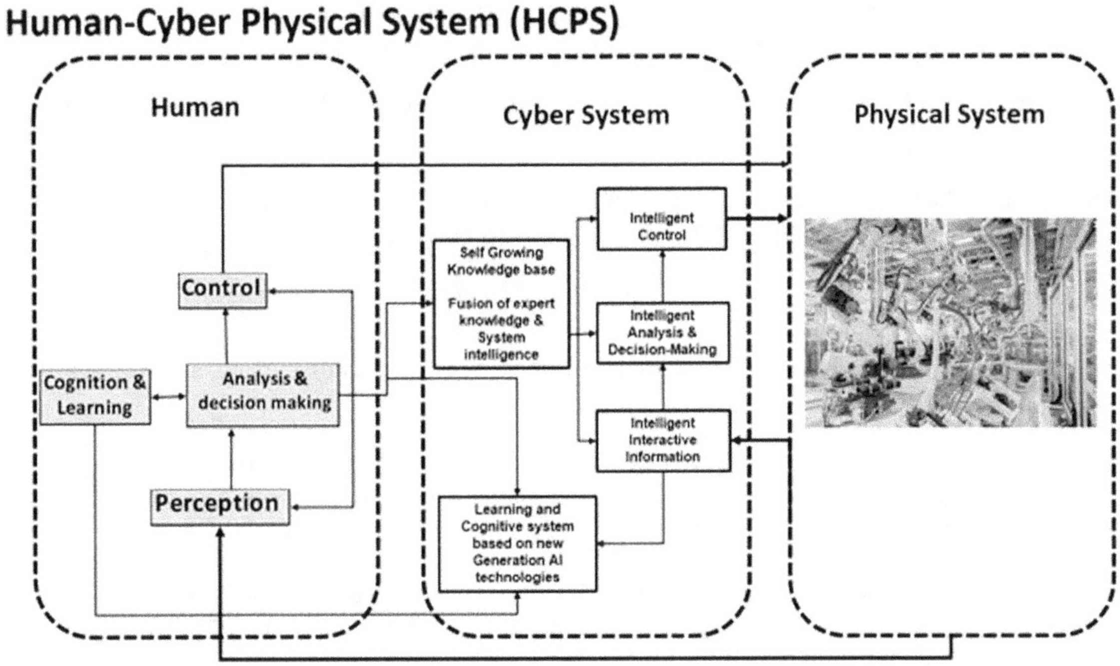

CONCLUSION

In new generation intelligent manufacturing systems, human-centric approach is going to be a core theme for Industry, with human machine interaction be through physical and cognitive means. The concept of Operator 4.0 will provide the opportunity to integrate operators into the concept of smart factories and encourages the concepts of human-in-the-loop in cyber-physical systems. This further enhance the knowledge and understanding for the development of Human Cyber-Physical Systems for new generation automated systems. This cooperation of human and automation make stability in socio- technical system with smart automation and human-machine interfacing technologies.

ACKNOWLEDGMENT

We thank Technalia Research & Innovation, Spain for supporting the work.

REFERENCES

Borchellini, R., Cardu, M., Colella, F., Labagnara, D., Martinetti, A., Patrucco, M., . . . Verda, V. (2013). A Prevention through Design Approach for the Environmental S&H Conditions and the Ventilation System at an Italian Underground Quarry. *Chemical Engineering Transactions, AIDIC - The Italian Association of Chemical Engineering, 32*, 181-186. doi:10.3303/CET1332031

Clark, J., & Goulder, R. (2002, July). Human systems integration (HSI) ensuring design & development meet human performance capability early in acquisition process. *Program Management*, 88–91.

Colson, E. (2019). *What AI-Driven Decision Making Looks Like*. https://hbr.org/2019/07/what-aidriven-decision-making-looks-like

David R., João P. M. (2020), Interoperability Enablers for Cyber-Physical Enterprise Systems. *Enterprise Information Systems.* October 2020. DOI: 10.1080/17517575.2020.181508

David R., Ovidiu N., Johan S., Peter B., (2015). *Towards a Human-Centred Reference Architecture for Next Generation Balanced Automation Systems: Human-Automation Symbiosis*, Conference: Innovative Production Management towards Sustainable Growth, Service, Manufacturing, and Resilient Value Chain, Volume: Part II, AICT 460, Springer, pp. 556-566

David R., Johan S., Thorsten W., Ovidiu N. (2016). Towards an Operator 4.0 Typology: A HumanCentric Perspective on the Fourth Industrial Revolution Technologies. *International Conference on Computers & Industrial Engineering* (CIE46), pp. 1-11 Tianjin / China, ISSN 2164-8689

David R., Peter B., Ovidiu N., Johan S., (2016). The Operator 4.0: Human Cyber-Physical Systems & Adaptive Automation Towards Human-Automation Symbiosis Work Systems Conference Paper in *IFIP Advances in Information and Communication Technology*. DOI: 10.1007/978-3-319-51133-7_80

Fahey, R. (2013). *Human factors in information security management systems*. Academic Press.

Fontaine, O., Martinetti, A., & Michaelides-Mateou, S. (2016). Remote pilot aircraft system (RPAS): Just culture, human factors and learnt lessons. Chemical engineering transactions, AIDIC - The Italian Association of Chemical Engineering, 205-210. doi:10.3303/CET1653035

Henshel, D., Cains, M. G., Hoffman, B., & Kelley, T. (2015). Trust as a human factor in holistic cyber security risk assessment. *Procedia Manufacturing, 3*, 1117–1124. doi:10.1016/j.promfg.2015.07.186

Henshel, D., Sample, C., Cains, M., & Hoffman, B. (2016). *Integrating cultural factors into human factors framework and ontology for cyber attackers.* In D. Nicholson (Ed.), *Advances in Human Factors in Cybersecurity. Advances in Intelligent Systems and Computing* (pp. 123–137). Cham: Springer International Publishing.

International, S. A. E. (2019). *Standard Practice for Human Systems Integration (SAE6906)*. Warrendale, PA: SAE. https://saemobilus.sae.org/content/sae6906

International Council on Systems Engineering. (2010). *Systems Engineering Handbook Volume 3.1 Appendix M*. Author.

Ji, Z., Yanhong, Z., Baicun, W., & Jiyuan, Z. (2019). Human–Cyber–Physical Systems (HCPSs) in the Context of New-Generation Intelligent Manufacturing. *Engineering*, *5*(4), 624–636. doi:10.1016/j.eng.2019.07.015

Kaasinen, E., Schmalfuß, F., Özturk, C., Aromaa, S., Boubekeur, M., Heilala, J., ... Mehta, R. (2020). Empowering and engaging industrial workers with Operator 4.0 solutions. *Computers & Industrial Engineering*, *139*, 105678. doi:10.1016/j.cie.2019.01.052

Labagnara, D., Martinetti, A., & Patrucco, M. (2013). Tunneling operations, occupational S&H and environmental protection: A Prevention through Design approach. *American Journal of Applied Sciences*, *11*(11), 1371–1377. doi:10.3844/ajassp.2013.1371.1377

Lee, J., Bagheri, B., & Kao, H. A. (2015). A cyber-physical systems architecture for industry 4.0-based manufacturing systems. *Manufacturing Letters*, *3*, 18–23. doi:10.1016/j.mfglet.2014.12.001

Martinetti, A., Braaksma, A. J. J., & van Dongen, L. A. M. (2017). Beyond RAMS Design: Towards an Integral Asset and Process Approach. In L. Redding, R. Roy, & A. Shaw (Eds.), Advances in Through-life Engineering Services (pp. 417-428). Springer. doi:10.1007/978-3-319-49938-3_25

Martinetti, A., Chatzimichailidou, M., Maida, L., & van Dongen, L. A. M. (2018). Safety I-II, Resilience and Antifragility Engineering: A debate explained through an accident occurred on a Mobile Elevating Work Platform. *International Journal of Occupational Safety and Ergonomics*, *25*(1), 66–75. doi:10.1080/10803548.2018.1444724 PMID:29473459

Modular. (2018). In *Merriam-Webster.* Merriam-Webster Inc.

Reyes Garcia, J. R., Martinetti, A., Jauregui Becker, J. M., Singh, S., & van Dongen, L. A. M. (2019). Towards an Industry 4.0-Based Maintenance Approach in the Manufacturing Processes. In V. González-Prida Diaz & J. P. Zamora Bonilla (Eds.), *Handbook of Research on Industrial Advancement in Scientific Knowledge* (pp. 135–159). IGI Global; doi:10.4018/978-1-5225-7152-0.ch008

Satya Sachdeva (2019). *BI to AI Journey*, https://labs.sogeti.com/bi-to-ai-journey/

Scheifele, S., Friedrich, J., & Lechler, A. (2014), Self-configuring control system for a modular production system. *2nd International Conference on System-Integrated Intelligence: Challenges for Product and Production Engineering Flexible*, 1-8.

Singh, S., Galar, D., Baglee, D., & Björling, S. E. (2014). Self-maintenance techniques: A smart approach towards self-maintenance system. *International Journal of System Assurance Engineering and Management*, *5*(1), 75–83. doi:10.100713198-013-0200-7

Zhang, B., & Wang, J. (1994). Knowledge information and human function in manufacturing systems. *Chinese Journal of Mechanical Engineering*, *30*(5), 61–65.

Zhou, J. P., Li, Y., Zhou, B., Wang, J., Zang, L., & Meng, L. (2018). MengToward new-generation intelligent manufacturing. *Engineering*, *4*(1), 11–20. doi:10.1016/j.eng.2018.01.002

Chapter 8
Augmented Technology for Safety and Maintenance in Industry 4.0

Vikas Kukshal
https://orcid.org/0000-0001-9277-5097
National Institute of Technology Uttarakhand, India

Amar Patnaik
Malaviya National Institute of Technology, Jaipur, India

Sarbjeet Singh
https://orcid.org/0000-0001-7229-4050
Luleå University of Technology, Sweden

ABSTRACT

The traditional manufacturing system is going through a rapid transformation and has brought a revolution in the industries. Industry 4.0 is considered to be a new era of the industrial revolution in which all the processes are integrated with a product to achieve higher efficiency. Digitization and automation have changed the nature of work resulting in an intelligent manufacturing system. The benefits of Industry 4.0 include higher productivity and increased flexibility. However, the implementation of the new processes and methods comes along with a lot of challenges. Industry 4.0. requires more skilled workers to handle the operations of the digitalized manufacturing system. The fourth industrial revolution or Industry 4.0 has become the absolute reality and will undoubtedly have an impact on safety and maintenance. Hence, to tackle the issues arising due to digitization is an area of concern and has to be dealt with using the innovative technologies in the manufacturing industries.

DOI: 10.4018/978-1-7998-3904-0.ch008

INTRODUCTION

As also indicated by Reyes Garcia et al. (2019), the first industrial revolution started in the eighteenth-century by utilizing steam as the source of power. It brought significant changes in all the industries by and increased the performance of the industrial sectors to a large extent. The second industrial revolution took place when electric power was used to run the machines. In addition, the assembly lines were used in large industries for mass production. The third industrial revolution integrated information technology with the manufacturing process. There were several dynamic changes in the manufacturing system with the swift exchange of information. The fourth industrial revolution is the amalgamation of the manufacturing process with emerging technologies like the Internet of thing, big data and cloud-assisted processes (Muhuri, 2019). Industry 4.0 is the solution to the increasing demand for product variety existing in the market (Kuo et al., 2019). It includes the use of all cutting edge technology with the automation resulting in smart manufacturing that can adapt the hasty environment transformation (Galati & Bigliardi, 2019, Corallo et al., 2020).

As the advancement in technology is increasing day by day, the safety and maintenance of the entire system is a serious concern for all the industries (Scurati et al., 2018). The level of the issues pertaining to the quality of the product has increased manifold in the current industrial scenario. The intensity of the analysis and identification of safety measures is the key challenge in the industry 4.0. In addition, the standard policy of the governments prevailing in a country needs to be implemented in all the industries in order to avoid major accidents occurring in any workplace. Advancement in technology has also affected the environment and hence the environment protection is to be dealt with seriousness (Ceruti et al., 2019).

Safety and maintenance of the processes involved in the industry is the essential requirement for the growth of any manufacturing plant and new approaches are rising for taking into account not only technical but also societal needs (Martinetti et al., 2017; Martinetti et al. 2019). It includes the implementation of the safety measures and identification of the factors that can cause risk arising due to uncontrolled and unprotected data management (Kaczmarek & Gola, 2019). Despite several attempts towards the safety and maintenance adopted in the manufacturing industry in the past, there exist several cases reporting the breakdown of the network connecting the different processes. A similar scenario prevails with the maintenance of the various tools used in the manufacturing industries wherein highly skilled professionals are required to maintain the dynamic environment of the industry 4.0 as compared to the conventional manufacturing system (Sarmiento et al., 2020; Ivanov et al., 2018). The effective use of information technology in the manufacturing and processing industry needs special attention.

The chapter aims to highlight the processes used in industry 4.0 for the safety and maintenance and the related physical system with the cyber world. The work done to improve the safety and maintenance processes involved in industry 4.0 is discussed. The present chapter also focuses on the various challenges existing in the highly sophisticated and automated processes involved in Industry 4.0.

INDUSTRY 4.0: THEORETICAL BACKGROUND AND PROCESSES

Industry 4.0 is a new industry era wherein all the process involved in the system is digitalized. There is an interconnection of the various steps involved in the process with is a corresponding virtual demonstration. The process flow is mapped with the cyber world in an incessant manner (Monostori, 2014). The integration of the cyber world with the production system relates the information technology and the communication system with the different processes involved in the system. The complete system falls under industry 4.0 which connects the real and virtual process close to each other and forms the internet of things (IoT) (Colakovic & Hadzialic, 2018). The progress of the manufacturing system without networking cannot be visualized in the present scenario. The growth of the industries entirely depends on the extent of the implementation of sensors, wireless communication and the internet of things (IoT). The cyber revolution has reduced the gap between the human-machine interface (Kadir et al., 2019).

Industry 4.0 can bring a radical change in the production, quality and all the allied services by making use of intelligent information service. Augmented reality provides a path for the integration of information technology with the real word processes (Gattullo et al., 2019). The real-time problems can be easily dealt with augmented reality at any instant. The application areas of augmented reality are still very limited and have to be widely used in productivity applications. This function can be very advantageous for real problems over the conventional user interface (Paelke, 2014). To further enhance the efficiency of the industrial sector, the real and virtual are connected in the form of the internet of things (IoT). The use of IoT improves flexibility in terms of the volume, customization, interconnection of customer and supplier and sustainability in the production system (Shrouf et al., 2014). The industries can respond quickly to the change in the environment and act wisely to overall improve the productivity of the system.

The emerging use of automation along with the IoT has resulted in the terminology called smart factories. The smart factories make use of the cloud system to store the information and share the service and product among all the resources connected to it (Hidalgo et al., 2019). Cloud computing connects the manufacturing system using many servers through a network. It provides an integration to maintain the whole production system and resolve various issues arising in a system.

INDUSTRY 4.0: SAFETY AND MAINTENANCE

Safety and maintenance are the key concerns in all the industries that can have a significant impact on the productivity of any business sector. With the incorporation of the information technology with the existing production system in Industry 4.0, there have been multiple threats that can cause a huge loss in terms of the output of the whole process (Sarmiento et al., 2020). However, the Internet of things can be used to monitor the processes and help largely in monitoring and maintenance of the machine equipment. The information technology is a great asset for providing a significant contributions to the areas of safety and maintenance in Industry 4.0.

Process Safety: Methods and Principles

There are several ways to control the risk in the production system by utilizing the tools of the cyber-physical process system. The risk assessment and risk management is essential to reduce the occurrence of serious events that can hamper the activities of the industry (Lee et al., 2019). Moreover, as indicated by Borchiellni et al. (2013) and by Labagnara et al. (2013) a correct starting methodology for safety-related problems is the adoption of the a Prevention through Design (PtD) approach, helpful in several industrial context. Consequentially, the internet of things can play a critical role in the planning of safety and maintenance of the products during manufacturing as well as its complete life.

The use of the internet of thing in industries is an upcoming technology for connecting the data of all the processes and the machine through a network competent to interact and collaborate without any interruption. The use of information technology in the manufacturing system will provide an interconnection with all the processes of the manufacturing system (Junior et al., 2018). The relationship between the physical and the cyber world can prevent the happening of a large number of accidents in the industries. All the information provided from the sensors and actuators are stored in the cloud-based system and can be retrieved by the maintenance system whenever required (Alqahtani et al., 2019). Therefore all the data is processed and stored in the clouds to further prevent any safety-related issue.

Cybersecurity

Several activities are involved in the industry level that can be related to each other employing information technology. Industry 4.0 facilitates the easy access of all the information to the specific user so that the whole process can be optimized. All machine tools can be easily controlled from the remote location and maintained retrieving the information stored in the cloud system. Real time monitoring all the devices helps in keeping safety at all level of the industry. In the past few years, smart sensors are widely employed by the industries for monitoring and tracking the devices (Xu et al., 2018). The expert of a particular device can provide the information from the distant place resulting in keeping the whole system updated without wasting too much of time as well as money.

Maintenance Processes

It is significant to recognize the maintenance of the various component of the manufacturing system at a different level. All the process involved during the process needs to be updated in an industry in a regular interval of time for sustainable manufacturing system. The complete maintenance process can be divided into a number of the subprocess. The main function is to minimize the overall maintenance of the plant thus reducing the overall cost disbursed in producing a product. The maintenance is performed using real-time monitoring through the sensors, thereby processing and storing the information for its use through cloud computing (Sipsas et al., 2016). The proper and sensible implementation of the process will lead to the economic sustainability of the industry. The other maintenance process is related to the reduction in the usage of the industrial resources affecting the environment. The optimization of the various resources used in an industry will increase the efficiency of the process and reduce the effect of multiple wastes and emissions on the environment. The other aspect of the maintenance process is social sustainability which includes the plant and human safety (Paelke, 2014). The hazardous activities in the industry can be performed using the automatic machine that can be controlled and monitored by information technology. The process of e-maintenance certainly improves the life of the equipment and the maintenance of the processes involved in the industry.

Challenges in Safety and Maintenance

Industry 4.0 has brought a radical change in the entire manufacturing sector but still, there remain many challenges to be overcome before its successful implementation. Interaction between new technology and operators, such as Unmanned Aerial Vehicles, represents an example of the mentioned challenges (Fontaine et al., 2016). The information technology plays a major role in revolutionizing the industry and hence many key issues need to be resolved for acquiring its full benefit (Alaloul et al., 2019). One of the concerns related to information technology is its data management. A lot of data is generated during the whole process of industry 4.0 and hence the maintenance and safeguarding the data from the foreign threats is very important. Once there is an intruder in the system, it will affect all the processes and the loss incurred will be unrecoverable. This issue can be addressed with the help of a strong support system and a backup plan. The inefficient method of processing and analyzing the data obtained during the real-time operation is another area of concern. The data obtained from the acquisition system, need to be properly processed and analyzed before its storage and retrieval. Loss of any data during the entire process can severely affect the entire manufacturing system (Heritage, 2019).

Industry 4.0 supports the use of wireless communication for the flow of information and regulation of activities. It is very critical to preserve the communication system from damage so that work can be continued uninterrupted. The communication system adopted in the industry is exposed to natural calamities and hence can provide an obstacle in the continuous flow of information. Hence it is important to protect the complete wireless communication from any kind of harm caused to the system. Environment-related issues also need to be addressed while implementing industry 4.0 (Moktadir et al., 2018).

All the processes involved in industry 4.0 are exposed to the threat of cybersecurity. As all the devices are controlled with the help of information technology connected globally, it is alarming to prevent the intrusion of cyber-crime in the whole process (Lezzi et al., 2018). Therefore, it is very important to protect such activities by using strong firewalls so that no person can access and harm any component of the manufacturing plant. It is also imperative to formulate the global standards to minimize the risk of failure of any kind of system in industry 4.0. The standard should be followed globally to meet the safety and maintenance requirements of all the industries. The standard rules and regulations should be followed strictly for maintaining harmony in the era of globalization and avoiding any type of dispute. Implementation of measures for safety and maintenance in industry 4.0 can lead to sustainable manufacturing to a particular extent.

CONCLUSION

Industry 4.0 has revolutionized the whole process of production, quality and all the allied services by making use of intelligent information service. Industry 4.0 facilitates the easy access of all the information to the specific user so that the whole process can be optimized. However, there is a number of challenges that need to be addressed in order to reduce the occurrence of the safety of the manufacturing industry. New technology should be developed for maintaining and regulating the flow of information safely during the complete life cycle of a manufacturing plant. A stringent policy is required globally for the safety and maintenance of processes involved in industry 4.0. Implementation of measures for safety and maintenance in industry 4.0 can lead to sustainable manufacturing to a particular extent. The environment-related issues also need to be addressed while implementing industry 4.0 for making the entire process sustainable.

REFERENCES

Alaloul, W. S., Liew, M. S., Zawawi, N. A. W. A., & Kennedy, I. B. (2019). Industrial Revolution 4.0 in the construction industry: Challenges and opportunities for stakeholders. *Ain Shams Engineering Journal*.

Alqahtani, A. Y., Gupta, S. M., & Nakashima, K. (2019). Warranty and maintenance analysis of sensor embedded products using internet of things in industry 4.0. *International Journal of Production Economics*, *208*, 483–499. doi:10.1016/j.ijpe.2018.12.022

Borchellini, R., Cardu, M., Colella, F., Labagnara, D., Martinetti, A., Patrucco, M., . . . Verda, V. (2013). A Prevention through Design Approach for the Environmental S&H Conditions and the Ventilation System at an Italian Underground Quarry. *Chemical Engineering Transactions, AIDIC - The Italian Association of Chemical Engineering, 32*, 181-186. doi:10.3303/CET1332031

Ceruti, A., Marzocca, P., Liverani, A., & Bil, C. (2019). Maintenance in Aeronautics in an Industry 4.0 Context: The Role of Augmented Reality and Additive Manufacturing. *Journal of Computational Design and Engineering*.

Colakovic, A., & Hadzialic, M. (2018). Internet of Things (IoT): A review of enabling technologies, challenges, and open research issues. *Computer Networks*, *144*, 17–39. doi:10.1016/j.comnet.2018.07.017

Corallo, A., Lazoi, M., & Lezzi, M. (2020). Cybersecurity in the context of industry 4.0: A structured classification of critical assets and business impacts. *Computers in Industry*, *114*, 103165. doi:10.1016/j.compind.2019.103165

Fontaine, O., Martinetti, A., & Michaelides-Mateou, S. (2016). Remote pilot aircraft system (RPAS): Just culture, human factors and learnt lessons. Chemical engineering transactions, AIDIC - The Italian Association of Chemical Engineering, 205-210, doi:10.3303/CET1653035

Galati, F., & Bigliardi, B. (2019). Industry 4.0: Emerging themes and future research avenues using a text mining approach. *Computers in Industry*, *109*, 100–113. doi:10.1016/j.compind.2019.04.018

Garrido-Hidalgo, C., Olivares, T., Ramirez, F. J., & Roda-Sanchez, L. (2019). An end-to-end Internet of Things solution for Reverse Supply Chain Management in Industry 4.0. *Computers in Industry*, *112*, 103127. doi:10.1016/j.compind.2019.103127

Gattullo, M., Scurati, G. W., Fiorentino, M., Uva, A. E., Ferrise, F., & Bordegoni, M. (2019). Towards augmented reality manuals for industry 4.0: A methodology. *Robotics and Computer-integrated Manufacturing*, *56*, 276–286. doi:10.1016/j.rcim.2018.10.001

Heritage, I. (2019). Protecting Industry 4.0: Challenges and solutions as IT, OT and IP converge. *Network Security*, *2019*(10), 6–9. doi:10.1016/S1353-4858(19)30120-5

Ivanov, D., Sethi, S., Dolgui, A., & Sokolov, B. (2018). A survey on control theory applications to operational systems, supply chain management, and Industry 4.0. *Annual Reviews in Control*, *46*, 134–147. doi:10.1016/j.arcontrol.2018.10.014

Jasiulewicz-Kaczmarek, M., & Gola, A. (2019). Maintenance 4.0 Technologies for Sustainable Manufacturing-an Overview. *IFAC-PapersOnLine*, *52*(10), 91–96. doi:10.1016/j.ifacol.2019.10.005

Junior, J. A. G., Busso, C. M., Gobbo, S. C. O., & Carreão, H. (2018). Making the links among environmental protection, process safety, and industry 4.0. *Process Safety and Environmental Protection, 117*, 372–382. doi:10.1016/j.psep.2018.05.017

Kadir, B. A., Broberg, O., & da Conceição, C. S. (2019). Current research and future perspectives on human factors and ergonomics in Industry 4.0. *Computers & Industrial Engineering, 137*, 106004. doi:10.1016/j.cie.2019.106004

Kuo, C. C., Shyu, J. Z., & Ding, K. (2019). Industrial revitalization via industry 4.0–A comparative policy analysis among China, Germany and the USA. *Global Transitions, 1*, 3–14. doi:10.1016/j.glt.2018.12.001

Labagnara, D., Martinetti, A., & Patrucco, M. (2013). Tunneling operations, occupational S&H and environmental protection: A Prevention through Design approach. *American Journal of Applied Sciences, 11*(11), 1371–1377. doi:10.3844/ajassp.2013.1371.1377

Lee, J., Cameron, I., & Hassall, M. (2019). Improving process safety: What roles for Digitalization and Industry 4.0. *Process Safety and Environmental Protection, 132*, 325–339. doi:10.1016/j.psep.2019.10.021

Lezzi, M., Lazoi, M., & Corallo, A. (2018). Cybersecurity for Industry 4.0 in the current literature: A reference framework. *Computers in Industry, 103*, 97–110. doi:10.1016/j.compind.2018.09.004

Martinetti, A., Braaksma, A. J. J., & van Dongen, L. A. M. (2017). Beyond RAMS Design: Towards an Integral Asset and Process Approach. In L. Redding, R. Roy, & A. Shaw (Eds.), *Advances in Through-life Engineering Services* (pp. 417–428). Springer; doi:10.1007/978-3-319-49938-3_25

Martinetti, A., Chatzimichailidou, M., Maida, L., & van Dongen, L. A. M. (2018). Safety I-II, Resilience and Antifragility Engineering: A debate explained through an accident occurred on a Mobile Elevating Work Platform. *International Journal of Occupational Safety and Ergonomics, 25*(1), 66–75. doi:10.1080/10803548.2018.1444724 PMID:29473459

Moktadir, M. A., Ali, S. M., Kusi-Sarpong, S., & Shaikh, M. A. A. (2018). Assessing challenges for implementing Industry 4.0: Implications for process safety and environmental protection. *Process Safety and Environmental Protection, 117*, 730–741. doi:10.1016/j.psep.2018.04.020

Monostori, L. (2014). Cyber-physical production systems: Roots, expectations and R&D challenges. *Procedia CIRP, 17*, 9–13. doi:10.1016/j.procir.2014.03.115

Muhuri, P. K., Shukla, A. K., & Abraham, A. (2019). Industry 4.0: A bibliometric analysis and detailed overview. *Engineering Applications of Artificial Intelligence, 78*, 218–235. doi:10.1016/j.engappai.2018.11.007

Paelke, V. (2014). Augmented reality in the smart factory: Supporting workers in an industry 4.0. environment. In *Proceedings of the 2014 IEEE emerging technology and factory automation (ETFA)* (pp. 1-4). IEEE. 10.1109/ETFA.2014.7005252

Reyes Garcia, J. R., Martinetti, A., Jauregui Becker, J. M., Singh, S., & van Dongen, L. A. M. (2019). Towards an Industry 4.0-Based Maintenance Approach in the Manufacturing Processes. In V. González-Prida Diaz & J. P. Zamora Bonilla (Eds.), *Handbook of Research on Industrial Advancement in Scientific Knowledge* (pp. 135–159). IGI Global. doi:10.4018/978-1-5225-7152-0.ch008

Ruiz-Sarmiento, J. R., Monroy, J., Moreno, F. A., Galindo, C., Bonelo, J. M., & Gonzalez-Jimenez, J. (2020). A predictive model for the maintenance of industrial machinery in the context of industry 4.0. *Engineering Applications of Artificial Intelligence*, *87*, 103289. doi:10.1016/j.engappai.2019.103289

Scurati, G. W., Gattullo, M., Fiorentino, M., Ferrise, F., Bordegoni, M., & Uva, A. E. (2018). Converting maintenance actions into standard symbols for Augmented Reality applications in Industry 4.0. *Computers in Industry*, *98*, 68–79. doi:10.1016/j.compind.2018.02.001

Shrouf, F., Ordieres, J., & Miragliotta, G. (2014). *Smart factories in Industry 4.0: A review of the concept and of energy management approached in production based on the Internet of Things paradigm. In 2014 IEEE international conference on industrial engineering and engineering management* (pp. 697–701). IEEE.

Sipsas, K., Alexopoulos, K., Xanthakis, V., & Chryssolouris, G. (2016). Collaborative maintenance in flow-line manufacturing environments: An Industry 4.0 approach. *Procedia CIRP*, *55*, 236–241. doi:10.1016/j.procir.2016.09.013

Xu, L. D., Xu, E. L., & Li, L. (2018). Industry 4.0: State of the art and future trends. *International Journal of Production Research*, *56*(8), 2941–2962. doi:10.1080/00207543.2018.1444806

Chapter 9
Applying the Fuzzy Inference Model in Maintenance Centered to Safety:
Case Study – Bucket Wheel Excavator

Predrag D. Jovančić
Faculty of Mining and Geology, University of Belgrade, Serbia

Miloš Tanasijević
https://orcid.org/0000-0002-9629-1513
Faculty of Mining and Geology, University of Belgrade, Serbia

Vladimir Milisavljević
Faculty of Mining and Geology, University of Belgrade, Serbia

Aleksandar Cvjetić
Faculty of Mining and Geology, University of Belgrade, Serbia

Dejan Ivezić
Faculty of Mining and Geology, University of Belgrade, Serbia

Uglješa Srbislav Bugarić
Faculty of Mechanical Engineering, University of Belgrade, Serbia

ABSTRACT

The main idea of this chapter is to promote maintenance centered to safety, in accordance to adaptive fuzzy inference model, which has online adjustment to working conditions. Input data for this model are quality of service indicators of analyzed engineering system: reliability, maintainability, failure consequence, and severity and detectability. Indicators in final form are obtained with permanent monitoring of the engineering system and statistical processing. Level of safety is established by composition and ranking of indicators according to fuzzy inference engine. The problem of monitoring and processing of indicators comprising safety is solved by using the features that Industry4.0 provides. Maintenance centered to safety is important for complex, multi-hierarchy engineering systems. Sudden failures on such systems could have significant financial and environmental effect. Developed model will be tested in the final part of the chapter, in the case study of bucket wheel excavator.

DOI: 10.4018/978-1-7998-3904-0.ch009

INTRODUCTION

Maintenance engineering is an important concept within the life cycle of engineering systems. Furthermore, development of maintenance engineering follows the steps of system sciences achievements. One of the development goals is to identify the concept which describes the maintenance quality and quality of service in best way. There are numerous maintenance strategies, concepts and technologies currently in use, depending on type of engineering system, availability of trained personnel, tools, spare parts, logistics etc. However, initial issue is always selection of critical indicator or concept in terms of establishing maintenance actions and procedures. Reliability and maintainability are the first developed theoretical indicators of maintenance quality (Martinetti et al., 2017). Dependability and availability are the first developed concept in maintenance engineering. Advanced maintenance relies on risk and safety, based on numerous methods developed for monitoring condition of the system, related to risk and safety. Failure modes and effect analysis and Risk Priority Number are usually used method in that sense (Papic, 2009). These methods are almost based on experience. Systems whose failures bring potentially high risk (for example: Nuclear plants, Oil refinery, Mining industry in general), maintenance according to these two approaches are standard.

Safety as theoretical approach has no strict definition in systemic sciences. Some interesting suggestions for using Prevention through Design approach (PtD) have been suggested by Borchiellini et al. (2013) and by Labagnara et al. (2013) for mining and underground activities. This chapter provides definition of safety critical value, for later introduction of safety "alarm". Comprehensive character of safety as phenomena, as well as uncertainty, indefinitely, multiplicity, subjectivity and mutual over-lapping, is leading researchers to use fuzzy theory, which here represents mathematical and conceptual synthesis model (Yang, 2006). Theories of Fuzzy sets and Fuzzy logic is specifically used for description and proposition of indicators, i.e. their composition (Teodorovic, 1998) and identification at safety level.

Existing problems in identification of maintenance parameters, such as data collection and grouping, acquisition, and synthesis of partial indicators can be solved, in general, with recent technical solutions according to *Industry4.0* achievements. *Industry4.0.* integrated for the first time digital and cyber technologies into the all production levels to such extent that advanced technological and cognitive machines are automated for intellectual tasks, aside for physical ones (Reyes Garcia et al., 2019). Development of autonomous networks for execution of physical tasks, creates condition for mending imperfections without engaging human action. Practical elimination of failure risks is expected in near future through improved capability of detection of hidden anomalies, by cyber-physical technology. Advantages of *Industry4.0* could easily transcend the risks, having in mind physical capabilities of cyber system.

Developed fuzzy safety model will be tested in the final part of the article, in the case study of bucket wheel excavator. Case study will be of general type, with analysis of two catastrophic failures, occurred on lignite open cast mines of Electric Power Industry of Serbia. It will also include functional algorithm of maintenance related to safety.

CONCEPTS AND INDICTORS IN MAINTENANCE ENGINEERING

This part of the article provides an analysis of indicators and concepts used for description of maintenance engineering and quality of service, as well as life cycles of engineering systems. Indicators are of partial character, while the concepts are overall, i.e. of synthesis origin. Indicators are defined in detail, while the exact structure of concepts is mainly lacking.

Reliability and Maintainability

Reliability and maintainability are the most important theoretical indicators of engineering system' quality of service analysis. These are based on the probability theory and statistics, and they are in extensive use since late 1940-ies to date. Such indicators are representing the probability that the system is in operation and probability that it is returned to operating condition.

Reliability as the indicator of technical system behavior in operation and maintainability as the indicator of technical system behavior during the period of failure can be stated as traditional and the two most familiar concepts. Regard to time dependent systems, reliability is theoretically expressed as probability of operation without any failure during the period t, i.e. as the time function $R = f(t)$. Maintainability performance evaluation can be performed as probability function $M = f(t_0)$ determination for maintenance operation with duration t_0.

The time state picture of operation and maintenance is shown on Fig. 1. Up-time of the system can be divided into stan-by time – system is waiting to operate (t_{11}) and operational time (t_{12}). Down-time comprises of organization time (t_{21}), logistics time (t_{22}) and active repair time (t_{23}), which can be time for corrective repair (t_{231}) or time for preventive repair (t_{232}). Times (t_{21}) and (t_{21}) are related to failure identification, planning, administration, acquisition of spare parts and tools, engaging labour, etc.

Operational time of the system (t_{12}) are periods used for calculation of reliability function R. Active repair time includes repairs, assembly, disassembly, replacements and similar. Active repair time t_{23} refers to construction maintainability (not logistic) and it is used for calculating function $M(t)$. In this case it is $t_{12} = t$ and $t_{23} = t_0$.

Time state picture is not always of the same character. Example shown on Fig. 1 is just one possibility.

Neither analysis based to maintainability function nor based on mean time, do not give clear and sharp picture if the problem is in construction or logistic maintainability. The mean time in operation and the mean time in failure are indicators obtained on the base of statistics analyses.

Figure 1. Time state picture

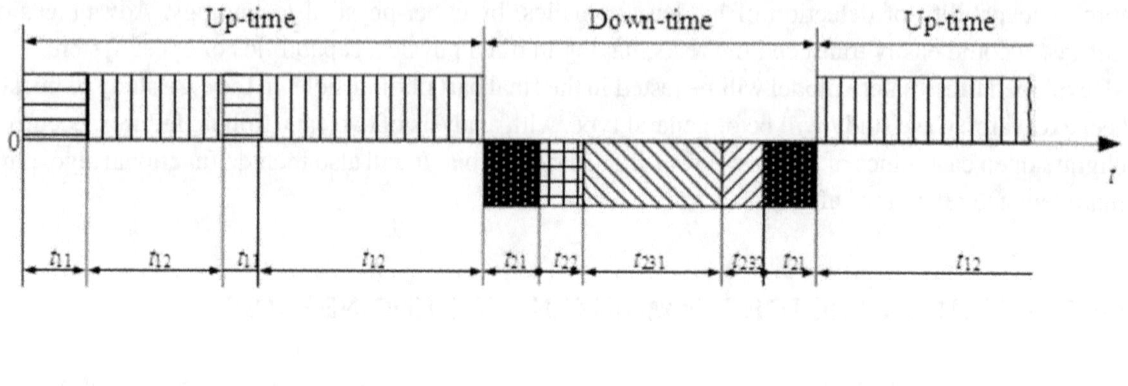

Availability and Dependability

Availability is comprehensive concept that includes entire service life of engineering system. Various definitions of availability can be found in references (Ivezic, 2019). In general, availability is a property of engineering system to execute required function in given conditions and in given time, assuming enough supply of necessary resources (Papic 2007). Availability can also be expressed as probability that engineering system will be available to operate or to start operation, at any given calendar time. It is a measure of engineering system's condition in its capability to operate within the performance limits of criterion function, in given time and given conditions (Erkoyuncu 2017). Availability is determined in relation to reliability function and maintainability function (Dhillon 2008).

Availability can be determined as a ratio of total (cumulative) up-time and sum of up-time and down-time (calendar time), and it can be expressed as:

$$A(t) = \frac{\sum t_{11}, t_{12}}{\sum t_{11}, t_{12}, t_{21}, t_{22}, t_{231}, t_{232}} \quad (1)$$

Equation (1) is frequently called operation availability and it is denoted as $A_o(t)$. Further on, achieved availability for down-time includes only active repair time for corrective and preventive maintenance $A_o(t)$.

$$A_a(t) = \frac{\sum t_{11}, t_{12}}{\sum t_{11}, t_{12}, t_{231}, t_{232}} \quad (2)$$

Inherent availability $A_i(t)$ (3) is obtained by including only active repair time for corrective maintenance:

$$A_i(t) = \frac{\sum t_{11}, t_{12}}{\sum t_{11}, t_{12}, t_{231}} \quad (3)$$

Such approach does not consider the reasons for increasing or decreasing of availability, i.e. it is not clear if the problem is in reliability, maintainability from the point of construction or logistical lacks.

Accurate availability function can be seldomly defined. More common is to calculate the availability by mean time between failures (MTBF) and mean down time (MDT) as:

$$A = \frac{MTBF}{MTBF + MDF} \quad (4)$$

If using exponential function of reliability $R(t) = e^{-\lambda \cdot t}$ and maintainability $M\left(t\right) = 1 - e^{-\mu - t}$, following relations are known:

failure rate: $\lambda = \dfrac{1}{MTBF} = const.$ and

maintenance rate: $\mu = \dfrac{1}{MDT} = const.$

Availability function $A(t)$ in this case is:

$$A(t) = \frac{\mu}{\lambda + \mu} + \frac{\lambda}{\lambda + \mu} \cdot e^{-(\lambda+\mu)\cdot t} \quad (5)$$

while the stationary value of availability is:

$$A = k_A = \lim_{t \to \infty} A(t) = \frac{\mu}{\lambda + \mu} = \frac{1}{1 + \frac{\lambda}{\mu}} = \frac{1}{1 + B} \quad (6)$$

Parameter k_A is called availability coefficient and it is calculated from A(t) for $t \to \infty$, i.e. when the availability value becomes stationary (Figure 2).

Figure 2. Relations between reliability R(t) and availability A(t)

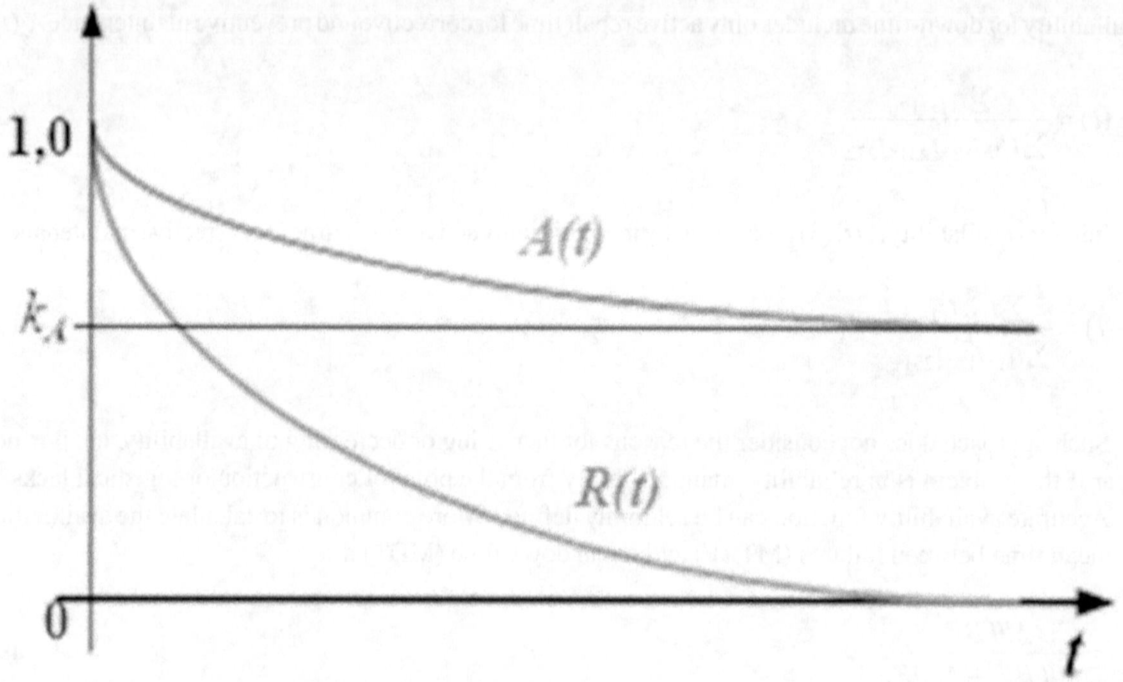

Reliability and availability functions are shown on Figure 2. It can be seen that availability requirement is stricter than requirement for reliability, $R(t) \leq A(t)$.

For realistic production and maintenance conditions of engineering systems availability is calculated by (1), (2) or (3), depending on available data. Analytical requirements are seldomly met to define the availability function in form of (4) and (5). For mentioned approaches (R, M, A), it is necessary to have comprehensive collection of data for periods of continuing operation, as well as data about times necessary for transition of engineering system in failure to system in operation. Determination of R, M and A as described above requires suitable IT structure for monitoring times t_i. In real-world scenario, this means that industrial process is supported by powerful IT sector, including monitoring and acquisition.

At the end of XX century, concept of dependability management (Strandberg, 1991) has been developed by the International Electrotechnical Commission (ISO-IEC 300), to provide an integrated approach to management and assurance of dependability, i.e., the reliability and maintainability performances of products and the performance of the maintenance support system. This concept is also well-known as International Standard IEC 300. This standard gives the explanation of dependability in expressive, linguistic manner without firm, formal determination. Dependability is only standardized, all-in concept, which describes technical systems from the point of: design, operations and maintenance (Ivezic, 2008). While in the same spirit, some authors have defined dependability slightly differently. In (Ebramhimipour, 2006) dependability was defined as total ability that has other measures such as: maintenance time, maintenance labor, maintenance frequency, reliability, availability. While in (Avizienis, 2001) dependability was understood as an integrative concept that includes the next attributes: availability, reliability, safety, confidentiality, integrity and maintainability. In cited articles dependability was defined as a concept of failure engineering, which includes reliability, safety, maintainability, security, risk and quality control. In contrast to dependability definition in ISO IEC standards, these definitions include similar concerning of reliability, maintainability, and availability. Availability contains reliability and maintainability, and this make trouble in partial analysis of dependability indicators and their further mixing. Standard IEC 300, among other things, states that performance of dependability includes availability, as its measure (Strandberg, 1991). Dependability is defined as «the ability to avoid service failures that are more frequent and more severe than is acceptable to the user(s)», that is an integrating concept that includes the following attributes: availability, reliability, safety, confidentiality, integrity, maintainability, etc. The problem with dependability concept is a lack of calculation model.

Some authors present effectiveness as overall concept of quality of service (Miodragovic, 2012). The effectiveness of a system can be defined as the probability that a system will operate successfully and reach required criterion function within the limits of allowed discrepancies for the given time period and for the given conditions (Papic, 2007). Effectiveness consists also reliability and maintainability plus functionality. Described three overall concepts (Availability, Dependability and Effectiveness) are inherently very similar and it is evident that essentially consists of reliability, maintainability and supportability in general (Ivezic, 2019).

RISK AND SAFETY

Risk and safety are separate group of overall indicators. Their importance is reflected in prediction of accidents and analysis of severity. Therefore, risk in this context is described by following performances: occurrence, severity and detectability. Occurrence can be identified as reliability. Risk is defined by Risk Priority (RPN) calculation (Petrovic 2014); hence its application is standardized. Safety is a concept complement to the risk, and there is no standardized implementation model (Wang, 1995). In general, safety can be described as distance from operating and critical value of considered performance of the system (Figure 3). Safety is larger as this distance increase. The issue in this approach is that both operating and critical values are represented by density function of distribution. There is also a possibility for over-lapping due to high deviation of the functions, regardless to distance between average values.

it is important to note that such over-lapping represents the risk. In case that equipment owner requires high safety – low risk, it will come at high costs. Failure modes and effect analysis (FMEA) is a method usually used for safety analysis (Bowles, 1995)). FMEA is an effective qualitative approach used to find all possible ways in which part of system can fail and the possible resulting effects on the total system and related personal. This model is based on experience in defining weak points. Also, it doesn't provide clear safety structure and relation of indicators used for description of safety.

As indicated by Martinetti et al. (2019) Safety is changing thorough the decades, and it is complex apprehension. It can be presented as "post festum" difference between operational and critical value of condition. Prediction requires in depth analysis of influencing parameters, where it is evident that dependability, probability of consequence, severity and possibility of detection are having impact. Several models were developed up to date for the purpose of identification of these parameters, which are based on long term monitoring and statistical processing. As it can be seen, myriad of parameters has impact on safety and intercorrelation between these parameters is unclear. Parameters are not complementing one another and cannot be arranged by common mathematics operations. Scheme shown on Figure 3 provides a safety outcome only by single parameter. For the safety analysis this approach is unsustainable. Safety should be approached through financial parameters, reduced reliability and structural integrity, environmental and personnel hazards.

MAINTENANCE STRATEGY

Main aspects of maintenance strategy are its concept, organization and technology. There are two approaches to the maintenance: Reliability Centered Maintenance (RCM) and Total Productive Maintenance (TPM). RCM engineering management makes decisions following detailed analysis and with thorough understanding of system properties, especially reliability properties. TPM engineering management makes its decisions based on permanent and comprehensive insight on current system condition, especially regarding execution of its tasks (Jovancic, 2012a).

Maintenance concept is one of the most important features of maintenance system. Overall maintenance quality is highly dependent on conceptual solutions. There are two basic concepts related to failure occurrence: Preventive Maintenance and Corrective Maintenance. Concept of preventive maintenance dictates maintenance action before the failure occurs, while the system is in the up-time. Purpose of preventive maintenance action is to prevent or postpone the occurrence of failure. On the other hand, concept of the corrective maintenance dictates maintenance action only if the failure occurs Jovancic, (2012a).

Corrective maintenance activities are performed only when the system is in the down-time. Most common approach is combination of these approaches, meaning that preventive maintenance is used on one part of the system and the corrective on the other.

There are two types of the preventive maintenance:

- First one is based on the information on reliability and empirical probability distributions of time between failures for whole system and its components.
- Second type is also based information on reliability, but also on permanent systemic monitoring of system's operation.
- Types of preventive maintenance procedures are shown in the lower part, while the dependence on time requirements of these procedures are shown in the upper part.

Figure 3. Relations between safety and risk

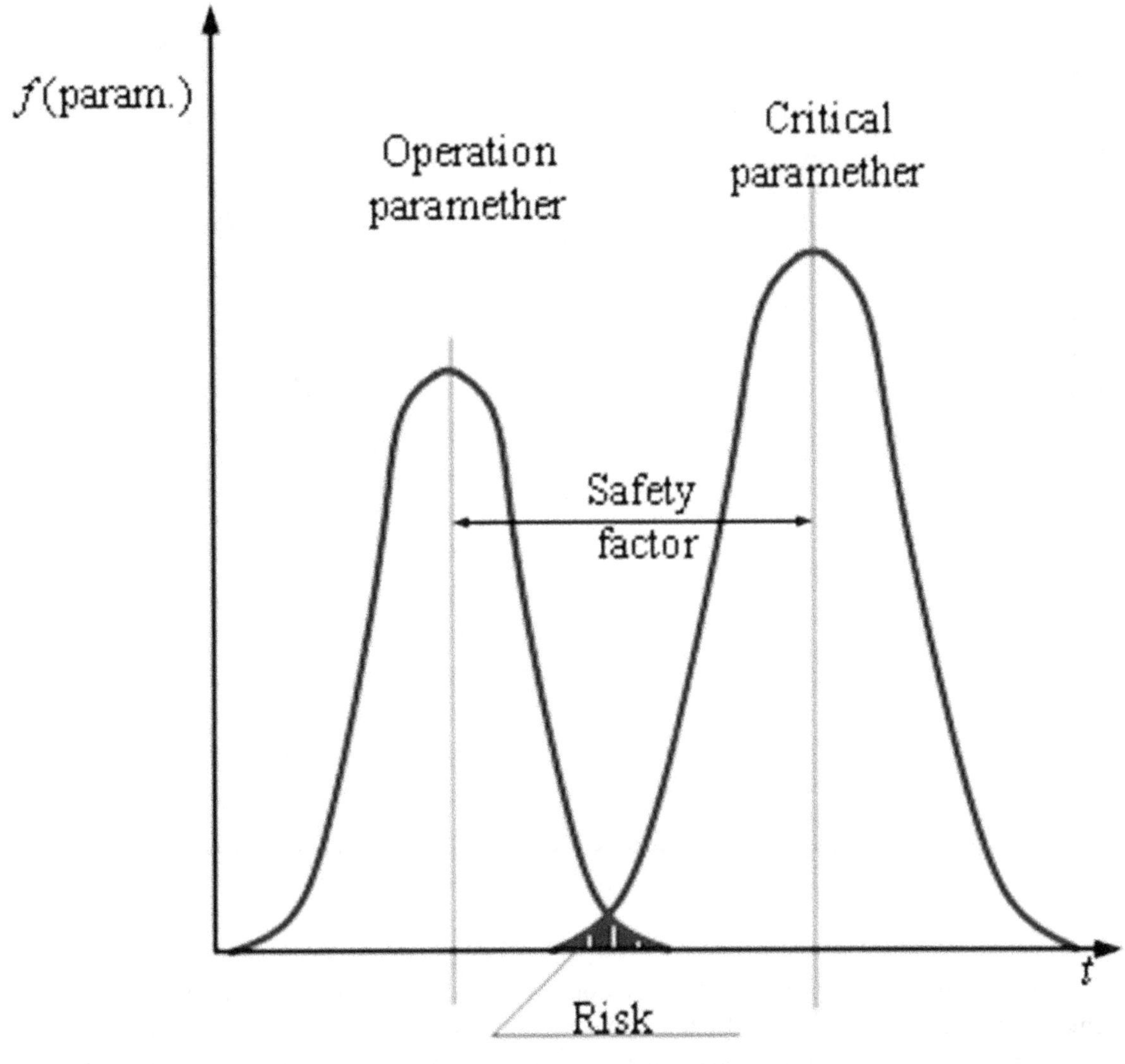

One approach to the preventive maintenance is Condition Based Maintenance (Jovancic, 2016). This approach is based on condition of the system, which is described by data acquired through condition monitoring. Monitoring of mechanical engineering systems are usually including tracking of fatigue indices, initial cracks, wear, overheating, noise, vibrations, etc. This approach is more cost efficient, in comparison to the typical preventive or corrective maintenance. Condition based maintenance is the best option in cases when the failure rate is constant and for prevention of unexpected failures. Requirements for this approach are:

- maintenance schedule, established according to numerous parameters and real-time data providing insight into the condition of engineering system and its components;
- advanced diagnostics techniques, for establishing relations between afore mentioned parameters.

Figure 4. Time period variations for identification of failure occurrence

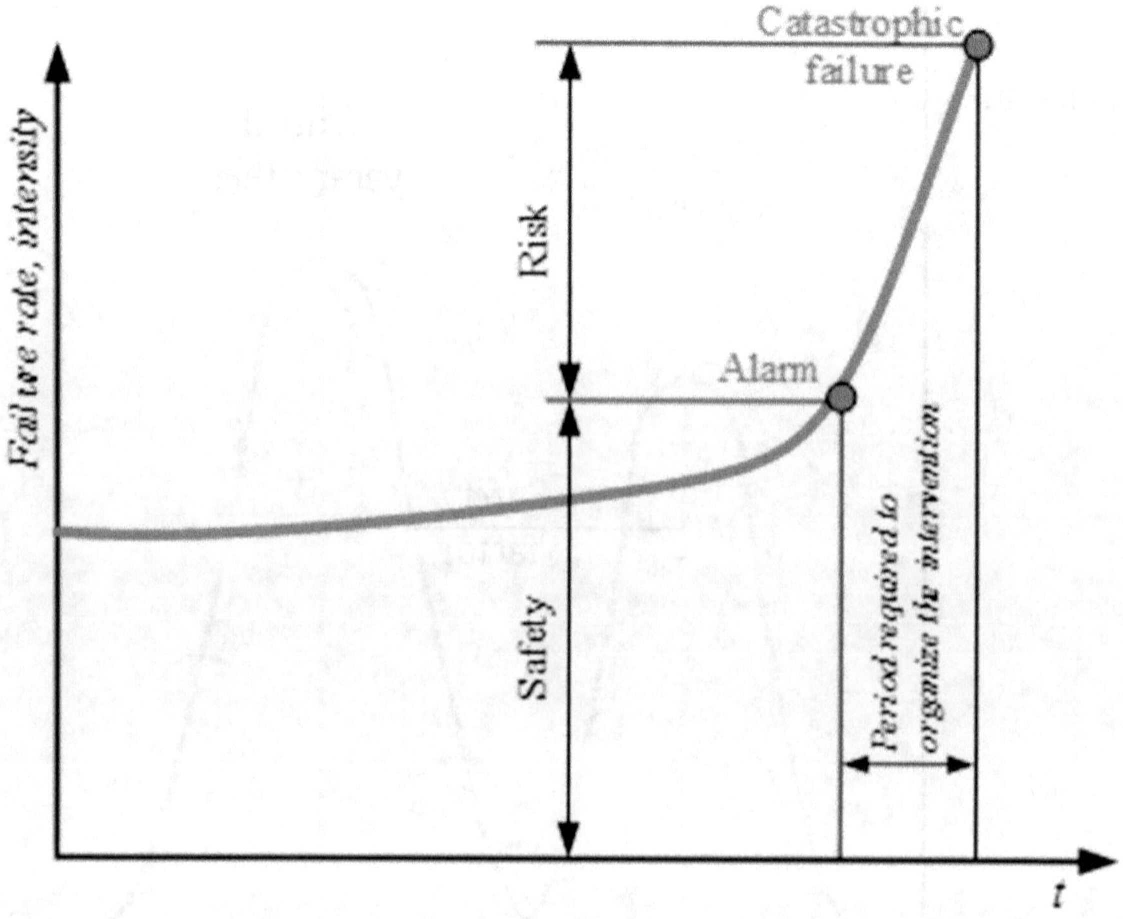

Corrective maintenance is mainly composed of unplanned activities, and therefore unplanned requirements in specific time interval. Hence, down-times cannot be foreseen, neither repair times. It is obvious that this approach isn't suitable for maintenance of important and expensive equipment, such is the case in mining industry (Karimnia, 2015). Corrective maintenance activities are demanding urgent action of the staff, causing the delays and changes in previously planned activities. This action must also be swift, with technical and technological understanding to repair failures and to return the equipment into the original performances.

Safety and risk centered maintenance (SCM) is approach to maintenance based on RCM (IAEA-TECDOC-658). Crucial difference is in boundary conditions, which are defining whether the strategy is successful or not. SCM analyze direct consequence of failure to the employees, environment, business results, structural integrity of the system, providing the opportunity to detect the failure in-time before causing the consequences. SCM requirements are supervising, management and logic (artificial intelligence) networks for monitoring the system. Therefore, SCM can be considered as an aspect of predictive maintenance.

APPLICATION OF PREDICTIVE MAINTENANCE 4.0 (PDM 4.0) IN MINING INDUSTRY

There are 5 asset management approaches to maintenance which can be applied in mining industry (Jovancic, 2016): reactive (corrective), preventive, condition based, predictive and regulated. Reactive maintenance is of no strategic importance for critical assets in mining industry, since most of the equipment operates in series connections. Failures are expensive on various levels, with impact on productivity and availability, spare parts can be very expensive, as well as labor and energy costs. These items are key success indicators, therefore, preventive maintenance results in improved reliability but even preventive maintenance may not be sufficiently efficient against unplanned down-times and avoidable expensive repairs. Calendar based planned maintenance is inefficient, since majority of failures (over 75%) are of stochastic nature.

Condition monitoring of mining machines is the initial step in acceptance of maintenance strategy. This approach usually incorporates following diagnostics techniques: vibration analysis, ultrasonic analysis, thermography, endoscopy, magnetic methods, spectrometry analysis of wear. Data can be acquired on-line or off-line, depending on importance of the machine or component. Predictive maintenance further improves condition-based maintenance by detecting modelled anomalies. It relies on on-line data acquisition combined with data analysis to evaluate reliability of the machine and establishing main cause of the failure.

Final maintenance level in accordance to the asset management principles and recent technological achievements based on intelligent systems, can be called regulated maintenance. This includes integration of big data, analytics, machine learning and artificial intelligence. Regulated maintenance system is cognitive system – it has capability to "think" and to operate with multiple heterogenous engineering systems enabling exchange of information for the final user. Regulated system is currently most advanced maintenance system, belonging to the *Industry4.0*. It is evident that last mentioned maintenance concept is most efficiency.

Concept of *Industry4.0* for mining industry equipment comprises of principles utilizing all capabilities of available technology (Jovancic, 2016):

- Integration of mining equipment into IoT (capability of different systems to operate together);
- Capability of digital technology to create virtual mining machines, based on available information (transparency);
- Digitalized mining equipment with artificial intelligence as a support to decision making process (technical assistance);
- Capability of digitalized systems to operate independently (decentralized decision making).

Industry4.0 integrated for the first-time digital or cyber technology into production to such an extent that high-tech cognitive machines are automating even intellectual tasks. Imperfections can be corrected as soon as detected, without human intervention.

Despite to the huge potential in the mining industry, there are still numerous obstacles for the full implementation of such solutions:

- Reliable machine to machine communication is still not at required level of performance and stability;

- IT security issues, especially when using on old machines;
- Uncertainty with IT omissions, regardless to capabilities of artificial intelligence;
- Insufficient training and experience of staff for implementation of *Industry4.0* in mining industry;
- Social issues related to labor redundancy caused by *Industry4.0*.

Further improvement in detection of omissions in cyber technology will reduce the risks of failures, and eventually the advantages of the *Industry4.0* principles would overcome the risks.

Purpose of modern mining systems is to neutralize potentially expensive and time demanding human errors, by using advanced diagnostic techniques and optimal maintenance strategy. To achieve this, it is necessary to develop advanced intelligent maintenance systems or technologies of smart maintenance. Principles of *Industry4.0* i.e. Predictive Maintenance 4.0 (PdM 4.0) are defining the future of engineering systems maintenance, including the maintenance of mining equipment. Machines capable for self-maintenance are better option if combined with condition monitoring, diagnostics, repairs planning in order to extend its operational life and improve performance. Target is to produce the self-maintenance system for mining equipment to enable machines capable for reconfiguration, compensation and self-maintenance, and PdM 4.0 is surely excellent starting point.

Concept of self-maintenance with autonomous repair module could have significant impact on labour costs and delays during maintenance. Further on, adaptive self-maintenance system could independently compensate failure of any sub-system by reconfiguration of parameters and minimize human intervention. Such systems could also identify source of the failure and estimate future trends. Self-maintenance based on PdM 4.0 could suggest accurate maintenance action, including capability to deal with errors in order to increase overall reliability. Self-maintenance systems would minimize preventive maintenance costs, optimize maintenance plans and reduce costs related to spare parts and resources.

Real-time access to operational data is the most important property for improving efficiency. Big data analytics in maintenance is the next level of maintenance strategy, which is incorporated into PdM 4.0. This approach would foresee future failures and delays and provide most efficient preventive measures, by using advanced big-data analytics on technical condition of equipment, its use, maintenance history, operating conditions, etc.

Such level of maintenance is feasible through implementation process comprising of following steps:

1. First step is acquisition of mining equipment real-time operational data and referencing them to the systems and staff. This is crucial step for enabling analysis, and the basis for digital transformation, such is the implementation of PdM 4.0.
2. Second step is developing solution for data storage and transformation of data into the information (putting the data into the context). In order to understand data analyst must know whether the equipment is running or not, for example. There is not much data value without context.
3. Third step is implementation of maintenance strategy based on contextual data – information. This includes establishing priorities on specific machines, systems and sub-systems and determination of conditions causing the failures and delays.
4. Fourth step is implementation of PdM 4.0. Intelligent system would enable optimization of operations and activities in combination with advanced analytical tools (condition and behavior pattern recognition in real-time causing the failure or delay).

Equipment condition and behavior monitoring in real-time leads to specific level of reliability. However, equipment will still malfunction due to unforeseen failures. PdM 4.0 also includes capabilities of artificial intelligence for permanent insight and detection of sources and anomalies for initially unforeseen failures. Therefore, PdM 4.0 would be capable to foresee previously unpredictable circumstances.

SAFETY EVALUATION MODEL

Safety evaluation comprises of two modules. First module consists of data processing, while the second module is synthesis of acquired data. Data acquisition and statistical processing of input data is related to partial safety indicators. The second module is represented by the mutual ranking of partial indicators and their composition. Recommended ranking method is Analytic Hierarchy Process (AHP) (Saaty 1980). Composition is performed according to the fuzzy inference rules (Zadeh, 1996).

Processing of Input Data

Records of time related occurrences for each partial indicator creates a data set, which can be processed by statistical tools and probability theory. The 2 Parameter Weibull Distribution (2PWD) is one of the most common distributions used in reliability and maintainability engineering since it attains numerous shapes for various values of scale parameter (η) and shape parameter (β) (Weibull, 1951). It can therefore be used to model a variety of data and real-life characteristics. Input data for the model are measured periods $t = t_1 \ldots t_n$ according to time picture of state. Based on these data, density function (7) can be calculated (Dhillon, 2008):

$$f(t) = -\cdot t^{-1} \cdot e^{-\left(\frac{t}{}\right)} \tag{7}$$

Fuzzy model operates with fuzzy numbers as input values. Features of fuzzy numbers, fuzzy sets and fuzzy arithmetic in general are well known and will not be described in detail here (Klir, 1995). Fuzzy numbers are defined in coordinate system of membership function (μ) and classes, as unit measure of indicator (j). Any integer > 0 can be accepted for j. Larger values are contributing to the accuracy of calculation. In case of $j = 5$ the fuzzy number (FN) can be presented as:

$$FN = (\mu_{(j=1)}, \ldots, \mu_{(j=5)}) \tag{8}$$

where $\mu = 0 \ldots 1$

Next step of input data processing is fuzzification of time dependent cumulative function i.e. transformation of density function in fuzzy shape. Fuzzification is a mapping procedure of distribution density function from coordinate system $f(t)$-t into coordinate system μ-j, where j is ranging from t_{min} to t_{max}.

Below is an example of mapping procedure, for clarification purpose. Let's assume it is necessary to calculate reliability (R) for a given machine, where it is recorded up-times are: $t_{min} = 187$ and $t_{max} = 921$. Parameters of 2PWD are: $\beta = 1{,}24$; $\eta = 824{,}43$; $\overline{T} = 773{,}43$. Coordinates of intersection points of curve *f(t)* and ordinates for each *j = 1 …5* can be read from Figure 5. Finally, assessment of R is obtained in the form (9):

$$R = (0.91_{(j=1)}, 0.88_{(j=2)}, 0.59_{(j=3)}, 0.34_{(j=4)}, 0.20_{(j=5)}) \quad (9)$$

Figure 5. Mapping process of reliability density function

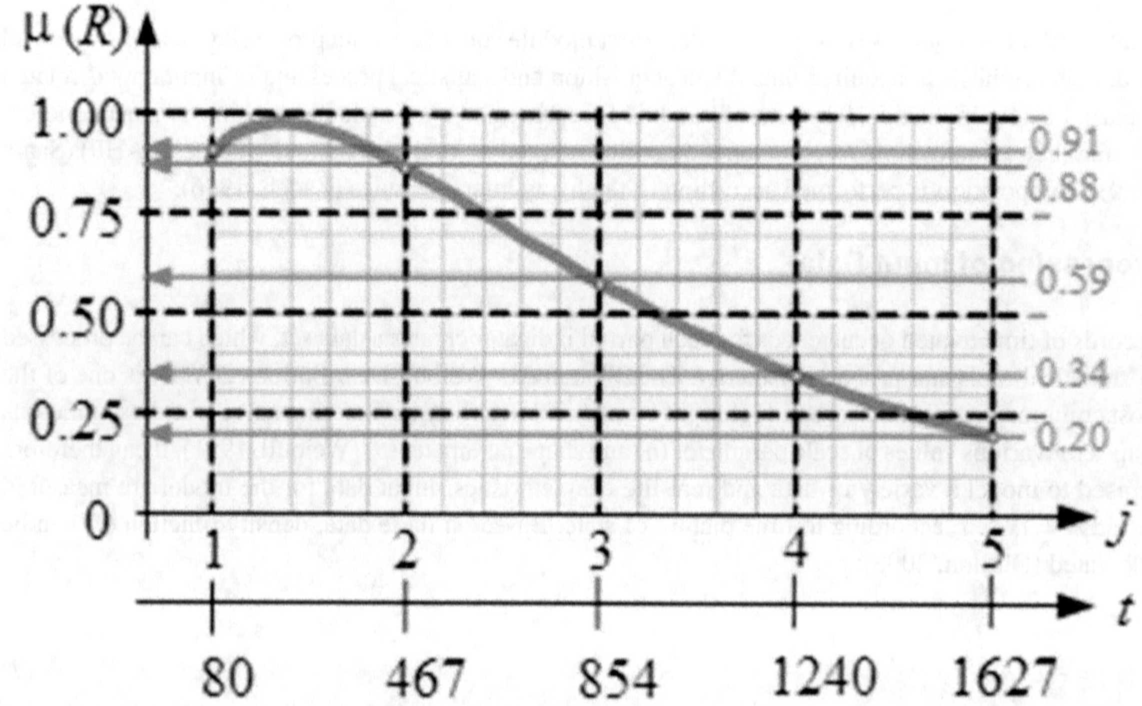

FUZZY INFERENCE ENGINE IN SYNTHESIS MODULE

Safety (further on denoted as – S) is defined as comprehensive concept (umbrella term) consisting of following phenomena – partial indicators: reliability, maintainability (constructive, design aspect and logistic aspect), consequence, severity and detectability. First two indicators can be joint into one – dependability (RM); while the probability of consequence and its severity also can be joint into single indicator (CS); leaving detectability by itself (Dt). Therefore, fuzzy module has two levels of synthesis. First level is synthesis of reliability and maintainability into RM indicator and synthesis of consequence and severity into CS indicator. Second level is synthesis of RM, CS and Dt indicators into S. Structure of influence indicators on safety is given on figure 6.

Figure 6. Structure of safety

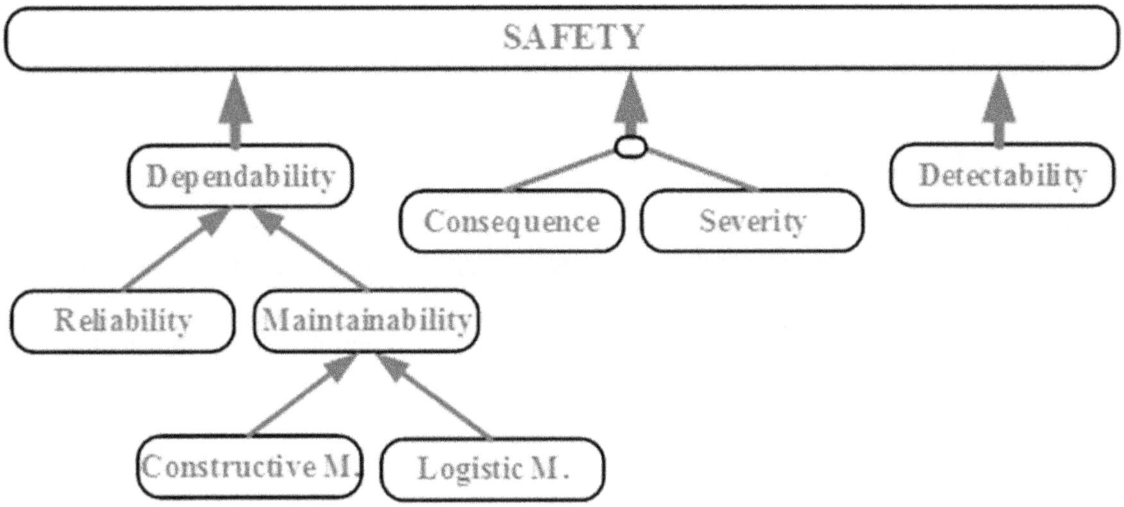

Synthesis is done according to the fuzzy composition rules, commonly represented as "IF-THEN" rules. Specific assessments are using particularly developed composition models, while the most frequently used compositions are *Max-min* and *Min-max*. *Max-min* composition (10) provides synthesis assessment through application of representative partial assessment, which is defined as the best possible one among the worst expected particular outcomes. This composition is used for representing phenomena such as safety of dependability (Tanasijevic, 2013), (Wang, 2000). *Min-max* composition is fuzzy model which provides synthesis assessment as representative partial assessment as the worst one among the best expected particular outcomes. It is commonly used for interpretation of risk (risk priority number) (Petrovic, 2014). Description of max-min composition is provided below in steps *i* to *vii*.

$$S = \max\{\min(P_1, P_2, \ldots, P_m)\} \tag{10}$$

Synthesis indicator is denoted with S, while partial indicators with Pi. Synthesis indicators in this case are: S, RM and CS, while partial indicators are:

- In case of S = S: $P_{i=1}$ = RM, $P_{i=2}$ = CS, $P_{i=3}$ = Dt;
- In case of S = RM: $P_{i=1}$ = R, $P_{i=2}$ = M, $P_{i=3}$ = L;
- In case of S = CS: $P_{i=1}$ = Cn, $P_{i=2}$ = Sv;

1. Let's define m fuzzy numbers $P_1, P_2, \ldots, P_m$ with membership function μ and class $j = 1$ to n:

$$P_{i=1} = (\mu_{P1(j=1)}, \ldots, \mu_{P1(j=n)});$$

$$P_{i=2} = (\mu_{P2(j=1)}, \ldots, \mu_{P2(j=n)}); \tag{11}$$

…

$P_{i=\mathrm{m}} = (\mu_{\mathrm{Pm}(j=1)},\ldots, \mu_{\mathrm{Pm}(j=n)});$

2. Membership function can form $C = n^m$ combinations. Each combination represents one possible assessment of S.

$$S_c = [\mu_{R(j=1,\ldots,n)}, \mu_{C(j=1,\ldots,n)}, \mu_{D(j=1,\ldots,n)}], \text{ for each } c = 1 \text{ to } C \tag{12}$$

If considering only values meeting the criteria $\mu_{Pi\ (j=1,\ldots,n)} \neq 0$, resulting outcomes o will be ($o = 1$ to O, where $O \leq C$). Each outcome has appropriate values (*iv*) and (*v*) defining for further calculation.

3. Value of J_c is calculated for each combination c meeting the criterion that it is an outcome, which is rounded as integer as follows:

$$J_c = [(w_{Pi=1} \cdot j_{(\mu Pi=1)c}) + \ldots + (w_{Pi=m} \cdot j_{(\mu Pi=m)c})] \tag{13}$$

Where:

- w weighting factor of specific partial indicator on the synthesis indicator (AHP method), obtained by ranking of all partial indicators providing: $\sum w = 1$;
- j_c class of the fuzzy number (8) for considered membership function and given combination c, where: $j_c = 1,\ldots, n$;

4. Minimal value of μ_{Pi} in vector S_c (12) is determined for each outcome, as follows:

$$MN_o = \min\{\mu_{Pi=1(j)o}, \ldots, \mu_{Pi=m(j)o}\}, \text{ for each } o = 1 \text{ to } O \tag{14}$$

5. Outcomes are grouped according to the values J_c. Number of such groups can be from 0 to n.
6. Each group of outcomes (*vi*) is searched for maximal value MX among identified minimums (*v*). Maximum corresponding to j^{th} value is calculated as:

$$MX_j = \max\{MN_1, \ldots MN_o, \ldots, MN_O\}_{Jc}, \text{ for each } j = o \tag{15}$$

Assessment of analysed engineering system in relation to synthesis indicator is finally obtained in form corresponding to the equations (8) and (9):

$$S = (MX_{j=1}, \ldots, MX_{j=n}) = (\mu_{S(1)},\ldots, \mu_{S(j)},\ldots, \mu_{S(n)}) \tag{16}$$

Equations (16) gives an assessment in relation to membership function and class.

In order to establish critical level of safety alarm (figure 4) it is necessary to reduce the fuzzy equation (16) to its final value. This procedure is called defuzzification. Center of mass point calculation (17) is commonly used (Ying-Ming, 2009).

$$Z = \frac{\sum_{j=1}^{n}(\mu_j \cdot j)}{\sum_{j=1}^{n}\mu_j} \quad (17)$$

In case of (9) this equation should be applied as follows:

$$Z_R = (0.91 \bullet 1+0.88 \bullet 2+0.59 \bullet 3+0.34 \bullet 4+0.20 \bullet 5)/(0.91+0.88+0.59+0.34+0.20)=2.33 \quad (18)$$

Centre of mass of fuzzy number R, on scale from 1 to 5, is 2.33.

Overall safety of multi hierarchy engineering systems is obtained as composite safety assessment of units at lower level of hierarchy. This can be expressed as:

$$S_t = \max\{\min(S_1, S_2, \ldots, S_N)\} \quad (19)$$

Where: S_t - safety of upper level of engineering systems hierarchical structure;

$S_{1..N}$ - safety of lower level engineering systems hierarchical structure;

CASE STUDY

Bucket Wheel Excavator (BWE) is a machine with inherent high risk (Tanasijevic, 2016). This machine is unique by its hierarchy structure, high price, high costs of unplanned breakdowns, high costs of technological processes, permanent requirement for ever higher production rates, extremely complex operating conditions and potential hazardous consequences for environment. Decades of operating these machines resulted in respectable operating and maintenance levels despite the difficult conditions, resulting in high time and production utilizations. Time utilization of all analyzed BWEs (31 machines operating in Electric Power Industry of Serbia – EPS) was 40%, while production rate utilization was 36%. Occurrence of failures have an impact on these parameters, and the engineering staff is trying to prevent them. Stationary failure rate with constant frequency of failure occurrence, on all BWEs, is a circumstance with limitation. Operational life of BWE is determined by weakening of critical load bearing components of steel structure. Catastrophic failures on BWEs had an impact on structural integrity to the largest extent, with the damages measured in millions of Euros. Also, there were accidents of smaller scale, mainly related to the sub-systems of the BWE (drive of the bucket wheel, bucket wheel, track sub-system, rubber belt conveyors on BWE, etc.) with larger impact on production and lesser on the structural integrity of the machine (Bosnjak 2015). Major task of engineering was to increase the safety of BWE operation at open cast mine.

Successful reduction of risk can be achieved with comprehensive systematic approach, including numerous measures, starting with maintaining nominal operating conditions, required maintenance, condition checks, securing firefighting service, telecommunications, as well as elimination of internal weak points by reconstruction and modernization. BWEs are machines manufactured in individual production, hence they are underdeveloped in terms of structure. Therefore, bucket wheel excavator can be categorized as a high-risk machine.

Sudden failure of BWE can have numerous consequences, such as: lost production, catastrophic failure – losing static integrity of machine, injuries, environmental contamination. Two examples from Kolubara mining basin are provided further on. Bucket wheel excavator SRs 1200 24/4 had catastrophic failure in 1995, following the fire which started by heating and ignition of rubber belt (figure 7). Source was the blocked bearing and idler on the conveyor, as well as incompetent intervention during fire suppression. Refurbishment of this BWE was completed by 2003, which included complete replacement of superstructure. Total investment was around 7 million €, i.e. around 60% of the value of new machine. Second example is bucket wheel excavator SchRs 630 25/6 which had catastrophic failure after the fire initiated by welding in the zone of central pillar (figure 8). Fire caused the failure of supporting steel ropes, resulting machine collapse. This machine was also refurbished, but with some structural improvements to increase safety and maintainability. In both examples machine was damaged by fire. Fire in second example was caused by maintenance activity (welding) i.e. by outer source and in first example by inner source. It should be noted that fire shall be more intense when the machine has grease residue and coal dust sediment.

Figure 7. SRs 1200 after catastrophic failure

Figure 8. SchRs 630 after catastrophic failure

BWE consists of large number of components assembled in functional systems, whose proper operation is necessary for efficient work of BWE (Tanasijevic, 2012). Each machine has nine basic systems: system for digging, system for transport of materials, system for transport of excavator, system for boom lifting, system for slewing of superstructure, main steel structure, auxiliary structure, control system and energy supply system. Each of these systems has a whole set of subsystems. For example, system for transport of materials, consist of: drive unit, rubber belt, lubrication system, steel structure, drive and return pulleys and idlers. Further on, drive unit sub-system consists of: electric motor, coupling (hydrodynamics and elastic), mechanical brake, gear-box and locking assembly (between hollow shaft of gear-box and shaft of drive pulley) (figure 9). It is evident, that BWE is real multi hierarchy machine, which makes maintenance complex task.

Figure 9. Hierarchical structure of BWE, with detail of drive unit

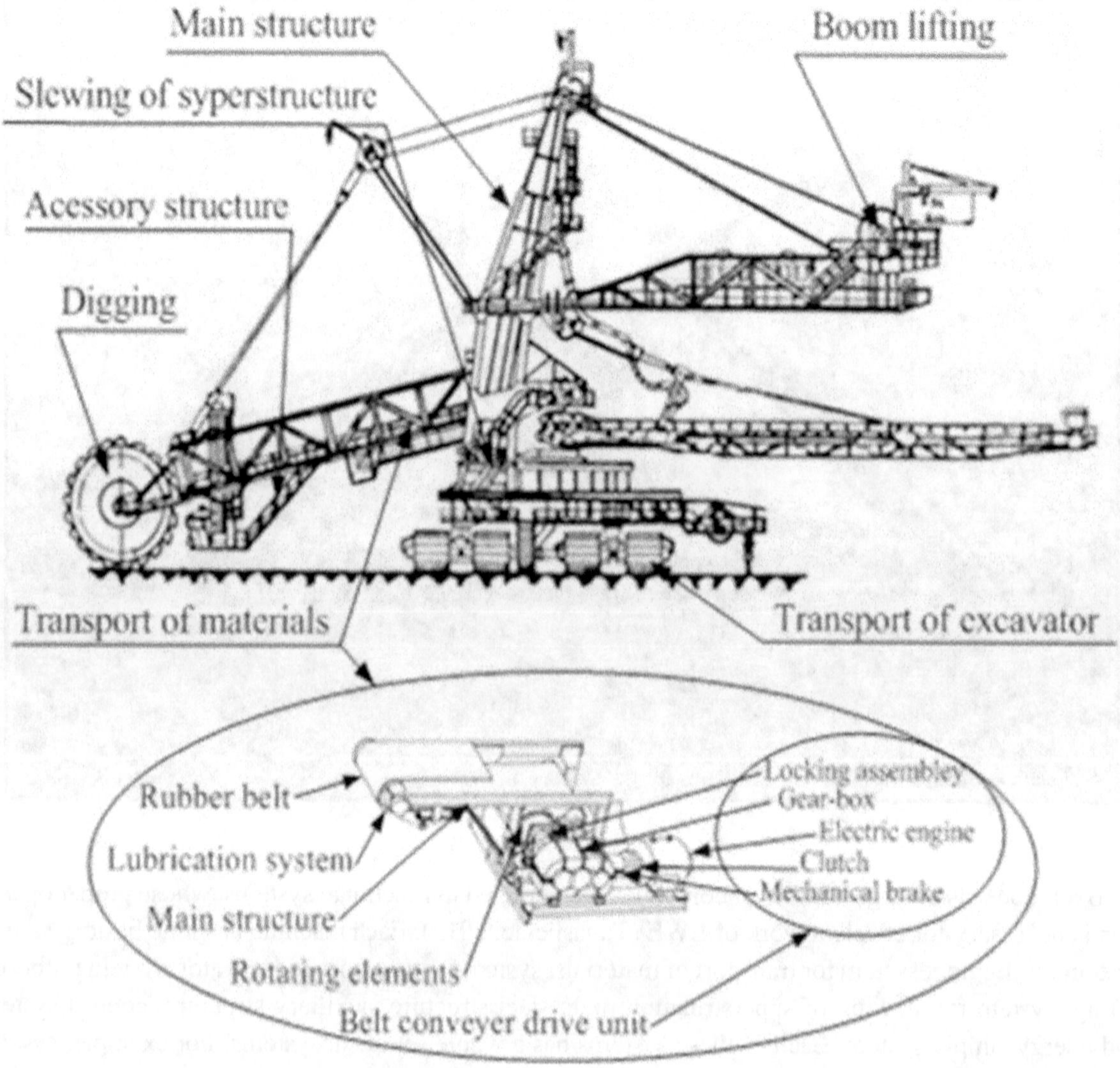

There are numerous advantages of modern technical systems for simple monitoring of partial indicators. Operation with big data, on-line monitoring and acquisition, are resources available through existing monitoring systems. Therefore, these are the basis of weak points investigations, reliable safety monitoring and maintenance decision making. In order to apply suggested maintenance model according to safety it is necessary to equip each weak point of the BWE with system for monitoring, data acquisition and data recording/storage. Weak points can be structural components of the BWE or parts of procedures and routines executed during operation and/or maintenance. Inclusion of more BWEs into the network would make the model more objective by making the history of each partial indicator more complete and more objective. Importance of this is reflected by increased accuracy in establishing boundary values for j_{min} and j_{max}.

Monitoring of partial indicators and definition of function $f(t)$ requires following records/data:

1. time of occurrence and down-time period;
2. reason for down-time occurrence;
3. time required for repair work; time state picture (Fig. 1);
4. logistics requirements for repair and maintenance (labor, machines, tools, spare parts, etc.);
5. level of administrative support;
6. types, number and level of consequential occurrences.

Specific task is monitoring of cause and source of the down-time, which in general can be triggered by:

- By mechanical failure: fatigue, corrosion, wear and rarely static fracture;
- Electrical failure;
- Technological down-time;
- Weather conditions.

Currently available reliability monitoring systems are commonly challenged by down-time categorization for large and complex engineering systems. Accurate identification of down-time causes frequently required comprehensive expertise, and often remains unidentified. All of this once the consequences occurred, means that detectability was at low level.

Regarding first of above mentioned catastrophic BWE failures potential causes for bearing failure could be fatigue of rolling elements, poor assembly, penetration of moisture and dirt into the bearing and insufficient greasing. Recent *Industry4.0* systems are providing myriad of applicable solutions for condition monitoring of components, detection of unforeseen situations – above recommended values of operational parameters, and similar. Online infrared camera would signal that there is a problem with bearing. Application of such online monitoring systems would increase level of detectability and result in more reliable history which could be statistically processed for detection of weak points. In this case, one of the reasons is underperformance of firefighting department. Procedure in this circumstance is to hook the burning rubber belt to the bulldozer and to pull it away from the BWE, and it is not performed in time. Safety in presented case was low, mainly due to the low level of reliability, maintainability, detectability and severity.

Cause in the second presented example is mainly low maintainability and high severity. This problem could be avoided if repair location were more accessible and more isolated from flammable components, even more if these were load bearing components of the excavator. Firefighting department underperformed in this case also. Such maintenance actions have inherited low safety, thus requiring larger resources for firefighting.

Conceptual algorithm for safety centered maintenance is shown on figure 10. Each weak point, either structural or procedural, would be equipped by safety "alarm". Critical value of safety would be defined by ranking of partial indicators and by management policy in the mine, according to restrictions. In doing so, restrictions can be financial, legal, in relation to standards, according to environment, health of employees, working parameters ... Partial indicator assessment, as well as the safety, can be defined based on permanent monitoring of partial indicators for given weak points of the engineering system.

Figure 10. Safety centered maintenance, BWE application algorithm

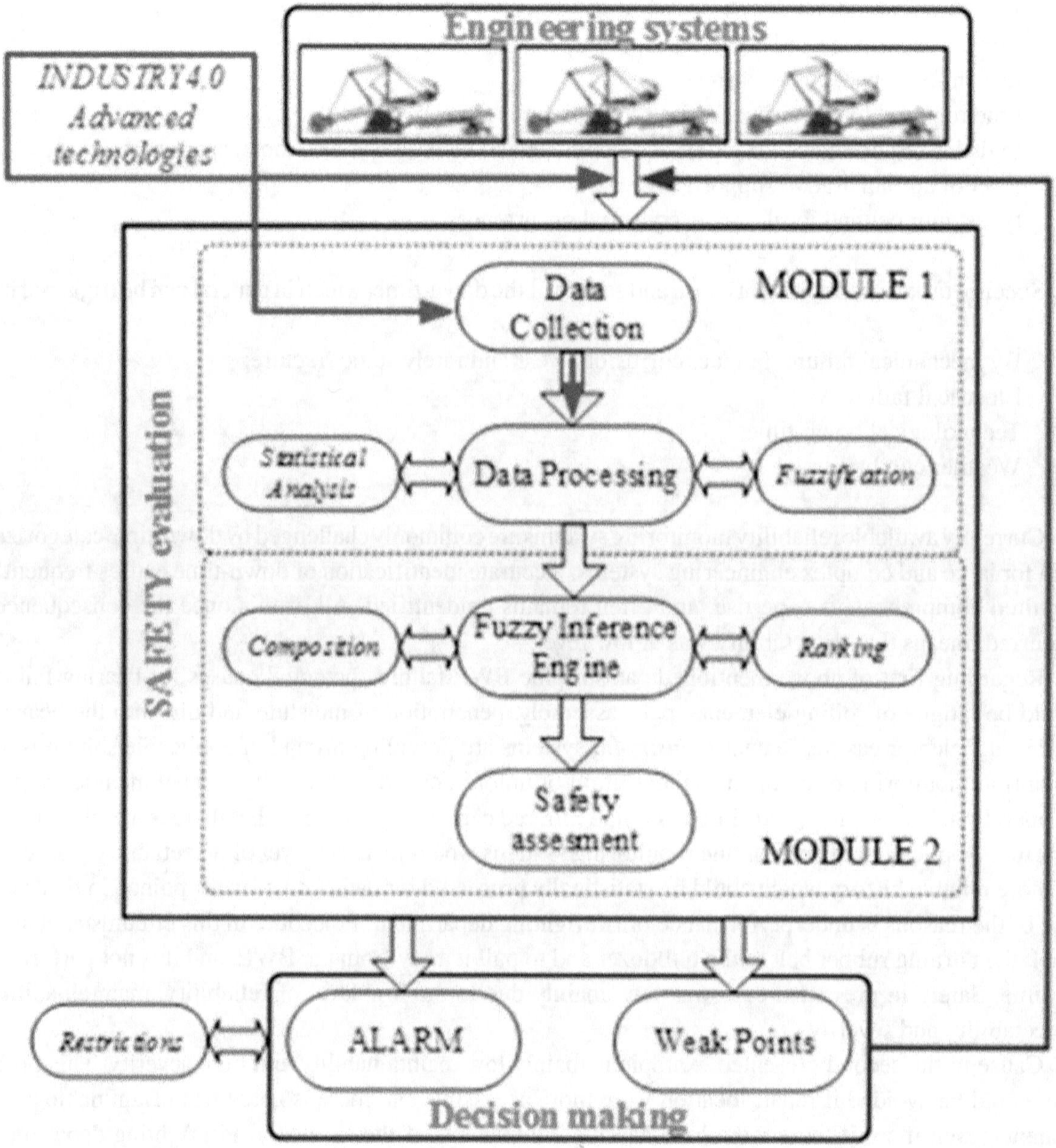

CONCLUSION

The quest for optimal maintenance strategy is the topic in maintenance engineering, both in theory and in practice. Maintenance parameters are often conflicting (such as risk and costs). This problem is especially evident for complex, multi hierarchical systems. Suggested solution to this problem, presented in this article, is based on maintenance centered safety. This approach accounts for: weak points investigations, failure analysis in sense of reliability and maintainability, severity and consequences probability. Mentioned parameters are synthetized into safety concept by using adaptive fuzzy inference model. Input data, represented by behavior history of the engineering system through time depending cumulative function, are fuzzified in appropriate manner. Fuzzy inference engine with ranking is most suitable screening model, which transforms input data into unique outcome – cumulative safety of engineering system. Applied logic for assessing the system's safety is to identify the best outcome among the worst ones.

Second common issue in practical maintenance engineering is monitoring, acquisition, data recording and processing. This article also reviews solutions for this problem within framework of the *Industry4.0*.

The paper specifically analyses the Bucket Wheel Excavator. This engineering system is typical representative of complex engineering systems, operating in difficult working conditions. Article provides two case studies of catastrophic failures of this machine, with analysis of reasons for reduced safety.

REFERENCES

Avizienis, A., Laprie, J.-C., & Randell, B. (2001). *Fundamental Concepts of Dependability*. Research Report No 1145, LAAS-CNRS.

Borchellini, R., Cardu, M., Colella, F., Labagnara, D., Martinetti, A., Patrucco, M., . . . Verda, V. (2013). A Prevention through Design Approach for the Environmental S&H Conditions and the Ventilation System at an Italian Underground Quarry. *Chemical Engineering Transactions, AIDIC - The Italian Association of Chemical Engineering, 32*, 181-186. doi:10.3303/CET1332031

Bošnjak, S. M., Savićević, S. D., Gnjatović, N. B., Milenović, I. L. J., & Pantelić, M. P. (2015). Disaster of the bucket wheel excavator caused by extreme environmental impact: Consequences, rescue and reconstruction. *Engineering Failure Analysis*, *56*, 360–374. doi:10.1016/j.engfailanal.2015.01.002

Bowles, J. B., & Pelaez, C. E. (1995). Fuzzy logic prioritization of failures in a system failure mode effects and criticality analysis. *Reliability Engineering & System Safety*, *50*(2), 203–213. doi:10.1016/0951-8320(95)00068-D

Dhillon, B. S. (2008). *Mining Equipment Reliability, Maintainability and Safety*. Springer. doi:10.1007/978-1-84800-288-3

Ebramhimipour, V., & Suzuki, K. (2006). A synergetic approach for assessing and improving equipment performance in offshore industry based on dependability. *Reliability Engineering & System Safety*, *91*(1), 10–19. doi:10.1016/j.ress.2004.11.008

Erkoyuncu, J. A., Khan, S., Eiroa, A. L., Butler, N., Rushton, K., & Brocklebank, S. (2017). Perspectives on trading cost and availability for corrective maintenance at the equipment type level. *Reliability Engineering & System Safety*, *168*, 53–69. doi:10.1016/j.ress.2017.05.041

IAEA-TECDOC-658, Safety related maintenance in the framework of the reliability cantered maintenance concept, International Atomic Energy Agency.

International Electrotechnical Commission. International standard - Dependability management ISO-IEC 300. https://www.iec.ch/

Ivezić, D., Tanasijević, M., & Ignjatović, D. (2008). Fuzzy Approach to Dependability Performance Evaluation. *Quality and Reliability Engineering International*, *24*(7), 779–792. doi:10.1002/qre.926

Ivezić, D., Tanasijević, M., Jovančić, P., & Đurić, R. (2019). A Fuzzy Expert Model for Availability Evaluation. *Proceedings of the 2019 20th International Carpathian Control Conference (ICCC)*. 10.1109/CarpathianCC.2019.8766031

Jovančić, P. (2016). Asset management and condition monitoring on maintenance of mining equipment lignite mines. *13th International symposium Continuous Surface Mining ISCSM 2016, 12-14 September 2016, Belgrade – Serbia, Metropol Palace, Conference Proceedings*, 197-207.

Jovančić, P., Ignjatović, D., & Tanasijević, M. (2012a). Modern concepts of mining equipment maintenance. In *10th International Symposia Mechanization and Automation in Mining and Energetics MAREN2012*. University of Belgrade.

Jovančić, P., Ignjatović, D., & Tanasijević, M. (2012b). Proactive monitoring system for basic mining equipment at open pit mines of Electric Power Industry of Serbia. *XXI International Congress on Maintenance and Asset Management – EUROMAINTENANCE 2012, May 14-16, 2012 – Sava Centar, Belgrade, Serbia, Conference Proceedings*, 622-630.

Karimnia, H., & Bagloo, H. (2015). Optimum mining method selection using fuzzy analytical hierarchy process-Qapiliq salt mine. *Iran. Int. J. Min. Sci. Technol.*, *25*(2), 225–230. doi:10.1016/j.ijmst.2015.02.010

Klir, G. J., & Yuan, B. (1995). *Fuzzy Sets and Fuzzy Logic, Theory and Applications*. New York: Prentice Hall.

Labagnara, D., Martinetti, A., & Patrucco, M. (2013). Tunneling operations, occupational S&H and environmental protection: A Prevention through Design approach. *American Journal of Applied Sciences*, *11*(11), 1371–1377. doi:10.3844/ajassp.2013.1371.1377

Martinetti, A., Braaksma, A. J. J., & van Dongen, L. A. M. (2017). Beyond RAMS Design: Towards an Integral Asset and Process Approach. In L. Redding, R. Roy, & A. Shaw (Eds.), Advances in Through-life Engineering Services (pp. 417-428). Springer. doi:10.1007/978-3-319-49938-3_25

Martinetti, A., Chatzimichailidou, M., Maida, L., & van Dongen, L. A. M. (2018). Safety I-II, Resilience and Antifragility Engineering: A debate explained through an accident occurred on a Mobile Elevating Work Platform. *International Journal of Occupational Safety and Ergonomics*, *25*(1), 66–75. doi:10.1080/10803548.2018.1444724 PMID:29473459

Miodragović R., Tanasijević M., Mileusnić Z., Jovančić P., (2012) Effectiveness assessment of agricultural machinery based on fuzzy sets theory. *Expert Systems with Applications, 39*(10), 8940–8946.

Papic, L., Aronov, J., & Pantelic, M. (2009). Safety based maintenance concept. *International Journal of Reliability Quality and Safety Engineering*, *16*(6), 533–549. doi:10.1142/S0218539309003563

Papic, L., & Milovanovic, Z.N. (2007). *Systems Maintainability and Reliability*. The Research Center of Dependability and Quality Management DQM.

Petrović, D. V., Tanasijević, M., Milić, V., Lilić, N., Stojadinović, S., & Svrkota, I. (2014). Risk assessment model of mining equipment failure based on fuzzy logic. *Expert Systems with Applications*, *41*(18), 8157–8164. doi:10.1016/j.eswa.2014.06.042

Reyes Garcia, J. R., Martinetti, A., Jauregui Becker, J. M., Singh, S., & van Dongen, L. A. M. (2019). Towards an Industry 4.0-Based Maintenance Approach in the Manufacturing Processes. In V. González-Prida Diaz & J. P. Zamora Bonilla (Eds.), *Handbook of Research on Industrial Advancement in Scientific Knowledge* (pp. 135–159). IGI Global; doi:10.4018/978-1-5225-7152-0.ch008

Saaty, T. L. (1980). *Decision Making: The Analytical Hierarchy Process*. New York: McGraw-Hill.

Strandberg, K. (1991). IEC 300: The Dependability Counterpart of ISO 9000. *Reliability and Maintainability Symposium*, 463-467. 10.1109/ARMS.1991.154481

Tanasijević, M. (2016). A fuzzy-based decision support model for evaluation of mining machinery. *Full Papers Proceeding of International Conference ''48th International October Conference on Mining and Metallurgy*, 15-18.

Tanasijević, M., Ivezić, D., Jovančić, P., Ćatić, D., & Zlatanović, D. (2013). Study of Dependability Evaluation for Multi-hierarchical Systems Based on Max–Min Composition. *Quality and Reliability Engineering International*, *29*(3), 317–326. doi:10.1002/qre.1383

Teodorović, D., & Vukadinović, K. (1998). *Traffic Control and Transport Planning: A Fuzzy Sets and Neural Networks Approach*. Boston: Kluwer Academic Publishers. doi:10.1007/978-94-011-4403-2

Wang, J. (2000). A subjective modelling tool applied to formal ship safety assessment. *Ocean Engineering*, *27*(10), 1019–1035. doi:10.1016/S0029-8018(99)00037-2

Wang, J., Yang, J. B., & Sen, P. (1995). Safety analyses and synthesis using fuzzy sets and evidential reasoning. *Reliability Engineering & System Safety*, *47*(2), 103–118. doi:10.1016/0951-8320(94)00053-Q

Weibull, W. (1951). A statistical distribution function of wide applicability. *Journal of Applied Mechanics - Transaction ASME, 18*(3), 293–297.

Yang, J.B., Wang, Y.M., Xu, D.L., & Chin, K.S. (2006) The evidential reasoning approach for MADA under both probabilistic and fuzzy uncertainties. *European Journal of Operational Research, 71*(1), 309-343.

Ying-Ming, W., & Kwai-Sang, C. (2009). Risk evaluation in failure mode and effects analysis using fuzzy weighted geometric mean. *Expert Systems with Applications*, *36*(2), 1195–1207. doi:10.1016/j.eswa.2007.11.028

Zadeh, L. A. (1996). Fuzzy Logic = Computing with words. *IEEE Transactions on Fuzzy Systems*, *4*(2), 103–111. doi:10.1109/91.493904

Chapter 10
The Support From Industry 4.0 to the Management of Change (MoC)

Micaela Demichela
https://orcid.org/0000-0001-5247-7634
Politecnico di Torino, Italy

Gianfranco Camuncoli
ARIA srl, Italy

ABSTRACT

Dealing with maintenance activities in complex systems often configures the so-called management of change (MoC). MoC is a process for evaluating and controlling modifications to facility design, operation, organization, or activities—prior to implementation—to be sure no new hazards are introduced. Traditionally, MoC is related to technical changes. Safety implications from organizational changes have recently led to proposed integrated management of both types. An inadequate MoC is recognized as a recurring cause of accidents, often resulting in major accidents, mainly in the process industry. Despite this recognised criticality, the MoC workflow in many companies is still far from being mature and there are still evident shortcomings in its application that have to be compensated. The technical solutions composed by the enabling technologies within Industry 4.0 can offer a valid support to overcome the MoC shortcomings, as will be discussed within this chapter.

INTRODUCTION

Dealing with maintenance activities in complex systems often configures the so-called Management of Change (MoC). Management of Change is a process for evaluating and controlling modifications to facility design, operation, organization, or activities - prior to implementation - to be sure no new hazards are introduced. As discussed by Martinetti et al. (2017) maintenance and changes cannot be reconducted

DOI: 10.4018/978-1-7998-3904-0.ch010

only to the productivity and safety (RAMS factor), but new paradigms for taking into consideration supportability and environment need to be implemented to minimize the risks.

Regular maintenance usually does not require to be managed as a change, since it can be configured as a "in kind" modification, which substantially leaves the equipment exactly as it was prior to starting the work. Special maintenance activities instead fall within the changes to be managed.

Traditionally, MoC is related to technical changes. Safety implications from organizational changes have recently led to proposed integrated management of both types, based on Prevention through Design (PtD) approach, applied in the tunnelling work environment in Borchiellini et al. (2013) and in Labagnara et al. (2013).

An inadequate management of change is recognized as a recurring cause of many accidents, often resulting in major accidents, mainly in the process industry, as discussed by several authors. The Flixborough accident, occurred in the UK on 1.6.1974 (Mannan, 2012) is the representative case of a major accident happened because of a unperformed management of change. A massive explosion in a caprolactam manufacturing process following a release of hot cyclohexane virtually demolished the site, killed 28 persons and injured 36. The cause of the accident was traced back to the improper management of a change made to the plant for maintenance purposes. In particular, being the process carried on in 6 cascade reactors, when a leakage was discovered in the reactor n. 5, due to a crack of about 2 m in the reactor shell, it was decided to install a temporary pipe to bypass the leaking reactor to allow continued operation of the plant while repairs were made. In the absence of 28-inch pipe, adopted in the reactor cascade, a 20-inch pipe was used to build the bypass pipe for linking reactor 4 outlet to reactor 6 inlet. The effect of the thermal and pressure cycles inherent in the process brought to the pipe or bellows material to yield, initiating the accidental event. The plant modification was not designed, constructed, tested and maintained to the same standards as the original plant and this brought to the destruction of its technical integrity.

Levovnik & Gerbec (2018) reviewed different sources highlighting the links between MoC and accidents. Their analysis started from the contribution of Keren et al. (2002), that performed a survey, on how often changes occur in the industry, and found among the respondents the rate between 1 and 37 changes per annum per ten employees (with an average at about 10). Then they consulted the two main accident databases related to process safety. Lessons learnt from the ARIA (Analysis, Research and Information on Accidents) database of the French Bureau for Analysis of Industrial Risks and Pollutions, consisting at the time of almost 49,000 events, mention examples of 28 events where MoC was reported deficient/absent. The review also found that 8 out of 15 closed investigations by CSB, the US Chemical Safety and Hazard Investigation Board, in the period from 2015 till the end of 2017, explicitly list MoC issues among the causes.

In Han Siong et al. (2017) a total of 630 chemical process industry related accidents cases were reviewed. Preliminary result shows how the contribution of MoC failure to the accidents analysed was found to be 9.1%, the 6th behind other 5 process safety management elements: process hazards analysis (17.7%), operating procedure (17.6%), employee participation (11.5%), training (11.3%) and mechanical integrity (10.1%). Among the MoC failures, requesting system change demonstrated the highest percentage contribution of 45.6% in MoC typology failure percentage. Breakdown/failure system change contributes for the 30.9% followed by temporary system change, for 10.1%, administrative system changes for 8.1% and organisation system change for 5.4% as summarised in Figure 1.

Figure 1. MoC typology failure percentage, according to Han Siong et al. (2017)

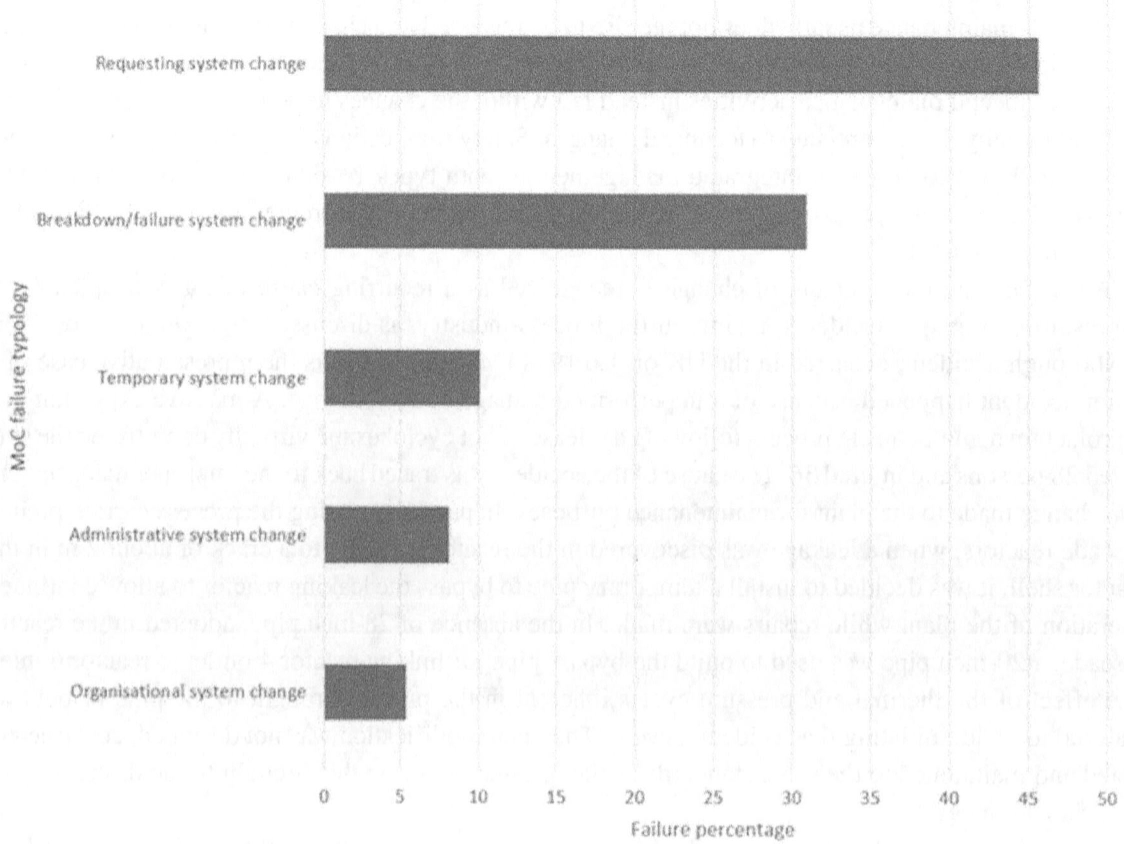

Management of Changes is one of the elements of safety management systems (SMS), implemented to control risks of different origin. The MoC process is a workflow that regulates the way an change is conducted: it must be a company approved procedure that guarantees the MoC process contains the necessary elements and is properly executed. An effective MoC work process should include a review and authorization process for evaluating proposed adjustments to facility design, operations, organization or activities prior to implementation to make certain that no unexpected new hazards are introduced and that the risk of the existing hazards to employee, the public, or the environment is not increased. MoC should also include steps to ensure that potentially affected personnel are informed of the change and that pertinent document, as procedures, technical drawings, and so on, are updated.

Despite the criticality related to a incorrect management of changes is well known, the MoC workflow in many companies is still far to be mature and there are still evident shortcomings in its application that have to be compensated. The technical solutions composed by the enabling technologies within Industry 4.0 can offer a valid support to overcome the MoC shortcomings, as will be discussed in the following paragraphs.

This chapter is thus organised as follow. In the following section a taxonomy of changes is proposed, highlighting the potential shortcoming in their management. Then, the structure of a management of change procedure is discussed, highlighting the issues related to its implementation. A section dedicated to the Industry 4.0 solutions that could support the MoC process will follow, based on literature case studies. In the final section, some conclusions are drawn.

MOC TAXONOMY

The definition of a taxonomy of the changes occurring in a production plant is recognised to be a not easy task, because of the multiple pressures, internal and external, that can drive the need for a modification, as discussed in Gambetti et al. (2013). According to Clarke (2019) the changes occurring in a production plant and that should be managed through a MoC can be grouped in clusters; within this chapter 3 main classes have been identified: hardware changes, software changes, qualified as organisational changes and miscellaneous. The proposed taxonomy of changes is shown in Figure 2.

Figure 2. Taxonomy of the changes

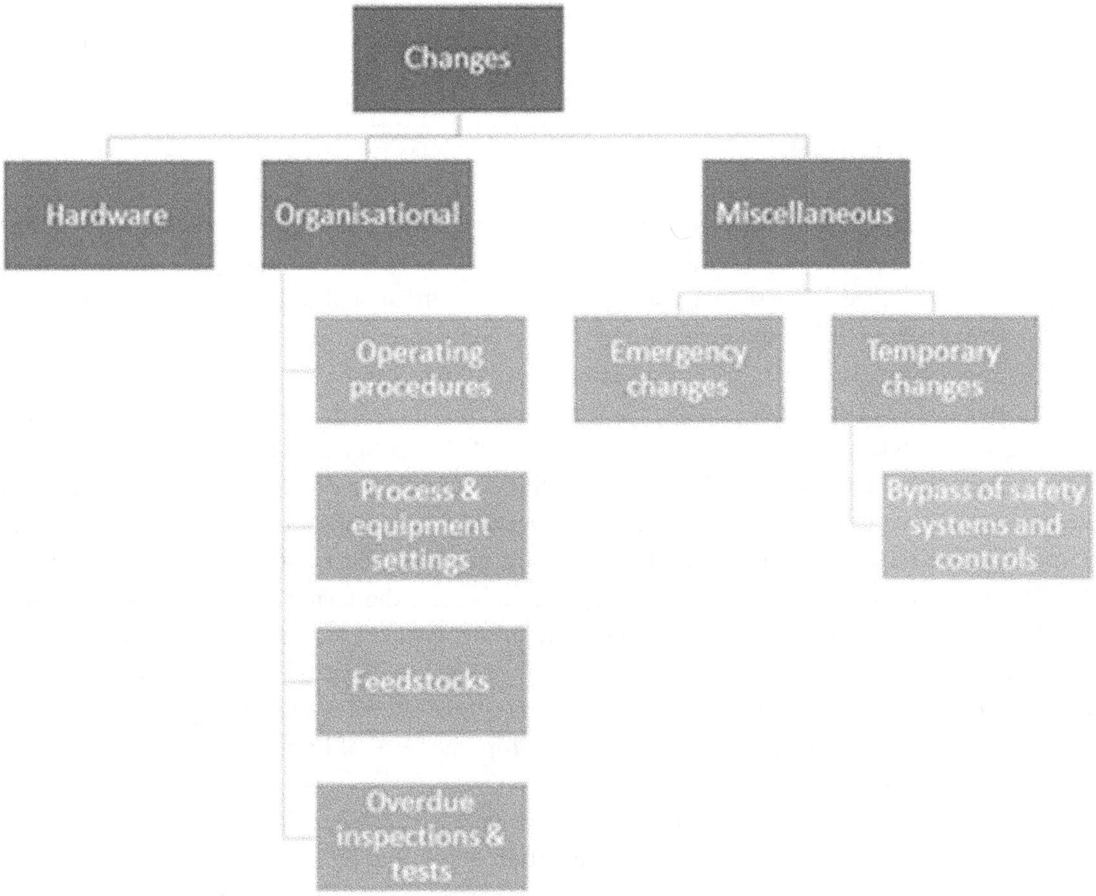

Hardware Changes

Any change in technology or process, or major change to equipment, being performed for maintenance or overhaul, is a significant change. The main issue is in this case to be able to identify the changes requiring a MoC or the "replacement in kind", that in general terms can be associated to a regular maintenance. As exemplified in the introduction, also upgrades to materials, particularly process piping to eliminate potential corrosion exposures, are hardware changes that should be subject to MoC, to avoid events as Flixborough accident.

Despite hardware changes are those easier to be recognised, a list of what is intended for technical change is shown, as extracted from Gambetti et al. (2013):

- Plant and equipment modification at start up
- Minor material, plant and equipment modifications made during maintenance
- Temporary material, plant and equipment modifications
- Process Modifications
- Introduction of New Tools
- Gradual Changes
- Modifications made to improve the environment.

Organisational Changes

Changes to systems, procedures and other management controls are considered as "software changes" that, as the analysis of accidents highlighted, are as critical as the hardware ones. There are in fact several accidents that were triggered by organisational change or where they played a relevant role. One of the most representative was the Exxon Longford explosion and fire, occurred in September 1998 in Longford, Alberta, where a massive release of hydrocarbon vapours from a brittle fracture in an occurred, caused by an excessive differential temperature between the fluids (Mannan, 2012). The investigation report following the accident highlighted the criticality of the Exxon decision to relocate engineering staff away from the plant to the Head Office about 200 km from the site, concluding that the absence of engineering oversight was a key contributor to the incident. Organisational changes are recognised to be more difficult to control and are often managed through a process outside of the MoC system.

Among organisational changes, the most representative are the changes to *operating procedures*, being them standard or emergency ones. Many organisations review operating procedures at given intervals, or when a change in the process itself requires a review. Typically, the procedures are revised every year for emergency operating procedures and every three years for standard operating procedures. If changes are made, these should be subject to an appropriate level of hazard analysis, although this is not frequently observed.

Another group of organisational changes are those to *process or equipment settings*. Many organisations maintain controlled lists of alarm set points, often within the distributed control system (DCS), which can only be changed after appropriate review and analysis. Similarly, for trip or emergency shut down (ESD) settings, which in many cases are PLC dedicated systems separated from the DCS. Clarke (2019) highlights other management controls that appears to be not well protected as the previous, as the setting of integrity operating windows (IOWs) which are usually developed in conjunction with inspection and/or process engineering. IOWs are intended to define the limits within which the process can be allowed to fluctuate for mechanical integrity, operational or product quality reasons. Extended operations outside of the optimum process conditions, even if "safe" from a process upset perspective, can often have damaging effects on equipment on a longer term. A typical example is reformer and/or furnace tubes, which suffer rapidly accelerated creep damage if exposed to temperatures even slightly above recommended operating levels, and this can drastically shorten tube life as a result. These parameters often can be changed from the DCS by the panel operator with only nominal referral.

Another typical situation refers to the *changes to feedstock*, that can cause serious corrosion problems. This is another example of a change that is frequently not managed appropriately: in fact, approval for feedstock changes is often given by the procurement function, with no involvement of operations or Inspection and Process Engineers to validate the suitability of the feedstock for the plant.

Still under the organisational change category stands the *overdue inspection and test activities*. These are particularly crucial when applied to safety critical elements (SCEs). Overdue inspections and tests usually occur because of operational constraints and could be acceptable only in case an accurate technical judgement that the plant will be safe to operate in the modified circumstances.

Miscellaneous

Under this category can be classified those changes, hardware or organisational, that are characterised by specific attributes. This category includes Emergency changes and Temporary changes.

Emergency changes includes those changes, being them both hardware or organisational, made under emergency conditions. They should be managed under MoC, but it must allow for deferred approval processes while still providing adequate oversight. As an example, if the Plant Manager has to approve a MoC by signing a hard copy of a form, operations personnel might avoid using the MoC system when the Plant Manager is not present. It is therefore of critical importance to define the controls, risk assessment processes and approval levels that are in place outside of normal working hours, what changes can be authorised on an emergency basis, and there should be a requirement to revert to the normal approval process within a defined (usually 24 hours) period.

Temporary Changes. Temporary changes are another type of intervention which frequently is not managed through MoC, also in this case they can be hardware or organisational changes. A typical example are hoses, which are commonly found throughout process units to facilitate various draining activities. Despite the hose is placed only temporary, often it remains in place for several months. Thus, temporary changes should have an associated finite time limit and should be subject to the same approval process as a permanent change. Once this time limit is exceeded then it should be managed as a permanent change. One or more extensions for a temporary change can be allowed, with appropriate risk assessment and approval for each extension. A particular temporary change is the bypass of safety systems and/or controls, that can be managed within or outside a MoC procedure. It requires in any case a robust risk assessment and an appropriate approval process, with a dedicated section on mitigations that will be in place while the system is unavailable. An example of a mitigation is to ensure that no hot work is authorised in the plant if one of the fire pumps is out of operation. Bypass authorisations also need to have a start and end date, with approvals for this period only. The potential weakness in bypass management is the chance that interrelated safety or control functions could be isolated or bypassed, leading to a cumulative reduction in layers of protection.

Han Siong et al. (2019) summarised the results of their analysis of the weaknesses in the management of changes that initiated or contributed to some accidents. Table 1 summarises the main failure factors for each typology of changes identified.

Table 1. Main failure factors in MoC

Typology	Drivers	Main failure factors
Requesting system change	· Modification to achieve higher production rate or better product quality · Change of production type with existing operating facilities and system · Operating control system change · Process line change · Start up and shutdown system change · Complete system change involving equipment, instruments, procedures, organisation, process · Setting higher production output without equipment / instruments upgrade	· Lack of Management involvement · Inadequate Procedure · Inadequate risk assessment · Lack of Competency · Resources limitation · Human factor · Inadequate Tools and Equipment · Lack of Supervision · Lack of Communication · Cost control
Breakdown/failure system change	· Equipment breakdown / failure change · Piping / vessel or high corrosion effect change · Sudden / urgent shutdown operating change · Change to prevent safety issue	· Lack of Management involvement · Inadequate risk assessment · Inadequate procedure · Cost control · Tools & Equipment · Pressure and stress (time constrain)
Temporary system change	· Temporary bypass normal operating system to keep operation process running with part of the system / equipment taken out for service or replacement · Temporary interlock bypass · Temporary safety protective devices bypass · Temporary change on equipment material and/or chemicals	· Lack of Management involvement · Inadequate risk assessment · Inadequate Procedure · Tools and Equipment
Administrative system change	· Changing SOP / work flow for operational and safety issue · Changes in establish training method · Change in operation parameter, limit, control · Change procedure from hard to soft copy	· Lack of Management involvement · Inadequate Procedure · Inadequate risk assessment · Inadequate Training
Organisation system change	· Manpower (work force / reduction / work distribution) · Contractors / vendors change · Human behaviour change (physical, emotional) · Cost saving implementation · Restructuring (competency) · Policy change · Stakeholder change · Business unit change · Realignment of audit function	· Lack of Management involvement · Inadequate Procedure · Human factor · Lack of Competency · Cost control · Pressure and stress

MOC WORKFLOW

Management of Change (MoC) is a process for evaluating and controlling modifications to facility design, operation, organisation, or activities. It is one of the most important elements of Safety Management, whose structure and contents are defined in a number of technical standards (among others the recently released ISO 45001:2018 Occupational health and safety management systems -- Requirements with guidance for use that at the clause 8.1 requires the implementation of a MoC system).

Once the taxonomy of the changes has been established and adopted in a formal procedure within the plant management system, a managing procedure must also be established. Several procedure flows have been proposed in the literature and adopted in the companies in the last decades. Among the more recent, the more complete, considering both technical and organisational changes was proposed by Gerbec in 2017 and further developed in Gerbec and Levovnik (2019). Figure 3 shows the flow chart proposed, that in the original methodology is complemented with forms and check lists to for its practical implementation.

Figure 3. Flow chart of MoC workflow (extracted from Gerbec, 2017)

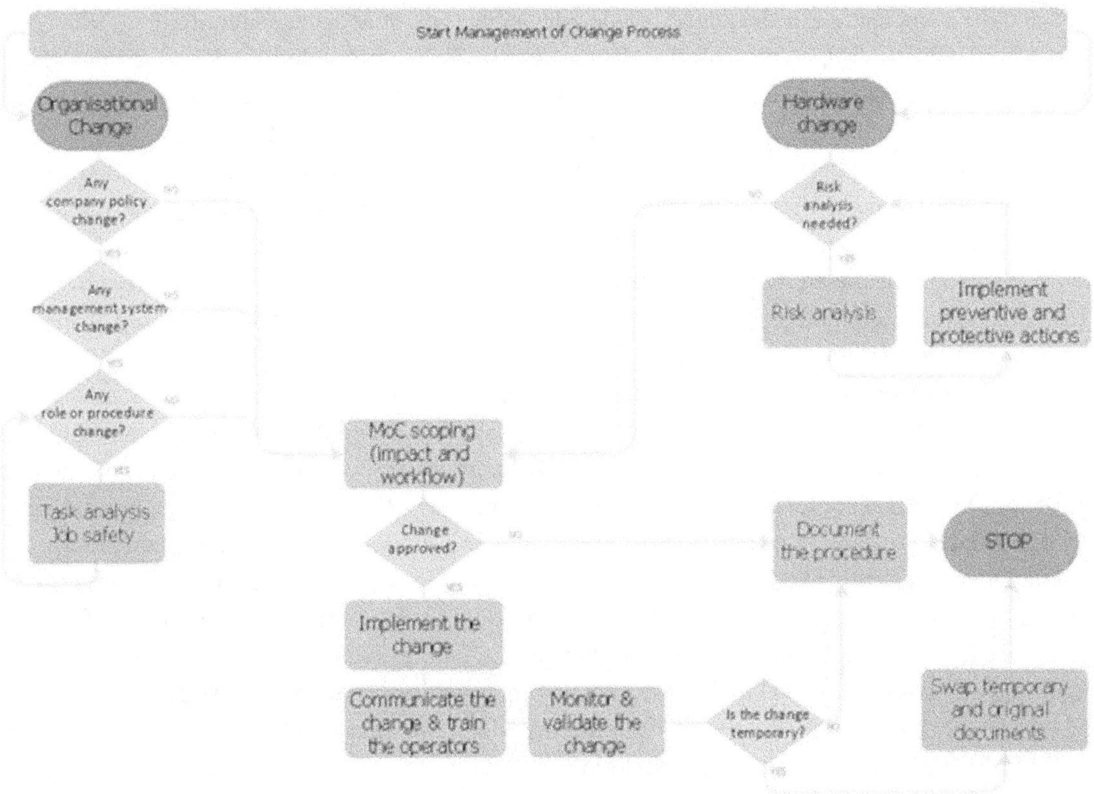

The MoC can be implemented through a set of paper files which passes through all members of the authorisation flow; software-based MoC systems are becoming increasingly common, as described e.g. in Cao et al. (2013). The major advantage of these systems is the possibility to hold a project at a "gate" until all aspects are signed off and reviewed. Another significant advantage is the ability of the system to generate KPIs which can quickly demonstrate the health of the MoC programme.

As discussed in the taxonomy section, the MoC process should not be too difficult or too easy to implement, to avoid bypasses of the procedure. Similarly, approval levels need to be effective and commensurate with the change being made and flexible enough to manage emergency changes.

An appropriately detailed safety review is a critical step in the MoC process. It enables the expertise within an organisation (and outside), to be deployed to investigate a change, and what mitigating actions or improvements need to be implemented to guarantee the safety both at the time of the change and after its implementation. The safety and/or hazard review that could be deployed are the traditional methodologies for risk assessment, widely used, adopted across several domains.

As widely discussed in the literature, it exists a proportionality between the level of risk of a system and the level of detail of the required risk assessment, as reviewed in Demichela & Baldissone (2019). Thus, the techniques for hazard identification and risk assessment, as Hazard and Operability (HAZOP) analysis, Preliminary Hazard Analysis (PHA), What-if analysis, Failure Mode and Effects Analysis (FMEA) or the more advanced analysis integrating the logical-probabilistic and phenomenological analysis, as in Demichela et al. (2017) and Baldissone et al. (2017), will have to be chosen depending on the type and complexity of the change to be implemented. As for traditional risk assessment applications, it is of importance the composition of the team who decides what type of analysis is required.

The era of industry 4.0, as discussed in Cimini et al. (2020), promoted the development of smart work environments characterised by relevant socio-technical interactions between humans and machines, increasing the system complexity, but also the availability of data to interpret it. Even as or maybe more than before, there should be a requirement for personnel from various departments to participate to the safety review, depending on the type of change: typically, a representation from operations, engineering, control, maintenance, inspection and HSE would be require. Mostly important, there must be a responsible person delegated to ensure all recommendations are followed to closeout, including the requirement for any further analysis highlighted during the safety review, and the inclusion of any proposed changes into the scope and documentation of the change.

Following the change, it is important to validate it, against procedures, documentation and technical drawings. If differences should emerge, these need to be addressed and the "to do list" tracked to ensure adequate completion. Following hardware changes, the appropriate technical drawings, documentation and files such as P&IDs, logic drawings and asset registers need to be updated, as do the possibly potentially large numbers of associated procedures. Following the completion of the change and all associated documentation changes, it is then necessary to give training to staff on the change and all associated procedural changes that have resulted from it. It is important to ensure that this training is effective and documented. It is also important that the new changes are incorporated into the organisation's training manuals and any competency validation assessments for operators, maintenance staff and any other affected personnel. This activity is termed "scoping the MoC". As discussed in Hoff (2013), the quality of MoC scoping depends largely on the methodology used, with different sites using anything from no scoping at all, guesswork approaches, checklist approaches, to more sophisticated asset-based scoping.

In the end, any recommendations made following the safety review, pre-start-up review or generated during the MoC process have to be completed. The MoC can then be formally closed out as completed.

HOW INDUSTRY 4.0 CAN SUPPORT MOC?

As discussed in Cimini et al. (2020) and by and by Reyes et al. (2019) Industry 4.0 is a complex technological system that encompasses several technologies (e.g., Cyber-Physical Systems, Internet of Things, Robotics, Big data, etc.), whose implementation allows the development of intelligent work environments, composed by devices able to exchange information, perform actions, and control each other. At the basis of the ongoing Industry 4.0 paradigm, there are foundational innovations involving not only machines but also sensors, work pieces, and IT systems connected throughout the entire value chain. Standard Internet-based protocols enable real-time interaction between devices, thereby supporting data analysis to avoid and predict failures, as well as reconfiguration and adaptability capabilities. Different technologies are identified as the pillars of Industry 4.0. The most cited technologies, labelled in literature as Enabling Technology of Industry 4.0, have been considered and reported in Figure 4, with a short definition.

Figure 4. Industry 4.0 enabling technologies (extracted from Cimini et al., 2020)

Technology	Definition
Cyber-Physical Systems	CPS integrate physical processes with computation capabilities and are able to operate in changing environments, maintaining robust behaviour against unexpected conditions and failures
Internet of Things (IoT)	Internet of Things integrates various devices equipped with sensing, identification, processing, communication, and networking capabilities.
Big data analytics	Big data is characterised by volume, variety and velocity (the 3 Vs), and it requires new techniques of data processing and analysis.
Cloud technologies	Cloud computing is related to the ICT-infrastructure that allows ubiquitous access to data from different devices.
Cybersecurity	Cybersecurity is the set of technologies, tools and processes to guarantee the security of networks, devices and the large amount of data collected, stored and communicated via IoT.
Virtual reality	Virtual reality is a computer interface that allows the user to be fully immersed in an experimental situation, i.e. a virtual environment, enabling the looking, moving and interaction of users in a world that is like the real one.
Augmented reality	Augmented reality (AR) allows the creation of a virtual environment in which humans can interact with machines using devices able to recreate the workspace.
Smart sensors	Smart sensors are traditional sensors embedded with intelligence capabilities, i.e. on-board microprocessors, which can be used for processing, conversions, calculations, and interfacing functions.
Simulation	Simulation provides a digital representation of products and processes, in order to identify in advance potential issues, avoiding cost and resource wastes in production.
Additive manufacturing / 3D printing	Additive Manufacturing (AM) consists in a cluster of technologies that enable to produce small batches of products with a high degree of customisation by adding rather than removing material from a solid block.
Advanced robotics	The evolution of traditional robots opened the way to new collaborative solutions of robots (i.e. Co-Bots) that are able to work together with humans in a safe and efficient way. Moreover, embedded intelligence in robots can allow them to learn from human activities, improving their autonomy and flexibility.
Energy saving technologies	Energy saving technologies include the monitoring and optimisation systems that allow reducing the energy consumption in manufacturing.
Horizontal and vertical integration	Horizontal integration refers to the creation of a global value network through the integration and the optimisation of the flow of information and goods between company, suppliers and customers. The vertical integration, instead, is the integration of functions and departments of different hierarchical levels of the single company creating a consistent flow of information and data.
Multi-agent technologies	Multi-agent systems (MAS) are an organised set of agents that represent the behaviour of objects of a system, capable of interacting and negotiate among them to achieve individual goals.

It has to be noticed that the enabling technologies included in Figure 4 are not independent one each other, but usually are well integrated to maximise their effectiveness, e.g. the IOT generates usually a very large amount of data that requires Big Data Analytics to be exploited; Smart Sensors require Simulation to allow their full potential to be exploited for process and/or machines control and safety.

In the follows, without any claim of being complete, relevant enabling technologies are reviewed with reference to the MoC, identifying significant applications that could support the MoC workflow to minimise the potential issues highlighted in the previous paragraphs.

Cyber-Physical Systems

A Cyber-Physical Systems (CPS) is a system that combines computational and physical components, allowing the system to interact with the real world. Moreover, CPSs can integrate networking capabilities, a feature that is a cornerstone of interconnected and autonomously operating manufacturing systems (Eckhart & Ekelhart, 2018). Increased digitization and connectivity open up new attack vectors that may put an organization's assets at risk but also endanger human life. This is especially important for industrial control systems—a subset of CPSs— where safety has been the main focus. Depending on the criticality of the running systems, testing in the operating environment could be not recommended. The setup and maintenance of test environments on the other hand, is expensive and time consuming, often leading to incomplete and outdated environments.

Eckhart & Ekelhart (2018) have proposed a framework, named CPS Twinning, to build and maintain fully functional digital twins of CPSs, with an approach to automatically generate the virtual environment from specification, taking advantage of engineering data exchange formats.

The term "Digital Twin" describes the use of holistic simulations to virtually mirror a physical system. Adopting such a concept could enable operators to monitor the production process, test changes in a virtual, isolated environment, and to further strengthen the security and safety of CPSs. A data-driven digital twin system allows integrating process monitoring, diagnosis, and optimized control into a cooperative framework, to reduce severe fluctuations and guarantee safe controls for industrial processes.

After the generation and configuration of digital twins, two modes of operation become available, i.e., replication and simulation. In simulation mode, the digital twins run independently of their physical counterparts. Similar to virtual commissioning, this mode allows users to analyse process changes, test devices or even optimize manufacturing operations. The replication mode on the other hand, mirrors data from the physical environment. Possible data sources to mirror in the virtual environment are logfiles, network communication and sensor measurements from the physical environment. Monitoring the physical process in replication mode and then switching to simulation mode allows to investigate specific states. This approach may provide insight regarding the root cause that led to an unexpected behaviour or a failure.

He et al. (2019) acknowledged that for complex fault scenarios, the process diagnostic performance of digital twins is still insufficient and requires to rely on online instruments to locate a specific fault.

For the benefit of MoC, within other advantages, digital twins can be leveraged to perform system tests and simulations. Real devices can be tested by first connecting them in the virtual environment. For instance, if the real device is connected as a replacement of the existing digital-twin PLC, the operator could observe the behaviour of the new PLC inside the virtual environment. Moreover, experimenting with configurations in a virtual environment provides the possibility to detect problems and incompatibilities early, without costly setups. Another relevant use case is to detect mismatches of the real environment and the maintained specification.

IOT

Specific application of IOT systems, consisting in the integration of different devices equipped with sensing, identification, processing and networking capabilities, has been already proposed for MoC.

In the work of Cao et al. (2013) the authors designed and developed a management of change platform based on web and RFID (radio frequency identification devices), whose purpose was to support the MoC workflow, minimising the potential shortcomings in the process. To reach its purpose the proposed system was made of different modules:

A change process management module used for change application, review and examination. According to the time characteristic, change activity can be classified into two categories: emergency change and general change, and the last into permanent change or temporary change. In the change file, there is the detail (time, site, equipment characteristic, etc.) about the change activity, the adopted plan and safety precautions. The module contains the relevant information of a change activity, and can provide original data for change notification, real-time monitoring and database update. It also provides the radio frequency identification interface, which can identify data about file, equipment, and process variable message imported from RFID.

Then a Database about change history is foreseen to record the detail about the change activities in the plant, which can be queried by project numbers, equipment name, station name, responsible person or identifier to get the historical change records.

A Risk analysis module is also foreseen to support the applicants and auditors in evaluating the change risk using a checklist risk analysis method. This module provides basic information for more detailed risk assessment methodologies (e.g. HazOp sessions), in the cases in which they are required.

In the end a change notification and real-time monitoring module allows the staff entering the area in which the change is occurring be advised about the change by wireless network and RFID reading devices.

To achieve effective supervision in change procedure, the system can include video surveillance and regular reports to the manager on a defined time interval.

Since one of the critical aspects of the MoC workflow in the authorisation step, in line with the company requirements, the system includes four roles, including staff, department management, safety and security management and system management. Every role has its specific module permission.

Other IOT based solutions, instead, despite not developed specifically for MoC can give a relevant support in its correct management.

As an example, Park et al. (2017) proposed a dynamic proximity sensing a processing for smart-work zone identification and management, through the adoption of Bluetooth Low Energy (BLE) based proximity sensing applied to interaction scenarios between ground workers and equipment in various dynamic conditions. More recently, the It is the case of the adoption of Bluetooth Low Energy (BLE) based sensing and alert system. More recently, the work of Arslan et al. (2019) starts from the consideration that unsafe worker movement behaviours are one of the major reasons of construction site fatalities resulting in serious collisions with site objects and machinery. The authors thus developed a system named 'WoTAS' (Worker Trajectory Analysis System). A real-time Bluetooth Low Energy (BLE) beacons-based data collection and trajectory pre-processing subsystem is built for extracting multifaceted trajectory characteristics and stay regions of the workers that will help in recognizing the important regions in the building for categorizing the worker movements. An ontology-based model 'STriDE' (Semantic Trajectories for Dynamic Environments) is applied which tracks the evolution of moving and changing building objects and outputs semantic trajectories. For extracting insights from the semantic trajectories, the Hidden Markov Model (HMM) is used to describe the object behaviour in time. Using the HMMs, a set of trajectories belonging to a stay region is analysed by categorizing the worker movements into different states. In the end, the output of the Viterbi algorithm is visualized using a BIM model for identifying the most probable high-risk locations involving sharp worker movements and rotations.

The developed 'WoTAS' system was developed to help safety managers in monitoring and controlling building activities remotely in dynamic environments by understanding the worker movements for improved safety management in day-to-day building operations. Understanding the worker movements will contribute towards reducing the chances of near-miss incidents on sites which have the potential to cause serious accidents.

A similar tool can directly help during the construction yards that usually accompany larger changes, and indirectly, in monitoring the change operations in order to guarantee lay-out and position constraints to be respected. Another important use could be in the phase of change risk assessment, in case risk sources, fixed and mobile, can be easily foreseen in advance and then risk-based planned.

Furthermore, the adoption of distributed and multi-purpose sensors and the increasing capability of elaborating their data, as will be discussed in the following paragraph, will also help in gaining insights in the operator interface and interaction with the work environment and its hazards, that during changes are often less controllable, allowing a better control of unsafe acts and unsafe conditions, recognised as precursors of occupational accidents, as discussed in Comberti et al. (2015) and Baldissone et al. (2019).

Big Data Analytics

There have been few scientific studies on solving a specific problem using big data method in the field of safety management, despite, as stated in Goel et al. (2017) there exist huge amounts of data that is being generated related to process safety at different levels-regulatory agencies, industry consortiums, and plant/facility. Based on data science, this raw data needs to be subjected to the process of pre-processing or cleaning in order to organize it into information. The data in the form of information or developed databases is then available for use by analysts to extract value from it and support decision-making. This procedure has been schematised as in Figure 5.

Figure 5. Framework for safety big data (extracted from Goel et al., 2017)

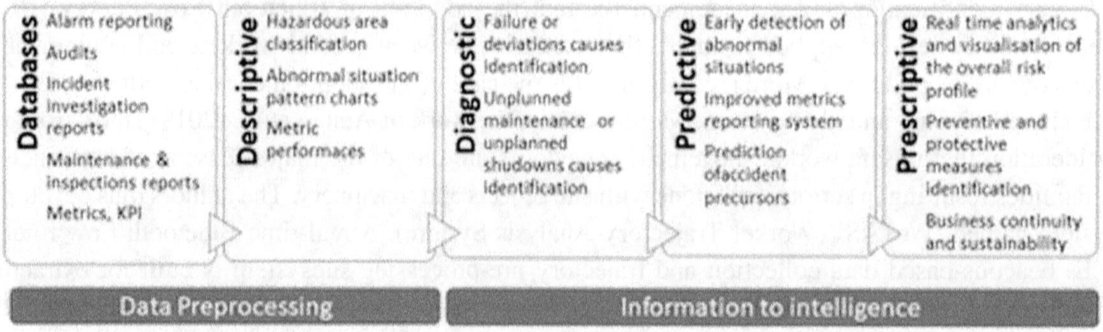

After the first phase of collection of process safety databases, the following steps can help in supporting process safety decision making:

- **Descriptive Analytics:** Deals With determining what happened and converting the data into information such as pattern charts or histograms.
- **Diagnostic Analytics:** Refers to data presentation to understand why something happened or underlying causes for undesirable situations or events.
- **Predictive Analytics:** Refers to developing models on existing datasets to extract information on what will happen or predict future trends.

- **Prescriptive Analytics:** Refers to support decision-making or what should be done by use of advanced analytics.

Implementing a framework for safety big data will have multiple advantages for the MoC, as:

- Dynamic evaluation of risk profile of a facility with the support of real time visualization.
- Safer and reliable operations by incorporation of insights from data analytics enabling optimal maintenance schedules.
- Improve monitoring by the introduction of new or revised metrics.
- Correlation development and use of detailed analysis (structured and unstructured data) to improve audits, incident investigations, hazard evaluation studies.
- Development of visualization dashboards for personnel from different levels within the organization, to communicate the changes.

A more specific application is described in Zhang et al. (2019), where the unstructured-data analysis of the hot-work permit-to-work text was investigated. Permits to work are tools used to control that critical activities, as hot works, that are usually part of larger maintenance or overhaul interventions, are authorised only in case all the relevant safety requirements are respected, in order to minimise the chance for incidents and accidents.

The benefit of adopting a computer-based management of the permits, due to their ineffective control on paper were discussed in Iliffe et al. (1999). Zhang and co-authors propose an advanced procedure to control them and the information they retain with a big data approach.

All hot-work recordings were manually grouped into semantic domains in the light of experts' experience. The data-driven reduction models of hot work number were proposed aiming at possible invalid hot work, long-time hot work, repetitive hot work and possible equipment defect. Based on the proposed reduction models, the authors mined the patterns of hot work and the unnecessary or high-risk hot work could be identified automatically.

The process was based on tools for the natural language processing (NLP), which drives the development of unstructured data analysis. Then the data-driven reduction models of hot work number were built to identify the needless hot work automatically: possible invalid hot work, characterised by the same work content being carried out in the same day; long-time hot work, being characterised by the time of hot work is longer than the average time of hot works in the company; repetitive hot work, being characterised by the same, repetitive times and discontinuous hot work; possible equipment defect, being characterised by different components of the same equipment requiring hot work frequently.

Other relevant applications have been discussed in Goel et al. (2017); of particular interest for the MoC could be the following.

The Dynamic Risk Mapping. Manufacturing facilities have been recognized as complex socio-technical system, having various subsystems and/or components, technical and human, that shows complex interactions, which result in changing operations environment, as previously acknowledged for e.g. aircraft systems (Fontaine et al., 2016). This affects the risk profile of the facilities and hence it is important to study the emergent behaviour of these interactions within the complex systems. There is a relatively small body of literature that is concerned with dynamic risk profiles due to emergent behaviour of complex process systems using big data analytics. The process unit system is reproduced as a system of layers; dynamic risk profile is obtained by the incorporation of the wealth of data generated in the facility from various sources such as historian, Computerized Maintenance Management System, operational data, Safety Management system. The evaluation of dynamic risk involves calculation of initiating event frequency (F1), operations hazard factor (F2), final probability of failure on demand (F3) to give the final risk as R= F1* F2* F3*C, where the consequence C is assumed to be fixed. Operations hazard factor and penalty factors for barriers can be retrieved on literature sources or from the real plant data, that could be assessed from safety culture surveys data, safety management system studies, audit reports and so on. In the original paper an example of a modified accident scenario on sugar dust explosion is considered to analyse and map the dynamic risk profile.

This type of dynamic risk profile analysis would support more informed operational decisions, even during MoC, improved maintenance plans, work execution strategies, and overall safer and more reliable operations.

Image Analysis. With the advancement in IoT, digitization and use of handheld devices, the recording and capturing of unstructured data (images and videos) has increased significantly in the manufacturing plants. During normal field visits, operators use these devices to enter the details related to the health of the equipment, instruments, process streams, etc. This has resulted in mobility and real-time insights of the plant for the managers and other operators. They can access this information through a dedicated application either on their phone or computer at any location. Image analysis can be used as a tool to assist and provide better insights for decision making by using images or videos. Another application of image analysis can be comparing two images captured at a different time to understand the physical changes occuring in the system, equipment or the instrument over a period of time. For this purpose, a dedicated application can be created, where images taken at different times can be compared to show the differences, which may or may not be visible during the routine operational visits. The opportunity of these tools can be exploited in the monitoring and validation phase of a MoC, comparing sections of the plant before and after the intervention.

Cloud Technology

The need for aligning documents underlying the plant safety with major changes, although quite simple in theory, is one of the most difficult elements to implement effectively in Safety Management, due to time and to expertise resources. Systematic methods are often applied just to major changes, such as those that occur when new equipment is added, when an interlock is bypassed, or a replacement is "not in kind". As discussed above, instead, the changes to be addressed should include all changes to processes, equipment, infrastructure, procedures and organisation. In order to follow up on all technical and organizational changes, documents have to be almost continuously reviewed. Consequently, a necessary condition for an effective safety management, and moreover of an efficient MoC system, is a good check by the safety manager on the risk assessment documents and technical documentation.

As discussed in Bragatto et al. (2010) A few forerunners of the idea of a "safety digital representation" can be found in the literature. The potential of the knowledge-based methods to capture experience and reuse it in other studies has been demonstrated, as the potential of the digital models coming from a computer-aided design system, for supporting hazard analysis.

Yoo et al. (2011) proposed the Drawing Information System (DIS). Piping and Instrumentation Diagrams (P&IDs) constitutes the basis of DIS, as, at least in process plants, they retain all the information of process design, process equipment and often also of operation. Such information is basically constructed in the bi-directional method that the information is exchanged in both directions in accordance with the potential risk analysis module.

The drawing management in the workplaces of equipment industry is generally conducted in the hard copy type of Master P&ID form. The change of processes and equipment is performed relatively often according to the safety and operation issue and market change. Therefore, it is necessary to perform the update of process information as soon as possible in order to prevent accident and operation error from the outdated process information. P&ID and other drawings, existing in the form of CAD file are converted into web images to allow multiple users viewing the drawings simultaneously through the DIS, while facilitating the update, through the uploading updated CAD files onto DIS.

During a MoC, the documents update is not limited to the technical drawings, but, as discussed in the previous paragraph many documents as procedures, risk analysis, etc. require updates following the change. The list of action items that need to be accomplished in order to successfully complete the MoC, including the list of documents to be updated, is termed "scoping the MoC". The quality of MoC scoping depends largely on the methodology used, with different sites using anything from no scoping at all, guesswork approaches, checklist approaches, to very sophisticated asset-based scoping. The work of Hoff (2013) provides specific guidelines on how to optimize scoping for small, medium and large MoCs, but above all demonstrates how asset-based scoping, based on local or cloud databases, is significantly better at creating a list of MoC action item, where the action, the type of action, the role, the reason for the action and the timing are all speciðed, without any omissions.

Augmented and Virtual Reality (AR/VR)

Augmented and Virtual reality have been already widely proposed for the training of maintenance operators and in case of emergency.

As discussed in Tatić & Tešić (2017), several applications exploiting augmented reality technologies have been developed for industrial purposes. The authors conducted a literature review, noticing two main categories of AR-systems aimed at application in industry: systems to support maintenance and systems to support training. Maintenance oriented systems usually offer virtual information, providing help in solving a specific problem or performing predefined steps in each operation, as the periodical check of a machine directly at the workplace. These systems can be viewed as an improved electronic version of manuals with built-in expert knowledge. The AR-based training systems are instead mostly focused on the educational component, providing virtual guidance combined with the real experience in the industry workplace and equipment. Tatić & Tešić, among others, propose an AR-based tool to ensure correct implementation of all the steps in the production or maintenance procedures, and to increase the occupational safety by an interactive verification procedure. Similar tools can be of great help in the implementation phase of the MoC.

Moving to virtual reality, Longo et al. (2019) propose a training system for emergency response training in industry. The overall solution encompasses a cooperative, experiential ad differentiated training strategy that is implemented and deployed through an immersive training environment based on Virtual Reality and distributed simulation. It has been designed and developed with the aim of enhancing preparedness skills for all the key roles that are part of the emergency response system including Emergency Managers and Emergency Team Members. Multiple interfaces and VR devices (such as head mounted displays) totally involve the EMs and immerse the ETMs in a realistic virtual training environment, which is built upon real procedures and realistic evolution of physical events in emergency scenarios. Large experimentations have been carried out on a real case study based on a wide set of KPIs. Experimental results have demonstrated that meaningful improvements can be achieved after repeated training sessions in terms of overall performances of Emergency Managers and Emergency Team Members as well as in terms of procedural compliance. Furthermore, the magnitude of the psychological pressure is analysed in terms of average heart rate for Emergency Managers (that cover a strategic role in emergency management) and Emergency Team Members (that are mostly field operators). Here research results show that Emergency Managers can improve their capability to manage stress. Aside from training outcomes, it is interesting to note that the proposed approach impacts not only on emergency response systems' performances, but it can have direct impacts on risk management practices in industries. Results of training sessions can provide considerable data and feedbacks that can support emergency response systems design and implementation, can highlight the need for actual emergency procedures (re)design process; enable greater flexibility and faster responses to comply with changes in regulations. Moreover, being conceived upon Industry 4.0 technologies and principles the proposed training system enables interoperability with other already available Industry 4.0 solutions. For instance, if an industry has a Cyber-Twin for remote monitoring and control the training system can provide information on people that are currently involved in training activities, the expected training duration, etc. Vice versa, whenever an accident occurs, data collected by IOT devices and available through the digital twin can feed into the training system to enable new training scenarios.

Furthermore Kwok et al. (2019), appreciated that conducting emergency drills in an actual environment is workforce and resources intensive. Hence, organisations often hesitate to conduct emergency exercises frequently. Because of the limited number of opportunities to conduct drills in a year, the content of the emergency drills often only focus on common cases and exclude rare cases. This constraint also restricts the members of crisis response teams from exploring and verifying new methods for tackling a crisis. In the proposed research, the authors uses information and communication technology (ICT), virtual reality (VR) and discrete-event simulation (DES) technologies to develop a hazard simulation system with the capability to recreate large scale and multi-agency emergency incidents that would be otherwise too costly, complex and dangerous to reproduce in the reality. This training method is called virtual collaborative simulation-based training (VCST).

Mitsuhara et al. (2019) concentrate their studies on evacuation training, as an important component of disaster education and survival. Evacuation training using a virtual reality (VR)-based disaster simulator that provides a highly immersive simulated evacuation experience (SEE) has attracted significant attention. To improve the training effect, a failure-enhanced evacuation training model based on Kolb's experiential learning theory was proposed. The model aims to purposefully induce participants to succumb to conformity bias and fail to evacuate during the first SEE because inactive evacuees (i.e., people who are not evacuating speedily or not starting their evacuation) are simulated in a VR-based disaster simulator. The participants are expected to overcome failure in the second SEE via reflection and conceptualization. Despite the study is still at an early stage, initial results indicated that failure-enhanced evacuation training can successfully improve the training effect.

Previous experiences were developed e.g. during the EU FP6 funded European programme VIRTHUALIS (2005–2010), that was aimed at the reduction of hazards in production plants and storage sites, one of the main end-users' practical safety issues that have been addressed was the control room operators training, proper alarm systems designing and teams' coping with emergencies, as described in Monferini et al. (2013). The objective was met through the development of an innovative methodology, which has merged HOFs-based knowledge and Virtual Reality (VR) technologies. The innovative character concerned mainly the new HOFs knowledge that was produced ad hoc in a simulated environment where operators had to cope with specific case studies. A Virtual Environment Lab has been developed that simulates the technical system behaviour and the performance of operator tasks in a realistic manner. Operators are able to manipulate process equipment and perform tasks in a virtual plant environment, while the risk analyst can conduct VR experiments that validate the assumptions made in paper work in the process of assessing critical human actions, deviations, error recovery and influential work conditions. The experimentations conducted with plant operators highlighted, through the analysis of user experiences, the need for an adaptation to users' needs (e.g. considering different levels of experience or the different capabilities in using senses).

From a MoC point of view, the above described experiences are worth of interest as they are, since it appears very difficult to train the operators and managers on single operations, but certainly having crews aware of the hazards and able to react correctly in case of accidents is fundamental in complex, high risk environment.

However, even though the use of AR/VR can be extremely beneficial to operations and training of technicians, still some disadvantages from an organizational perspective are present and they reduce at the moment the maximum benefit of these technologies and the overall pervasiveness in industrial sectors (Martinetti et al., 2019). As suggested by Martinetti et al. (2018), to build a complete AR/VR system applicable and flexible to several situations, some changes in companies' structure are needed. A dedicated team able to deal with the required digitalization of the physical asset is vital. Within this team, a job function, as AR/VR designer, should be created focused on maintaining available and reliable the digital system. He/she should work together in close cooperation with the CAD/CAM experts. Together they create, maintain, and adapt the augmented operations or virtual training environment based on changing work instructions, maintenance procedures, and feedback from technicians, safety engineers, and maintenance engineers. Their connection should be considered as a circular process where the experts delivers digital files and receives information relevant for design adaptations. This close cooperation should permit to optimize resources and working hours in the process, maintaining up-to-date the digital environment and achieving the aimed level of safety.

CONCLUSION

In this chapter, the management of change process has been reviewed versus the opportunities offered by Industry 4.0 enabling technologies.

The MoC process is recognised to be ineffective or under managed in many companies. This is often due to basic deficiencies, some of which have been highlighted within this chapter, such as poor documentation control, ineffectual approval regimes, poor hazard analysis, a lack of control on temporary and emergency changes, poor auditing and a lack of effective KPIs with which to monitor the health of the system. MoC, or rather a lack of it, has been implicated in many of the largest losses seen in the process sector over many years.

The Fourth Industrial Revolution, namely Industry 4.0, has emerged as a potentially disruptive force in the manufacturing landscape, with significant impacts on supply chains, business models, and business processes. A wide set of technologies, such as Internet of Things (IoT), Cloud computing, and Big data Analytics, are used to provide devices with connectivity, interoperability and intelligence capabilities. However, among the potential opportunities for organizations and supply chains to innovate and create strategic advantage, an underrepresented area is the one related to the change of the role of humans in manufacturing, where technology can greatly enhance the human-machine integration.

Operators, at any level, are increasingly required to perform complex activities, such as conducting data-driven decision-making processes, instead of physical and routine tasks. In this regard, big data and related technologies play a prominent, disruptive role in today's digital transformation that requires operators to extend their skill set.

This is particularly true in case of changes. In very complex and dynamic socio-technical systems, operators need to take decisions and work in conditions far from the regime ones, with outcomes, in case of errors that could affect the health and safety at the system start-up after modifications have been implemented.

As discussed in the previous paragraphs, the Industry 4.0 paradigm makes available tools and solution supporting the operators and decision-makers in managing the changes in a complete and safe way, minimising the threats to the health and safety of the operators and the safety for the environment and the asset.

REFERENCES

Arslan, M., Cruz, C., & Ginhac, D. (2019). Semantic trajectory insights for worker safety in dynamic environments. *Automation in Construction*, *106*, 102854. doi:10.1016/j.autcon.2019.102854

Baldissone, G., Comberti, L., Bosca, S., & Murè, S. (2019). The analysis and management of unsafe acts and unsafe conditions. data collection and analysis. *Safety Science*, *119*, 240–251. doi:10.1016/j.ssci.2018.10.006

Baldissone, G., Demichela, M., Camuncoli, G., & Comberti, L. (2017). Formaldehyde production plant modification: Risk based decision making. *Chemical Engineering Transactions*, *57*, 703–708.

Borchellini, R., Cardu, M., Colella, F., Labagnara, D., Martinetti, A., Patrucco, M., ... Verda, V. (2013). A Prevention through Design Approach for the Environmental S&H Conditions and the Ventilation System at an Italian Underground Quarry. *Chemical Engineering Transactions*, *32*, 181–186.

Bragatto, P. A., Agnello, P., Ansaldi, S. M., & Pittiglio, P. (2010). The digital representation of safety systems at "Seveso" plants and its potential for improving risk management. *Journal of Loss Prevention in the Process Industries*, *23*(5), 601–612. doi:10.1016/j.jlp.2010.05.013

Cao, G., Xu, C. A., & Shi, H. a (2013). Design of a management of change platform based on Web and RFID. In *1st CCPS Asia-Pacific Conference on Process Safety 2013* (pp. 784-792). Qingdao, China: AIChE.

CCPS. (2013). *Guidelines for Managing Process Safety Risks During Organizational Change*. Hoboken, NJ: Wiley.

Cimini, C., Pirola, F., Pinto, R., & Cavalieri, S. (2020). A human-in-the-loop manufacturing control architecture for the next generation of production systems. *Journal of Manufacturing Systems*, *54*, 258–271. doi:10.1016/j.jmsy.2020.01.002

Clarke, I. (2019). Common Failures in Management of Change. *Chemical Engineering Transactions, 77*, 439–444.

Comberti, L., Baldissone, G., Bosca, S., Demichela, M., Murè, S., Petruni, A., . . . Cencetti, S. (2015). Comparison of two methodologies for occupational accidents pre-cursors data collection. In *Safety and Reliability of Complex Engineered Systems - Proceedings of the 25th European Safety and Reliability Conference, ESREL 2015* (pp. 3237-3244). London: CRC Press/Balkema - Taylor & Francis Group. 10.1201/b19094-426

Demichela, M., & Baldissone, G. (2019). Process risk assessment: From the basics to new frontiers. In M. C. Leva, T. Kontogiannis, M. Gerbec, & O. Aneziris (Eds.), *Total Safety and the Productivity Challenge* (pp. 43–68). Abingdon, UK: Routledge. doi:10.4324/9781315108100-3

Demichela, M., Baldissone, G., & Camuncoli, G. (2017). Risk-Based Decision Making for the Management of Change in Process Plants: Benefits of Integrating Probabilistic and Phenomenological Analysis. *Industrial & Engineering Chemistry Research, 56*(50), 14873–14887. doi:10.1021/acs.iecr.7b03059

Eckhart, M., & Ekelhart, A. (2018). Towards Security-Aware Virtual Environments for Digital Twins. In *Proceedings of the 4th ACM Workshop on Cyber-Physical System Security* (pp. 61–72). Incheon, South Korea: Association for Computing Machinery. 10.1145/3198458.3198464

Fontaine, O., Martinetti, A., & Michaelides-Mateou, S. (2016). Remote pilot aircraft system (RPAS): Just culture, human factors and learnt lessons. *Chemical Engineering Transactions, 52*, 205–2010.

Gambetti, F., Casalli, A., & Chisari, V. (2013). Why sometimes may be neglected management of change? *Chemical Engineering Transactions, 31*, 553–558.

Gerbec, M. (2017). Safety change management – A new method for integrated management of organizational and technical changes. *Safety Science, 100*(Part B), 225-234.

Gerbec, M., & Levovnik, D. (2019). Management of technical and organizational changes in major hazard industries. In M. C. Leva, T. Kontogiannis, M. Gerbec, & O. Aneziris (Eds.), *Total Safety and the Productivity Challenge* (pp. 226–250). Abingdon, UK: Routledge. doi:10.4324/9781315108100-12

Goel, P., Datta, A., & Mannan, M. S. (2017). Application of big data analytics in process safety and risk management. In *2017 IEEE International Conference on Big Data (Big Data)* (pp. 1143-1152). Boston, MA: IEEE. 10.1109/BigData.2017.8258040

Han Siong, P., Chin, K. Y., Bakar, H. T. A., Ling, C. H., Kidam, K., Ali, M. W., ... Kamarden, H. (2017). The contribution of management of change to process safety accident in the chemical process industry. *Chemical Engineering Transactions, 56*, 1363–1368.

He, R., Chen, G., Dong, C., Sun, S., & Shen, X. (2019). Data-driven digital twin technology for optimized control in process systems. *ISA Transactions, 95*, 221–234. doi:10.1016/j.isatra.2019.05.011 PMID:31109723

Hoff, R. (2013). MoC scoping—Ensuring that MoC action items are correctly and completely described. *Journal of Loss Prevention in the Process Industries, 26*(3), 499–510. doi:10.1016/j.jlp.2012.07.007

Iliffe, R. E., Chung, P. W. H., & Kletz, T. A. (1999). More Effective Permit-to-Work Systems. *Process Safety and Environmental Protection*, *77*(2), 69–76. doi:10.1205/095758299529839

Keren, N., West, H. H., & Mannan, M. S. (2002). Benchmarking MoC practices in the process industries. *Process Safety Progress*, *21*(2), 103–112. doi:10.1002/prs.680210205

Kwok, P. K., Yan, M., Chan, B. K. P., & Lau, H. Y. K. (2019). Crisis management training using discrete-event simulation and virtual reality techniques. *Computers & Industrial Engineering*, *135*, 711–722. doi:10.1016/j.cie.2019.06.035

Labagnara, D., Martinetti, A., & Patrucco, M. (2013). Tunneling operations, occupational S&H and environmental protection: A Prevention through Design approach. *American Journal of Applied Sciences*, *11*(11), 1371–1377. doi:10.3844/ajassp.2013.1371.1377

Levovnik, D., & Gerbec, M. (2018). Operational readiness for the integrated management of changes in the industrial organizations – Assessment approach and results. *Safety Science*, *107*, 119–129. doi:10.1016/j.ssci.2018.04.006

Longo, F., Nicoletti, L., & Padovano, A. (2019). Emergency preparedness in industrial plants: A forward-looking solution based on industry 4.0 enabling technologies. *Computers in Industry*, *105*, 99–122. doi:10.1016/j.compind.2018.12.003

Mannan, S. (2012). *Lees' Loss Prevention in the Process Industries* (4th ed.). Oxford, UK: Elsevier.

Martinetti, A., Braaksma, A. J. J., & van Dongen, L. A. M. (2017). Beyond RAMS Design: Towards an Integral Asset and Process Approach. In L. Redding, R. Roy, & A. Shaw (Eds.), Advances in Through-life Engineering Services (pp. 417-428). Cham, Switzerland: Springer.

Martinetti, A., Costa Marques, H., Singh, S., & van Dongen, L. (2019). Reflections on the Limited Pervasiveness of Augmented Reality in Industrial Sectors. *Applied Sciences (Basel, Switzerland)*, *9*(16), 3382. doi:10.3390/app9163382

Martinetti, A., 't Hart, K., Damgrave, R., van Dongen, L. A. M., Turkenburg, R., & Nouwens, A. (2018). There is no spoon: Applying virtual reality for maintenance training of rolling stock technicians. *International Journal of System of Systems Engineering*, *8*(4), 398–415. doi:10.1504/IJSSE.2018.094563

Mitsuhara, H., Tanimura, C., Nemoto, J., & Shishibori, M. (2019). Failure-enhanced evacuation training using a VR-based disaster simulator: A comparative experiment with simulated evacuees. *Procedia Computer Science*, *159*, 1670–1679. doi:10.1016/j.procs.2019.09.337

Monferini, A., Konstandinidou, M., Nivolianitou, Z., Weber, S., Kontogiannis, T., Kafka, P., ... Demichela, M. (2013). A compound methodology to assess the impact of human and organizational factors impact on the risk level of hazardous industrial plants. *Reliability Engineering & System Safety*, *119*, 280–289. doi:10.1016/j.ress.2013.04.012

Park, J., Yang, X., Cho, Y. K., & Seo, J. (2017). Improving dynamic proximity sensing and processing for smart work-zone safety. *Automation in Construction*, *84*, 111–120. doi:10.1016/j.autcon.2017.08.025

Reyes Garcia, J. R., Martinetti, A., Jauregui Becker, J. M., Singh, S., & van Dongen, L. A. M. (2019). Towards an Industry 4.0-Based Maintenance Approach in the Manufacturing Processes. In V. González-Prida Diaz & J. P. Zamora Bonilla (Eds.), *Handbook of Research on Industrial Advancement in Scientific Knowledge* (pp. 135–159). Hershey, PA: IGI Global. doi:10.4018/978-1-5225-7152-0.ch008

Tatić, D., & Tešić, B. (2017). The application of augmented reality technologies for the improvement of occupational safety in an industrial environment. *Computers in Industry*, *85*, 1–10. doi:10.1016/j.compind.2016.11.004

Yoo, J. H., Lee, H. S., & Ko, J. W. (2011). Risk assessment and drawing information system based change management. *Korean Journal of Chemical Engineering*, *28*(2), 323–331. doi:10.100711814-010-0389-y

Zhang, G., Wang, Y., Mu, B., & Wang, T. (2019). Big Data and Safety Management Methods: The Reduction Model of Hot Work Number. In *2019 IEEE 4th International Conference on Big Data Analytics (ICBDA)* (pp. 140-143). Suzhou, China: IEEE.

Section 3

Chapter 11
Towards E-Maintenance:
An Exploration Approach for Aircraft Maintenance Data

Peter K. Chemweno
University of Twente, The Netherlands

Liliane Pintelon
KU Leuven, Belgium

ABSTRACT

Safety is an important concern for critical assets, such as aircrafts. E-maintenance strategies have long been explored for maintenance decision support and optimizing the operational availability of aircraft assets. Data-driven tools are an important influencer of day-to-day maintenance processes, which if optimally used may support practitioners to design more effective maintenance strategies. Recent trends show a correlation between e-maintenance strategies and enhanced use of data-driven tools for optimally managing technical assets. However, using data-driven tools for designing e-maintenance strategies is challenging because of aspects such as data-readiness and modelling-related challenges. This chapter presents a data-exploration approach for aiding root cause analysis of aircraft systems. The approach embeds algorithms for data preparation, text mining, and association rule mining and is validated in a use-case of maintenance of aircraft equipment, discussed in this chapter.

INTRODUCTION

Achieving safety during flight the mission is the highest priority of any aviation industry. As a result, the aviation industry usually tries to adopt schedules of maintenance activities, to optimize the operational lifetime of the aircraft, and minimize incidences of mechanical failures. A significant attention is usually focused on maintenance-related accidents, which according to a recent IATA (International Air Transport Association) document, 15% of reported aircraft accidents are attributed to maintenance-related factors (Chen et al. 2017). Additional factors influencing aircraft accidents includes meteorology, and human error related factors, which are largely mentioned as a leading factor contributor to accident events. It is

DOI: 10.4018/978-1-7998-3904-0.ch011

important to note that the report highlights a close correlation between maintenance-errors and equipment failure, which lead to accident events. Additional risk factors, which influences how maintenance interventions are carried out, includes the level and quality of compliance with standard operating procedures (SOP) while performing maintenance activities, training of maintenance crew and work-related factors such as fatigue. Furthermore, maintenance-related human errors tend to be influenced by communication challenges, work pressure, among other factors. To address these aspects, which influence the quality of maintenance activities for aircraft systems, IATA places considerable efforts developing and enforcing guidelines for maintaining aircraft systems, while emphasizing quality of the activities and compliance to standardized rules for maintenance.

Since aircrafts are equipped with sensors attached to the critical components of interest, often these sensors are used to record different attributes related to the aircraft condition. Consequently, this sensor information is useful for assisting maintenance technicians to monitor the performance of the critical components and derive diagnostic information to aid root cause analysis and plan maintenance activities. For example, sensors monitor attributes such as vibration signals and generate fault messages or error warnings. Once analyzed, the information aggregated from the sensors forms the basis of decision making on aspects such as planning maintenance interventions. Moreover, the sensor information is useful to support airlines to better allocate maintenance resources, such as technicians.

However, to a large extent, sensors and maintenance information collected from aircraft systems are not sufficiently used in decision making for maintenance. Rather, aircraft maintenance largely relies on standardized maintenance manuals, with insufficient attention focusing on maintenance data collected from aircraft sensors or repair information recorded by maintenance technicians. Some challenges limiting use of, especially repair information recorded by maintenance crew, pilots or quality assurance staff, is the textural form of the information. These limits use of traditional statistical analysis tools such as correlation analysis, which means that maintenance technicians cannot fully exploit maintenance records for decision support. This chapter proposes a data mining approach, which uses visual exploration, text mining, and association rule mining for deriving insights on patterns and correlation, which are useful for maintenance decision support, but embedded in data. Such patterns are useful for supporting decisions on aspects such as failure diagnosis and identifying the focal root causes of equipment failure. Importantly, the approach uses predictive analytics to identify critical failure patterns embedded in maintenance data and is useful for assisting maintenance planning and allocation of often scarce maintenance resources.

BACKGROUND AND LITERATURE REVIEW

Background on Aircraft Maintenance

Aircraft maintenance represents "all activities that involve repairing, cleaning, refueling, modifying and inspecting aircraft components" (Kroes et.al. 2013). It is often a very costly process, and safety-critical since sub-optimal maintenance may lead to accidents leading to loss of lives. Figure 1 summarizes stepwise, maintenance activities undertaken on the aircraft system. These steps follow the well-known Plan-Do-Check-Act cycle, which allows for continuous improvement of maintenance activities. For safety critical systems, Martinetti et al. (2019) mention examples of maintenance strategies usually implemented, including emergency or reactive maintenance, preventive and predictive maintenance. They also discuss

Figure 1. Aircraft maintenance cycle activities (adapted from Ackert and Shannon, 2010)

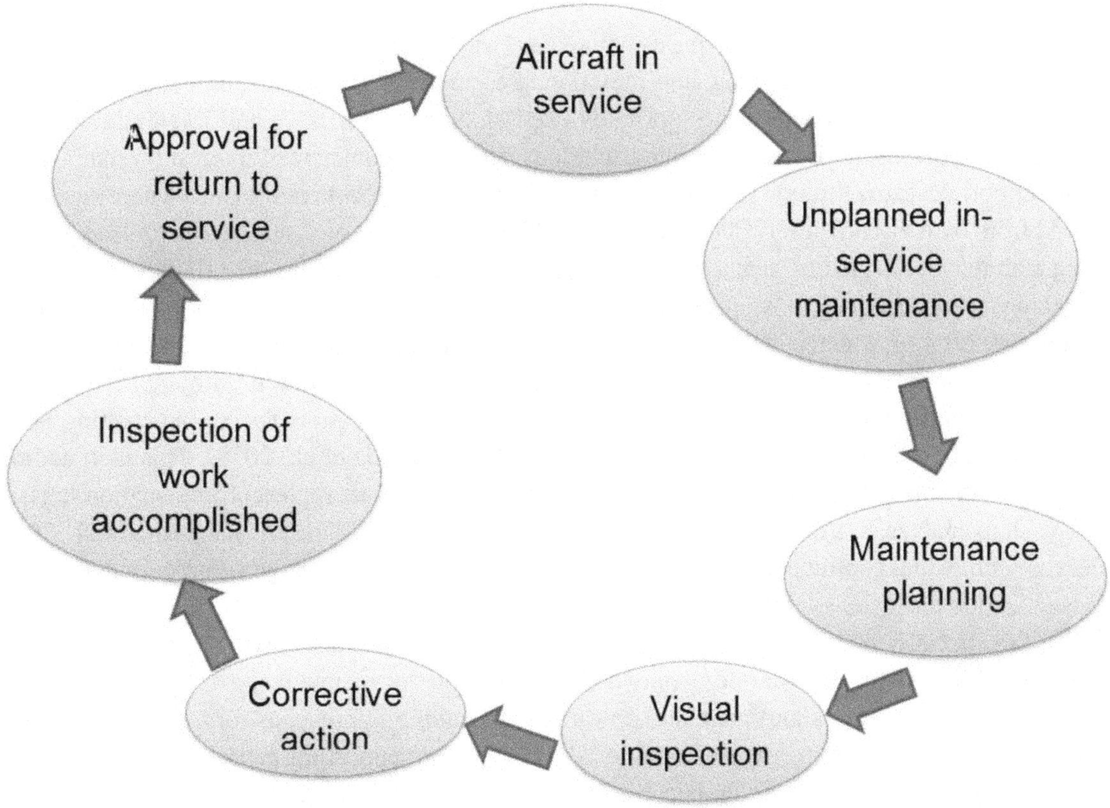

how data generated from critical components or sub-systems of a safety critical equipment can be used for supporting maintenance decisions.

For aircraft maintenance, data generated from sensors installed on critical components provides potential insights that could complement standardized maintenance protocols provided by aircraft manufacturers. For example, the Airbus aircraft uses the ECAM (Electronic Centralized Aircraft Monitoring) system to generate textural warning messages during different flights phases, such as take-off, flight and landing phases. These messages present opportunities for supporting technicians to perform diagnostics and root cause analysis. Young et.al. (2010) presented early work demonstrating how data mining can be useful for facilitating maintenance activities. For example, they describe how analytic tools for association rule mining can be useful to derive knowledge on potential association between failure modes and component failures. They further show how this knowledge is useful to complement preventive maintenance activities suggested by aircraft manufacturers and aviation regulators (e.g. FAA). Cost reduction is also another added advantage of proper analyzing aircraft maintenance data. For instance, a company can identify unproductive or unnecessary maintenance activities and eliminate them, hence save maintenance-related cost and remain competitive. Moreover, efficient maintenance strategies can be designed from the obtained knowledge.

From a business perspective, the advantage of using knowledge embedded in maintenance data for optimally allocating maintenance resources, such as technicians, through better planning can yield value in terms of cost savings.

Statistical Data Analysis and Data Exploration Approaches

Statistical analysis as a basis for supporting e-maintenance strategies is not new. Specifically, the analysis broadly involves "collection, analyzing, interpretation, presentation, and organization of data" (Fontaine et al., 2016). Two statistical approaches normally used includes, descriptive and inferential statistics, from which conclusions, for instance, from correlations between parameters of interest for maintenance decision support. Data exploration is an important complement to descriptive statistics where knowledge is derived from data visualization (Chemweno et al., 2016). Examples of visualization tools applied for deriving maintenance decisions such as interactive plots are discussed in Sun and Salgado (2017). Additional examples are described in Wang and Li (2017) where web-based data visualization is suggested as a decision support tool, which allows decision makers to gain quick insights on potential patterns hidden in data sets, for example recurrent equipment failure events for the aircraft system. Although visualization tools are easily understood by non-experts when limited information is displayed, interpreting highly detailed information is cited as an important challenge (Nimmagadda et al., 2018). This is in addition to potential challenges selection appropriate visualization tools to best represent information decision makers wish to analyze in order to support e-maintenance related activities (Olshannikova et al., 2015).

Because companies continually operate complex assets, which generate large volumes of data, simple data visualization tools are less effective to support design and implementation of e-maintenance strategies. This is especially relevant for companies wishing to transit to the Industry 4.0 paradigm, where big data and data analytics is expected to play an important role for maintenance decision support. For this reason, there is a more significant shift towards data mining approaches for decision support in maintenance (Chemweno et al., 2016; García et al., 2019). For supporting maintenance-related decisions for technical assets, Maquee, Shojaie, Mosaddar, and Management (2012) describes data mining as an appropriate process for discovering meaningful knowledge embedded in maintenance data sets, where usually, the derived knowledge is influenced by the domain in which the asset operates. Some examples of data mining approaches discussed in the study, includes, pattern recognition, predictive machine learning, among other tools.

Figure 2 shows several phases, based on the CRISP-DM (Cross Industrial Standard Process for Data Mining) proposed for data mining processes (Berry and Linoff, 1997). The first phase starts with decision makers clearly understanding requirements of a project, while considering the business perspective of the organization. For e-maintenance, this may involve situating e-maintenance as a strategic objective for improving asset availability. This understanding is necessary prior to defining which data to collect to design the e-maintenance programme. Borchiellinia et al. (2013) emphasize the need to involve multidisciplinary experts (maintenance, statisticians, etc.) in order to clearly understand and define objectives of the project at hand. This will support a company to understand which data is needed to support formulation of solutions which best fit the defined problem, here the e-maintenance strategy. For instance, mitigating risks of asset unavailability requires knowledge of risk assessment, for which, failure data is important for prioritizing safety-critical units of a system (Chemweno et al., 2016).

The next phases in the CRISP-DM approach, involves collection and preprocessing operation of maintenance data, with the pre-processing focusing on feature selection, dimension reduction, normalization, imputation of missing data and other related techniques to improve data quality (Chemweno et al., 2016). The pre-processing phase is followed by data analysis, validation and lastly, formulating e-maintenance strategies which best addresses asset management challenges defined during the first phase of the CRISP-DM approach. In this chapter, the CRISP-DM methodology is applied as a first

Figure 2. Cross industrial standard process for data mining (Berry and Linoff, 1997)

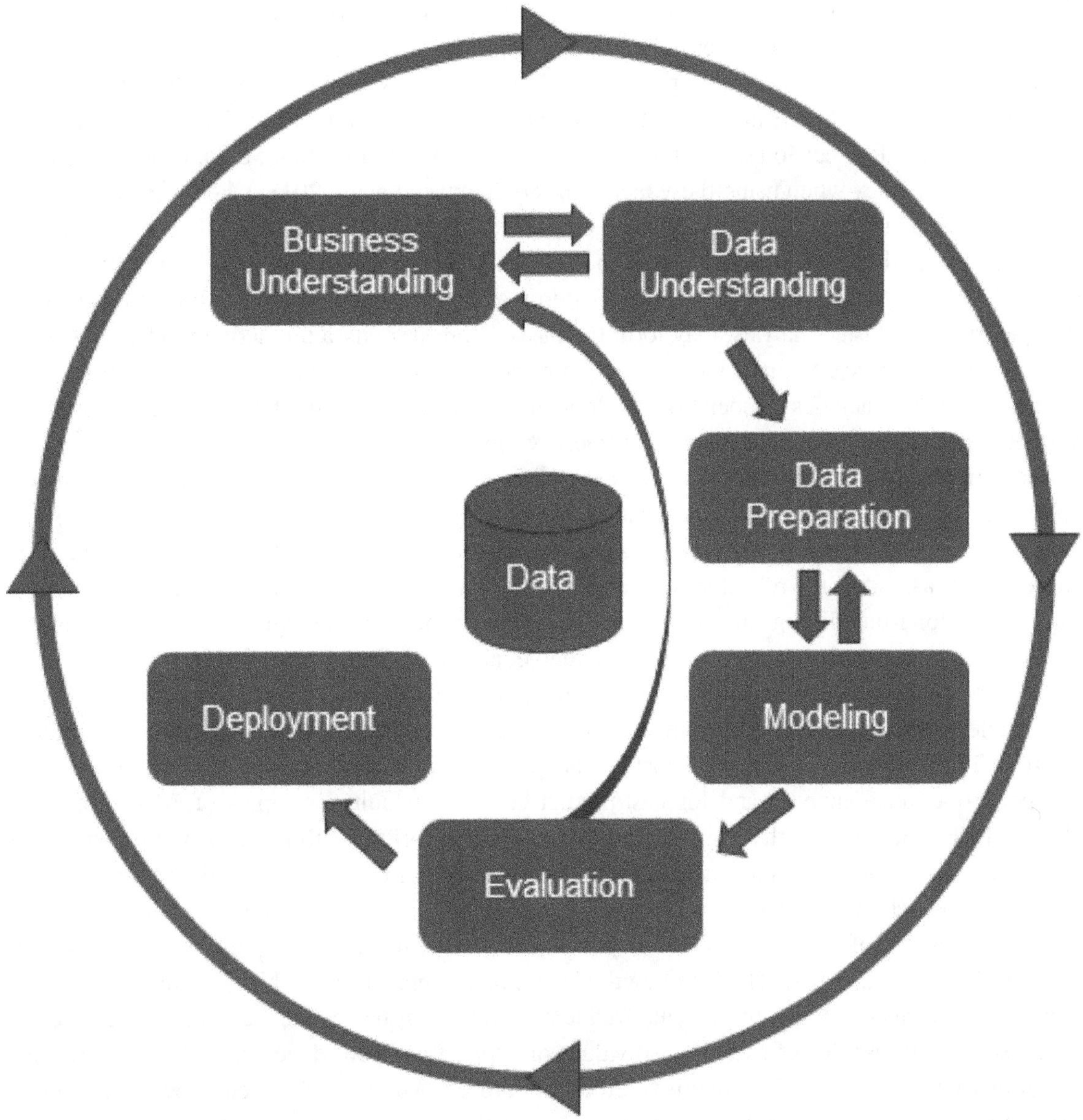

step to formulating e-maintenance strategy for an aircraft maintenance case. However, first, a review of data mining tools is discussed in the following sections.

Data Mining Techniques Applicable for E-Maintenance

Data mining techniques are classified into broadly, supervised and unsupervised learning methods. Supervised learning techniques/methods aims at predicting future observable events, using currently known attribute values. For instance, the methods can allow practitioners to infer future possible failure events based on historical failure patterns of an equipment. The predictive decision support of supervised learning relies on accurate mapping patterns (or data labeling) of historical data sets. Inference is then based on corelating mapped patterns to observable similar correlations in more extensive data sets of a similar equipment. Examples of techniques applied for data labeling includes; Decision Trees, K-Nearest Neighbor, Naive Bayes, Support Vector Machine and Neural Networks (Gerdes, 2019; Thai et al. 2019; Mack et al. 2016; Liu et al. 2018; Zhang et al. 2016). An important challenge of supervised learning techniques for decision support, is availably of historical maintenance data, which often affects the predictive accuracy when using these techniques (Chemweno et al., 2018; Chemweno et al., 2015). Moreover, several authors propose using unsupervised learning methods to map patterns in data, which is potentially useful for maintenance decision support, prior to training predictive supervised learning models (King et al., 2019). For instance, patterns of failure events derived using an unsupervised technique, such as cluster analysis, may form the basis of pre-training a predictive model to assess the effectiveness of specific e-maintenance strategy implemented by a company. Examples of unsupervised learning techniques includes Associations Rule Mining, Cluster Analysis, Regression, Classification, Anomaly Detection and Pattern Recognition (Gera & Goel, 2015).

Association Rule Mining (ARM) is cited as a useful method for extracting interesting relationships among a set of items. For instance, the relationship maybe between a failure mode such as vibration, and an actual failure event (such as a bearing failure). This approach defines two criterion, support and confidence, for assessing the robustness of relationship between attributes. Moreover, the rules derived using Association Rule Mining, are cited as useful for supporting root cause analysis or diagnosing faults of technical systems, such as an aircraft (Chemweno et al., 2016; Reuss et al., 2018).

The association rules appear in the form of an itemset, for instance, "vibration -> bearing failure". For example, here, a vibration failure mode is observed as a precursor of bearing failure (Turnbull et al., 2019). The advantage of the ARM method is its ability to extract robust association rules hidden to support root cause analysis and forecast/predict equipment fault. Examples of ARM approaches includes, the FP-Growth and the Apriori algorithms. The Apriori algorithm works by frequently scanning datasets and generating frequently occurring associated itemsets. However, the scanning process is often a bottleneck of this algorithm, because of large data storage needs and memory consumption, which slows mining of association rules, especially for large maintenance datasets (Djatna & Alitu, 2015; Rachburee et al., 2018). The FP-Growth algorithm attempts to solve this flaw and works by first scanning maintenance datasets to generate frequent itemsets. In the second step, all frequent itemsets meeting specific thresholds of user defined values of support and confidence, are filter out and robust association rules are generated, from this filtered-out itemsets. Moreover, frequently occurring itemsets meeting the user defined thresholds are visualized using the FP-tree, further aiding root cause analysis. Rachburee et al. (2018) combines both the Apriori and FP-Growth algorithms to mine failure association rules of automated teller machines, where they argue that using both approaches yields more robust rules and leads to a better trade-off between execution time, memory needs and validity of rules generated by both methods. Moharana et al., (2019) applies the Apriori algorithm to mine sequential patterns of maintenance activities and related spare part information from historical maintenance data.

Association Rule Mining methods work well with textural maintenance information. This makes the method suitable for extracting association rules from warning messages generated by automated systems such as the ECAM of the aircraft, which is usually in textural form. It is also helpful for mining associations from maintenance information recorded by aircraft crew, pilots and quality assurance staff as will be discussed further in Section 3. However, one disadvantage of ARM is the considerable time spent to pre-process data to a suitable form to extract association rules, using algorithms such as the Apriori and the FP-Growth algorithms. Recent research trends highlight the increasing use of text mining approaches as a means of supporting practitioners to efficiently pre-process textural data prior to extracting useful association rules for supporting e-maintenance strategies, for instance, online diagnostics and failure forecasting.

Text Mining Approaches

Text mining works by learning interesting patterns hidden in textural data, which could be harnessed to improve maintenance decisions, based on the knowledge obtained from the textural analysis (Aggarwal & Zhai, 2012). Algorithms for text mining are similarly useful for both supervised and unsupervised learning, since the algorithms can be linked efficiently to classification or association rule mining methods. Apart from pattern mapping, text mining is also useful for detecting outliers and analyzing trends hidden in maintenance data sets.

Figure 3. Text mining steps (adapted from Aggarwal & Zhai, 2012)

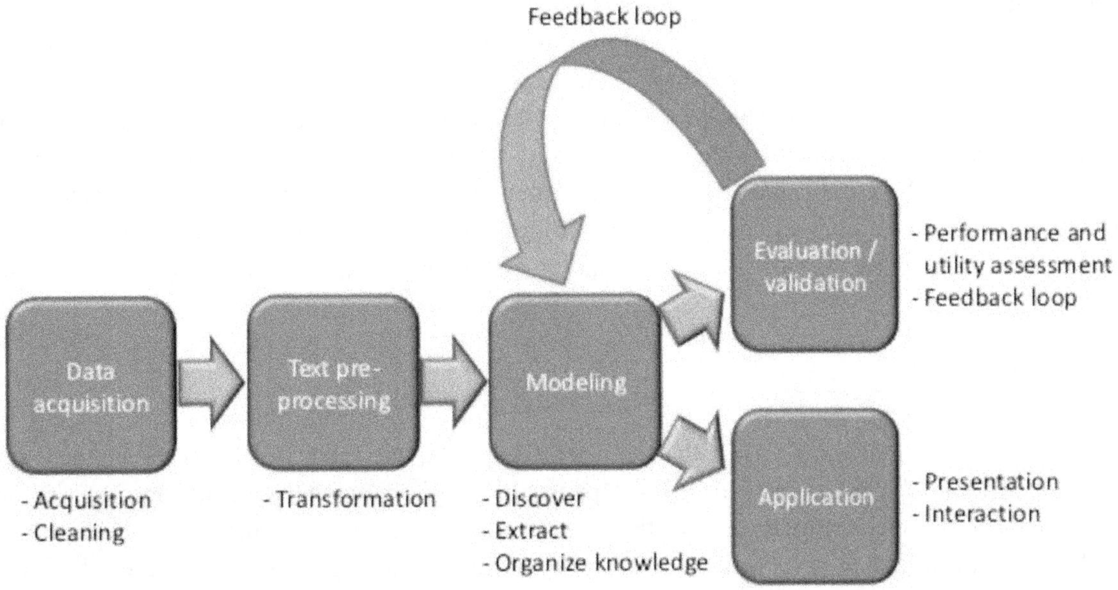

Figure 3 shows the text mining process, which involves two phases. The first phase focuses on pre-processing the textural data to a form which is suitable for analysis. The pre-processing involves filtering out redundant words, such as 'and' or 'the', which are usually less suitable for maintenance decision support, as they lack any meaningful attributes from a maintenance perspective, for an equipment. Additional pre-processing steps include tokenization (separating sentences into its component words) and removal of trivial words, which are not associated with important equipment attributes, again from a maintenance perspective.

After pre-processing, the next phase involves textural analysis which involves feature presentation. This phase presents the extracted words (features) in a format that is suitable for text mining using supervised and unsupervised learning algorithms. Rodrigues et al., (2012) describes an approach utilizing both text mining and neural networks to identify failure patterns hidden in textural failure logs, generated from a critical avionics equipment, of an aircraft system. They used the patterns as the basis of improving repair practices and enhancing diagnosis of the critical equipment.

The case study discussed in this chapter proposes a data mining approach following the CRISP-DM methodology and integrates text mining as an approach for pre-processing textural warning messages prior to extracting important association rules, using the FP-Growth algorithm. The methodology is discussed in the following sections.

CONCLUSIONS FROM REVIEW OF DATA MINING STRATEGIES FOR E-MAINTENANCE

From the literature review, several advantages of data exploration and mining techniques are highlighted, for pattern recognition and supporting e-maintenance strategies for aircraft systems. The motivation of the study discussed in this chapter, is to complement scheduled maintenance activities, largely carried out for aircraft systems. The proposed approach is important for supporting e-maintenance activities, and especially using maintenance data for supporting root cause analysis. As will be discussed in the next section, textural maintenance information used in this study, were largely recorded in an unstructured form, hence required considerable pre-processing effort prior to mining useful knowledge hidden in the datasets. Owing to the advantages of the FP-Growth, it was selected as an appropriate algorithm to support derivation of robust association rules between failure patterns and maintenance activities recorded for the aircraft systems. To support extraction of robust association rules, a text mining approach is integrated in our proposed approach. The methodology is discussed next section.

METHODOLOGY

The phases of the methodology start with an exploratory analysis from which an initial visualization of important attributes for maintenance decision support, derived from maintenance information recorded by maintenance technicians, pilots and quality assurance staff at an aviation airline. The next step focused on structuring the textural messages, where frequently occurring words, which are relevant for improving maintenance activities were extracted. This step is important as a pre-cursor to data mining in which, association rules were extracted from the textural maintenance information. Prior to applying the association rule mining approach, text mining was performed to pre-process the textural messages, such that important association rules from the perspective of maintenance decision support were generated. Lastly, a predictive analysis is carried out, using classification algorithms. A detailed explanation of these pre-processing activities and consequent text mining and data mining activities are discussed in the following sub-sections. For the study, the Rapid Miner software is used for the analysis.

Description of the Aircraft Maintenance Datasets

For this study, two separate datasets of maintenance information of aircraft systems of a case aviation airline were used. The first dataset reports maintenance-related attributes from the aircraft structure, while the second dataset reports maintenance information from the power plant system. Figure 4 shows an overview of the dataset, which includes description of failure events and associated maintenance interventions. Some of the maintenance-related attributes recorded for the two aircraft sub-systems, includes, the type of failure, date the failure occurred, etc.

Figure 4. Overview of maintenance-related attributes recorded the aircraft system

1	Acreg	Date	System_CODE	Description	Origin	Sta	Acc.Date	CSta	Action	Skill	MEL_Category	Mechanic	Inspector
2	AC6	1/6/2010	79	EOQ4-79-100511 IMPENDING BYPASS INDICATION F	OTHER	Station1	1/6/2010	Station1	DEFECT CLOSED I.A.W. ID30224 REQUEST.	4AT00		ID30183	ID30183
3	AC6	1/8/2010	61	ENG # 1 PROP BLADE # 1 FOUND PAINT/SEALANT I	MA	Station1	10/31/2010	Station1	ENG PROP.1 BLADE 1 PAINTED	SB1AS		ID30285	ID30184
4	AC4	2/26/2011	74	FAULT CODE 910.	MA	Station1	2/26/2011	Station1	REPLACED IGNITION EXCITER.	B1		ID30210	ID30210
5	AC6	1/11/2010	61	FUNCTIONAL TEST FOR THE BALANCING OF THE I	MA	Station1	9/11/2010	Station1	FUT FOR BALANCING PERFORMED.	4AT00		ID30142	ID30142
6	AC4	2/20/2011	75	FAULT CODE 908	MA	Station1	2/20/2011	Station1	ACC FAULT CODE REPLACED HBOV 2.7			ID30184	ID30184
7	AC2	2/18/2011	72	WO 1041262 ITEM 1 & 2 BSI OF COMBUSTION AND	MA	Station1	3/22/2011	Station1	ENGINE POWER ASSURANCE CHECK PERFC	4AT00		ID30117	ID30006
8	AC6	2/4/2010	61	RETORQUE OF PROPELLER BLADES REQUIRED	MA	Station1	6/22/2010	Station1	RE-TORQUE OF PROPELLERS BLADES PERI			ID30142	ID30142
9	AC2	2/13/2011	61	BOTH PROPELLERS RETORQUED I.A.W AMM 61-1(	MA	Station1	2/19/2011	Station1	PERFORMED, ALL PARAMETERS IN LIMITS	4AT00		ID30117	ID30117
10	AC4	12/2/2011	61	FUT FOR PROPELLER BALANCING HAS TO BE PEI	MA	Station1	2/13/2011	Station1	PERFORMED FUT FOR PROP BALANCING, IN			ID30142	ID30117
11	AC5	11/2/2011	61	DURING L CHECK FOUND SMALL DAMAGES IN CUF	MA	Station1	6/4/2011	Station1	PERFORMED	CAB		ID30216	ID30218
12	AC4	10/2/2011	61	POWERPLANT MESSAGE ON ED.FAULT CODE 224	MA	Station1	10/2/2011	Station1	CONFIRMED U/S GROUND BETA ENABLE SOL			ID30142	ID30142
13	AC6	2/7/2010	61	ON # 1 PROP BLADE FOUND DAMAGE (LIFTED ARE	MA	Station1	4/7/2010			4AT00		ID30285	ID30210
14	AC2	7/2/2011	74	ENG NO.2 FAULT CODE 910	MA	Station1	7/2/2011	Station1	REPLACED ENG NO.2 EXCITER BOX	TEC		ID30175	ID30175
15	AC1	5/2/2011	61	AFTER RAI OF BLADE 4 ON PROP No 2 RETORQUE	MA	Station1	6/2/2011	Station1	RETORQUE OF PROP 2 BLADE 4 AFTER RAI I	B1		ID30216	ID30216
16	AC1	5/2/2011	61	DUE TO BIRD STRIKE ON PROP 2 BLADE 4 ACC D	MA	Station1	2/13/2011	Station1	VISUAL INSPECTION OF PROP 2 PERFORMEI	B1		ID30216	ID30207
17	AC1	4/2/2011	61	BIRD STRIKE AT LDG ROLL, ENG #2 PROP BLADE ;	QA	Station1	4/2/2011	Station1	REMOVED BLADE ASSEMBLY AND BEARING /	B1		ID80008	ID30216
18	AC3	1/31/2011	61	DURING CRUISE AND DESCENT WITH PROPS AT 8 PI			1/31/2011	Station1	I.A.W. FIM 61-20-03-810-821 REV 36 PERFORM			ID30129	
19	AC1	1/29/2011	61	RESULTS FOR FUT FOR PROP BALANCING FOR P	MA	Station1	1/30/2011	Station1	FUT FOR PROP BALANCING PERFORMED VIE			ID30184	ID30184
20	AC2	1/27/2011	61	RESULTS FOR FUT OF PROP BALANCING ARE PR	MA	Station1	1/31/2011	Station1	FUNCTIONAL TEST FOR THE BALANCING OF			ID30199	ID30199
21	AC2	1/27/2011	77	NOT POSSIBLE TO PERFORM DOWNLOAD OF TH	MA	Station1	5/2/2011	Station1	DOWNLOAD OF DATA FROM EMU PERFORME	A		ID30131	ID30131
22	AC1	1/26/2011	77	DOWNLOAD DATA FROM EMU NOT POSSIBLE. CO	MA	Station1	1/29/2011	Station1	DOWNLOAD OF THE DATA FROM EMU PERFC	B1		ID30254	ID30226
23	AC1	11/1/2011	61	PROP#2 ON ALL BLADES FOUND MISSING PAINT O	MA		12/1/2011	Station1		TEC		ID30171	
24	AC1	11/1/2011	61	PROP. #1 ON 4 BLADES FOUND CRACK IN THE RA	MA		12/1/2011	Station1	REPAIR PERFORMED	TEC		ID30171	ID30171
25	AC1	10/1/2011	61	FUT FOR THE BALANCING OF THE PROPELLER RI	MA		12/1/2011	Station1	ITEM ALREADX EXISTS UNDER DE# 958354, D	B1		ID30171	ID30171
26	AC1	10/1/2011	61	FUT FOR THE BALANCING OF THE PROPELLER RI	MA		1/15/2011	Station1	PERFORMED READ OUT OF PMBS (ANVS FO	B1		ID30021	ID30021
27	AC2	6/1/2011	79	ENG#2 FAULT CODE READING REPORTS FAULT C	MA		9/1/2011	Station1	ENG#2 MAIN OIL FILTER REPLACED I.A.W. AM	B1		ID30200	ID30200

Table 1 summarizes a sample of the dataset. The first attribute, 'Acreg', describes the aircraft registration and specifies the asset the failure is recorded from. The second attribute, 'Date' describes the date the defective component of the aircraft was reported, while the attribute, 'Accdate' describes the date the defect was repaired. The 'description' attribute describes the nature of the reported defect, for example, damage of the aircraft door. The 'action' attribute describes the maintenance intervention carried out to rectify the reported defect, while the 'origin' attribute describes the persons responsible for recording the defect, for instance, maintenance crew, pilots, or from quality assurance personnel. The Minimum Equipment List (MEL) attribute describes information related to lead repair time required for defective components, and useful as a basis for determining the criticality of the type of failure event, based on the urgency needed to repair the failure to restore the aircraft to operational availability. Usually, the repair lead-time intervals are normally reported alongside the type of defect observed during the aircraft maintenance.

Apart from deriving insights on the urgency needed to repair a failure, the MEL specifies failure events allowable, which does not compromise the airworthiness of the aircraft. Failure events are categorized into classes. Class A failures in this instance, considers events which require immediate attention by maintenance technicians. This are usually critical failure events which influences the operational availability of the aircraft equipment. Class B failures, on the other hand, consists of defects which must be repaired within a period of 72 hours, while Class C defects should be repaired within a period of 240 hours. Lastly, Class D defects are the least critical and need to be repaired within a period of 2880 days (Kroes et al., 2013). Besides the MEL document, the Air Transport Association (ATA) adopts a convention for naming sub-systems of the aircraft equipment. Table 2 shows an overview, where the ATA 52, for example, describing the doors of the aircraft equipment.

Table 1. Summary of attributes

			Power Plant	**Structure**
Attribute code	Type	Description of the attribute codes	Missing Values	Missing Values
Acreg.	Numerical	Aircraft Registration number	3	1
Date	Numerical	Date defect was reported	0	0
System Code	Integer	Numbering system identifying aircraft components	0	0
Description	Textural	Description of the reported defect	1	0
Origin	Textural	Origin or source of the defect	0	0
Sta	Numerical	Station were defect was reported	120	92
AccDate	Numerical	Date a defect was closed	0	0
Csta	Numerical	Station a defect was closed	17	33
Action	Textural	Action performed on defect	27	49
Skill	Numerical	Mechanic or inspector skill	117	98
MEL Category	Numerical	Describe time for repair of defect component	386	275
Mechanic	Numerical	Unique identification of the mechanic	26	34
Inspector	Numerical	Unique identification of the Inspector	96	101

Table 2. Summary of ATA sub-systems defined for aircraft equipment

Air Transport Association (ATA) convention					
Structure			Power Plant		
ATA	51	General	ATA	61	Propellers
ATA	52	Doors	ATA	70	Standard Practices Engine
ATA	53	Fuselage	ATA	71	Power Plant
ATA	54	Nacelles/Pylons	ATA	72	Engine General
ATA	55	Stabilizers	ATA	73	Engine Fuel And Control
ATA	56	Windows	ATA	74	Engine Ignition
ATA	57	Wings	ATA	75	Engine Bleed Air
			ATA	76	Engine Control
			ATA	77	Engine Indicating
			ATA	78	Engine Exhaust System
			ATA	79	Engine Oil
			ATA	80	Engine Starting

The maintenance-related attributes recorded by maintenance crew, pilots, or quality assurance staff were largely captured in spreadsheets. Usually, as seen in the summary shown in Table 1, the information was in textural form and unstructured text, with missing values for specific maintenance-related attributes. Part of this, was because of absence of a robust data collection program at the airline. Especially, the missing information affected the MEL attribute, which specifies the observed defects, class, and repair information. This also meant that using the maintenance data as-is, for statistical analysis and data mining was rather challenging. This challenge motivated the methodology proposed in this chapter, to derive statistical insights from the data to support better maintenance of defective units of the aircraft.

Steps of the Methodology

The data mining approach developed for this study, considers five main steps discussed in the next subsections. The idea of the approach is to discover useful patterns embedded in the maintenance information recorded for the aircraft, to support a better data-driven maintenance planning approach. To achieve this goal, the RapidMiner© software is used for analysis. The first step, however, prior to discovering hidden patterns in the data, was carrying out a data exploration study discussed next.

Data Exploration

The objective of the exploratory analysis was to gain a better understanding of high-level relationships between maintenance-related attributes in the data. The analysis was an important preparatory phase where missing information was augmented with expert information. As discussed previously, the data recorded 13 attributes for both aircraft datasets. Figure 5 shows information of personnel who reported aircraft-related failure events or defects. For example, a large proportion of the aircraft power plant related defects were reported by the maintenance crew (MA), while carrying out scheduled inspections, or performing maintenance activities (59% of all recorded defects).

The next proportion of defects (20%) were reported by pilots (PI), while 18% of the remaining defects were reported by quality audit staff. The category, 'others' (18.18%) were reported, for instance, by the OEM representatives who did not immediately fit with maintenance crew, pilots or internal quality assurance staff. A similar reporting trend is observed for the aircraft structure, where 48% of all defects were reported by maintenance crew, while 37% of the defects were reported by pilots (PI). The remaining proportion of defects was reported by quality assurance staff (37% of the cases).

Figure 5. Origin of reporting of failure events

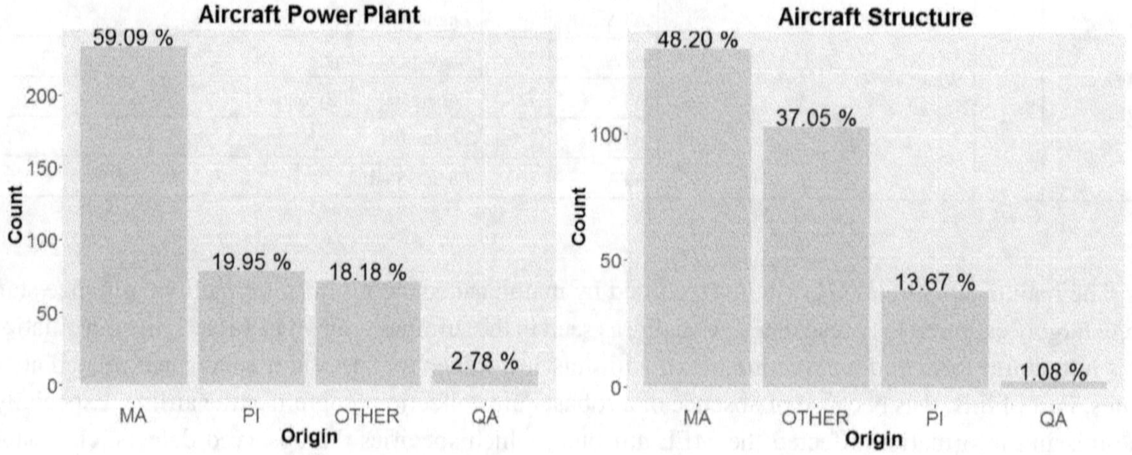

Figure 6 shows the distribution of repair lead time, in number of days taken until a defect was resolved. It was observed that the distributions were skewed to the right, meaning that most defects were repaired in less than 120 days, however, some defects took a much longer period to repair. Such a delay in repair is undesirable especially for MEL classes A, and B which need to be resolved immediately, and within 72 hours respectively.

Data Preparation

This step of the methodology was important to have quality and valid results, where pre-processing data was rather important. Several techniques are proposed for objectively pre-processing data, including, imputation, filtering, normalization, filtering among other approaches. Usually, the first step of pre-processing data involves augmenting inaccurate records, with assistance from maintenance experts. This is especially important for augmenting maintenance descriptions, with expert input. As an example, for the aircraft structure, the description below:

Prop#2 vibration level at cruise 850 0.104 ips due to balance solution more than **225 propeller blade retorque performed**.

Figure 6. Overview of distribution of repair lead-times

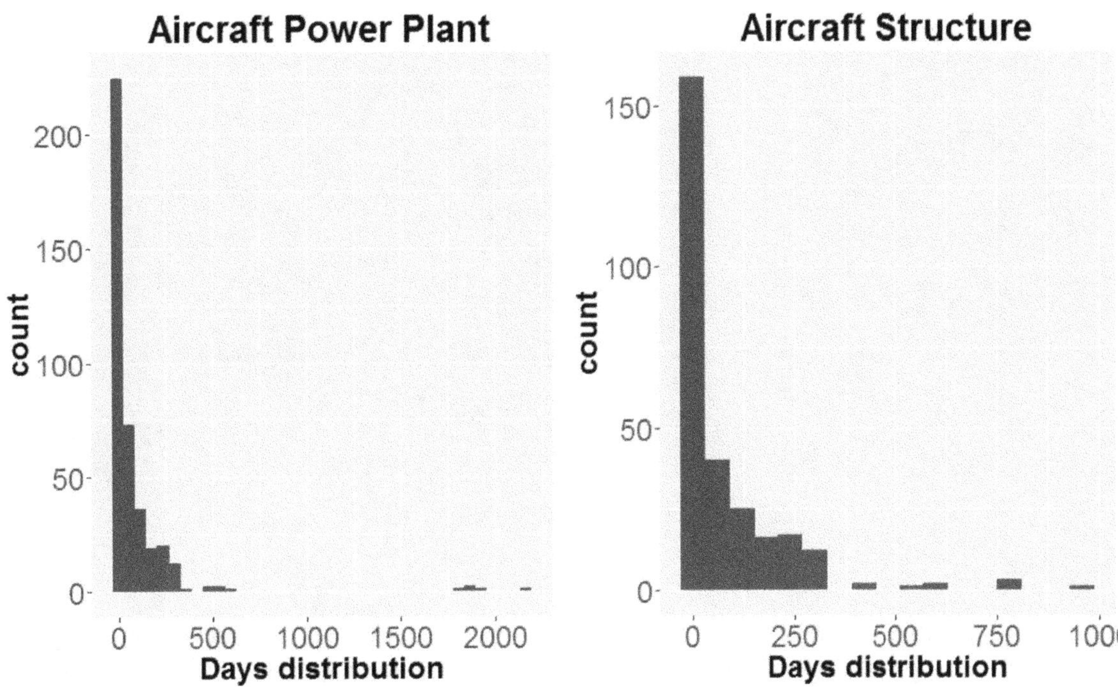

There was need to augment and separate the 'description' attribute from the 'action attribute', indicated by the bold statement. This separating is an important step to ensure that dissimilar attributes are recorded separately prior to carrying out the data mining step.

The second step involves filtering and removing generic terms and attributes, which are not considered useful for maintenance decision support. This step is important as a prior step to parsing the data, before applying association rule mining methods. For the same example (in italics above), generic terms which include, 'found' or 'performed', that contribute less to maintenance decisions, were filtered out because of low value for decision support, compared to words describing defects, such as 'broken' or 'misaligned'. Stop words which also do not contain meaningful failure information were also filtered out in this step. This includes words such as 'and' or 'the'. Imputation and filtering approaches were applied for omitting non-useful terms from textural attributes of the maintenance data.

The third step involves normalizing filtered text, and this step transforms the text, from upper to lower case. The lower-case format usually eases analysis of textural data in the data mining software used in this study. For this step, the 'transform case operator' function was used. For instance, transforming:

"PERFORMED RE-TORQUE OF THE BLADES" → "performed re-torque of the blades"

The fourth step involves text mining and complements the exploratory analysis step. From the analysis, it was observed that defects from several sub-systems of the aircraft were reported frequently, compared to other sub-systems. This includes systems such as the propeller of the aircraft power plant, and the doors, which forms part of the aircraft structure. To extract the most frequently reported defects, recorded in textural form, in the 'description' and 'action' attribute, text mining was performed.

Figure 7. Screenshot of pre-processing maintenance data

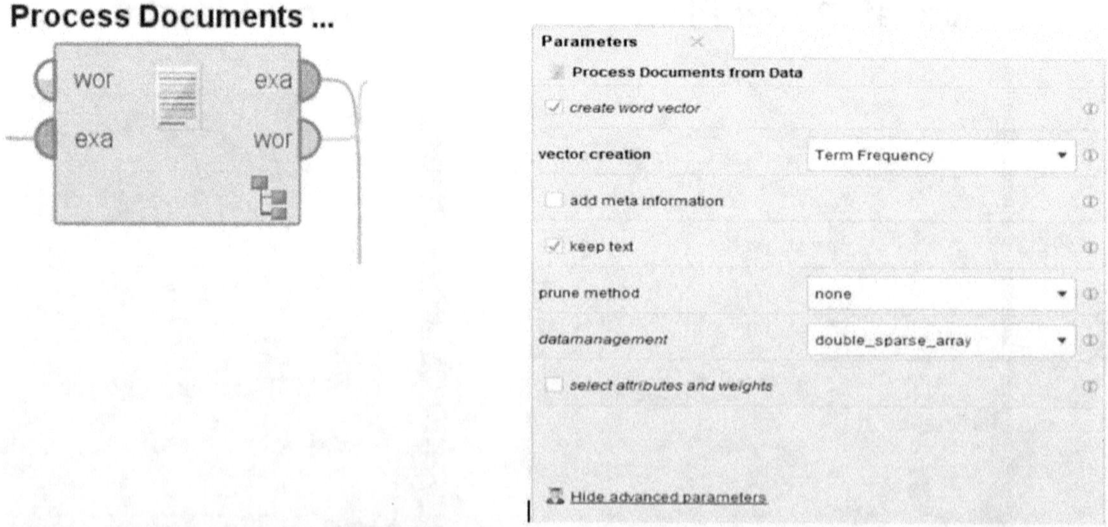

Figure 7 shows a screen shot of the operator, 'process documents from data' which helps pre-process textural information prior to text mining. There are several parameters which an analyst needs to specify for this operator, for example, 'vector creation' from which, the 'term frequency' is specified. 'Pruning' is an additional parameter, which specifies the length of words that should be ignored when forming a 'worklist'. Because our objective was to extract as many words as possible, when forming the 'wordlist', this parameter was not used for the text mining process.

For text mining, once the data was pre-processed, the first step involved 'tokenization' where sentences were separated into individual word elements (or tokens). For example, for the description below, initially one sentence, was split into two sentences:

FUT of prop balancing performed" ←→ "vibration level ok 0.038ips

Figure 8. Subprocesses under the text mining operator

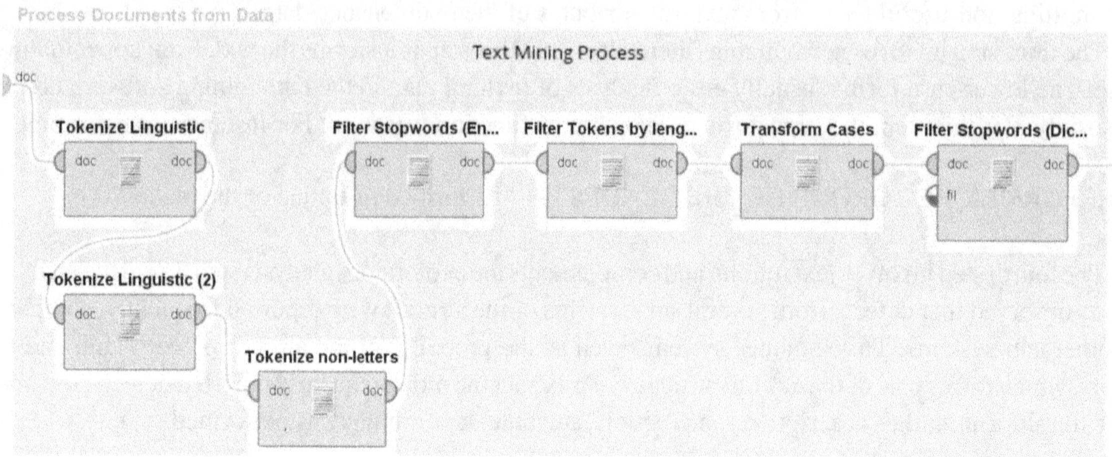

Association Rule Mining

In order to discover and quantify useful relationships between frequently extracted terms/words which describe the nature of defects, and maintenance actions performed on the defect, association rule mining was carried out. Specifically, this rule mining approach was used to find potential associations between maintenance actions on the equipment, recorded in the maintenance information. For this phase of the analysis, tokens extracted from text mining were used. Figure 9 shows the process steps as implemented in RapidMiner, up until tokenization, preparing the wordlist in columns, and inputting this list to the FP-Growth association rule mining algorithm.

Figure 9. Modelling steps from pre-processing to generating the wordlist

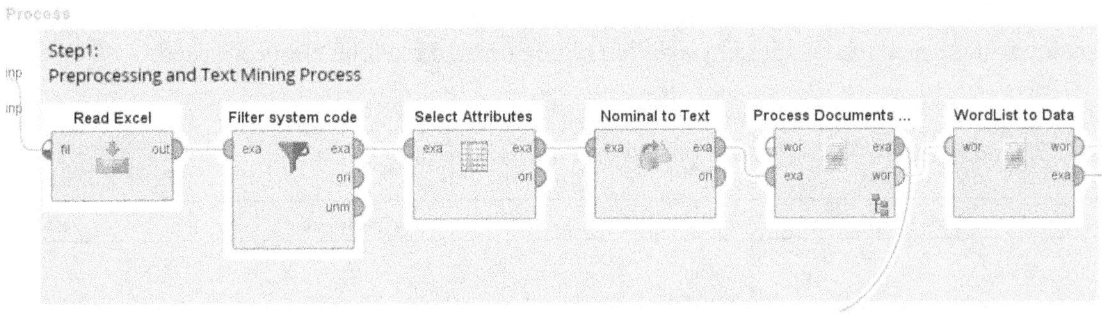

Care was exercised when setting the threshold values for confidence and support parameters, since values close to zero yielded too many association rules, while values approximating 1 yield too few rules, which were least useful for maintenance decision support. Figure 10 illustrates the steps of the association rule mining approach, which implements the FP-Growth association rule mining algorithm.

Figure 10. Implementation of the FP-Growth Association Rule Mining algorithm.

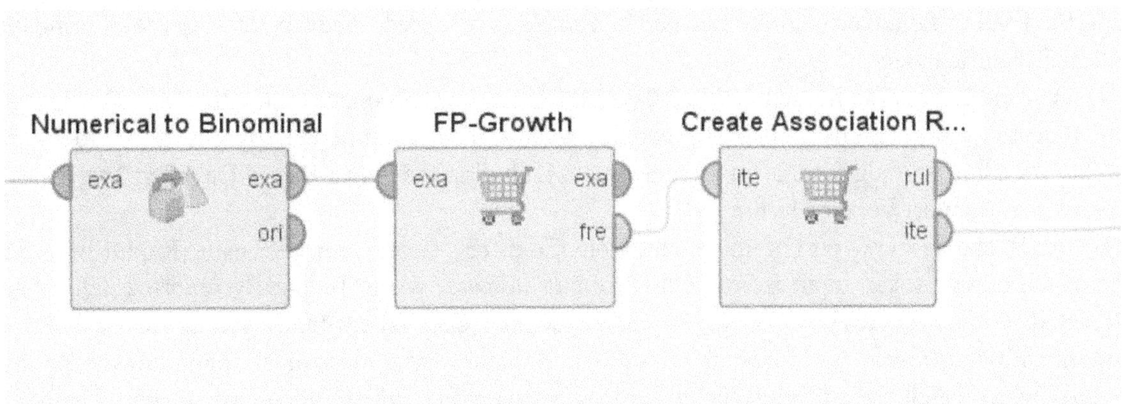

Classification Approach

The next step after association rule mining, was classification, which aims at predicting the level of criticality of the extracted frequently occurring tokens/words, which are associated with defect descriptions and maintenance interventions undertaken for the aircraft equipment. For this, a text classification approach is implemented, where we generate the level of criticality of the MEL defect based on the repair lead-time. As discussed previously, the criticality of a specific defect, is linked to the urgency needed to repair the defect.

The classifications illustrated in Table 3 below, differs from the standard practice discussed in Section 3.1, where class A defects require to be repaired within 24 hours. The case airline considers the criticality assessment classes as realistic, because of maintenance constraints, such as a delayed diagnosis of the defect, or few available maintenance technicians to address multiple failure. However, an important consideration here, is the effect of the criticality on the flight worthiness of the aircraft, for which, standard maintenance schedules required by aviation regulators and manufacturers are used.

Table 3. Criticality classification based on repair lead-time in days

Days	Criticality
0-2 3-5 6>	Urgent Medium Not Serious

Based on the "Guidance Material and Best Practices for Inventory Management document" published by IATA, the aviation industry aims to optimize their aircraft availability by having a well-defined inventory management system. This means a reliable supply of items is needed for maintenance activities, where spare part replenishment options suggested including stocking the parts in-house or leasing parts from other airlines. Hence, aircraft spare parts are almost always available when needed. Because of this fact, it is assumed that a delay in lead-time repair for classes illustrated in Table 3, is not because of lack of spare parts, but largely because of sub-optimal planning of maintenance activities. Hence, for this study, the level of failure criticality indicated in Table 3 is assumed to be a good measure of criticality level of a specific defect.

Thus, if the defect is addressed within 2 days, the defective component is classified as critical, while if the defect is addressed over a period exceeding 6 days, then it can be classified as non-critical. For the classification, four algorithms are implemented, including the Naive Bayes, Decision Tree, Neural Network and Support Vector Machine.

Figure 10 above, shows part of implementation of the classification analysis using RapidMiner. The process starts with token input from the text mining process, where frequently reported defects and associated actions were derived. For the classification analysis, two attributes were used, 'description' and 'criticality' attributes. Several operators were used to implement the analysis, including the *Set Role* operator, which labels the attributes that we are trying to predict, which in our case, is the criticality of the reported defects. A *split operator* was used to divide the token data into two sets, training and test datasets. A specification was made to split and use 70% of data as a training set and 30% as a testing set. A *sample (bootstrap)* operator is also to count how frequently a word occurs, but since there are numerous words, the bootstrap operator ensures that infrequently occurring words have a chance of being selected more than once. This avoids a situation where important, but infrequent words recorded in the datasets are filtered out, and not included for predicting decision support. For the classification, the Naive Bayes algorithm was used, implementing the steps shown in Figure 11.

Figure 11. Classification implementation in RapidMiner

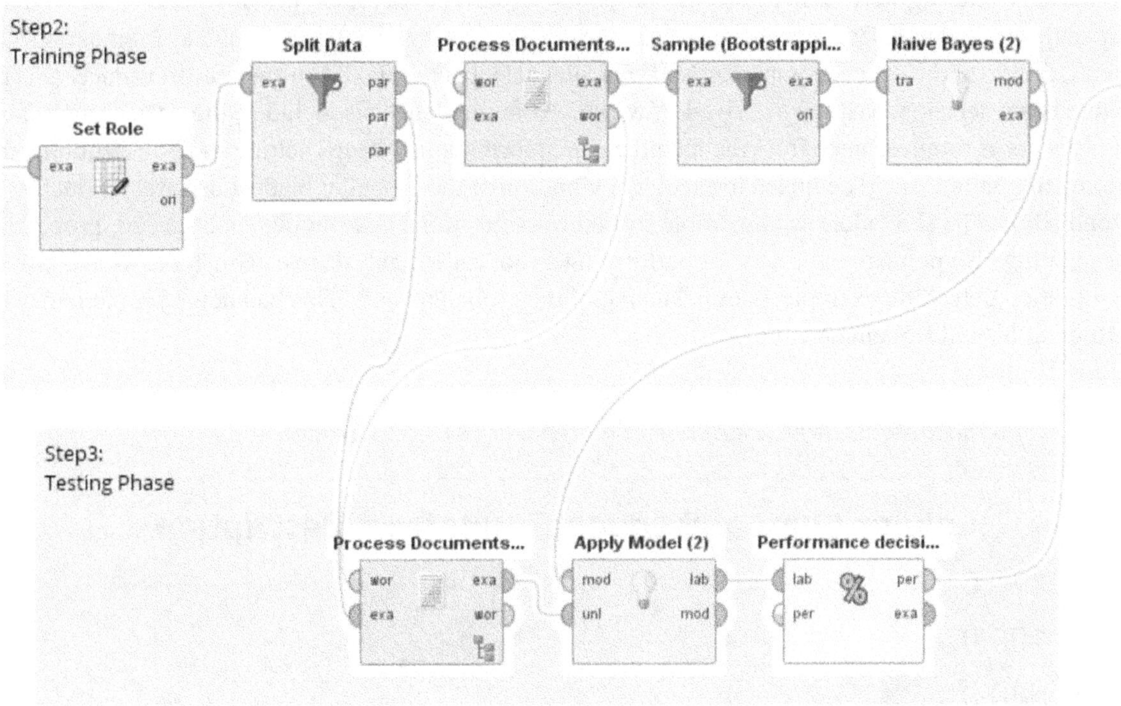

For the testing phase, the *Process Documents from Data* operator receives the 'wordlist' of frequently extracted words of defective components and their respective repair lead-times, based on the labeled data (training dataset). The *apply model* operator implements or executes the training step, while the *performance* operator measures the predictive accuracy of the selected criticality attribute. Results from the methodology are discussed in the next section.

RESULTS AND DISCUSSION

In this section, the results of the data mining methodology discussed in the previous section is presented. The objective is to derive insights from patterns embedded in the datasets of the aircraft equipment, which would support maintenance related decisions. From the results, important insights focus on among other decisions, spare part replenishment, prioritizing critical MEL defect classification, and furthermore, assist maintenance planning activities, for instance, optimal allocation of resources depending on the types and criticality of anticipated maintenance activities.

Insights from Text Mining for the Aircraft Power Plant System

The power plant system is perhaps one of the most critical sub-system of the aircraft equipment. Based on the defects reported for this sub-system, it is important to prioritize critical defects appearing in this system. Figure 12 shows a prioritized list of defects, based on frequently occurring terms extracted using text mining. The extracted terms mainly focused on 13 maintenance-related attributes of the ATA 61 (propeller system) discussed previously in Section 3.1. A total of 461 frequently occurring words of maintenance interest, were extracted for the propeller, with some specific types of defects reported more frequently than other defects. From the figure, one can see that critical defects include, engine propeller defects, failure of the de-icer component/sub-unit, and defect of the flexible ultrasonic transducer (FUT). Maintenance decisions that were derived from the text mining analysis includes, the need to audit how the propeller is torqued, since from the maintenance information, it keeps getting loose. Potentially, the re-torqueing failure could be linked to excessive vibration of the propeller blade, quantified as inch-per-second (IPS). The IPS failure is observable from the results of the text mining analysis (ips_propeller). The auditing also pointed to the need of performing a root cause analysis to establish the focal cause of the vibration defect, for example, poor balancing of the propeller shaft. This balancing problem may be an implicit insight from the analysis.

Figure 12. Frequent words from description of a propeller (ATA 61) system

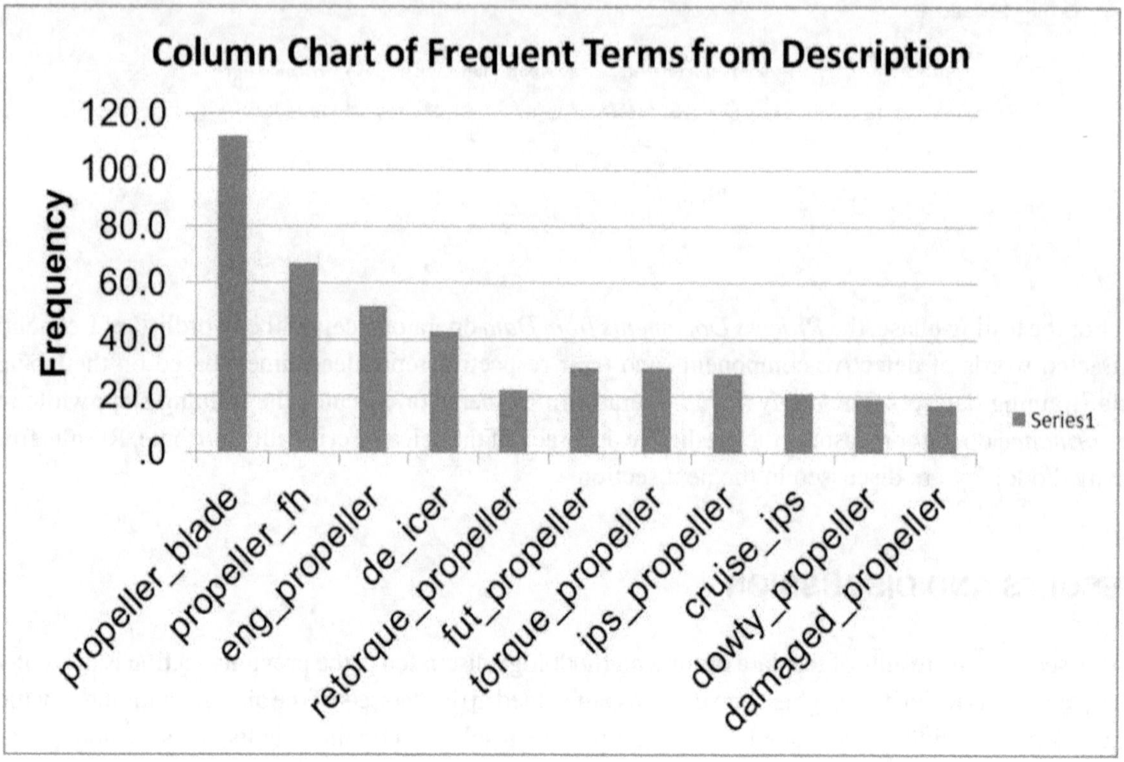

Text mining was also carried out to extract maintenance interventions performed on the propeller sub-system. These interventions were recorded under the 'action' attribute, which essentially reports maintenance activities carried out while repairing defects. Figure 13 shows a prioritized list of maintenance interventions, including balancing of the propeller blades, torqueing and re-torqueing interventions. Similarly, insights from text mining was observed for additional sub-systems of the aircraft, which includes, the aircraft structure (ATA 53).

Figure 13. Frequent words from actions performed on propeller (ATA 61) system

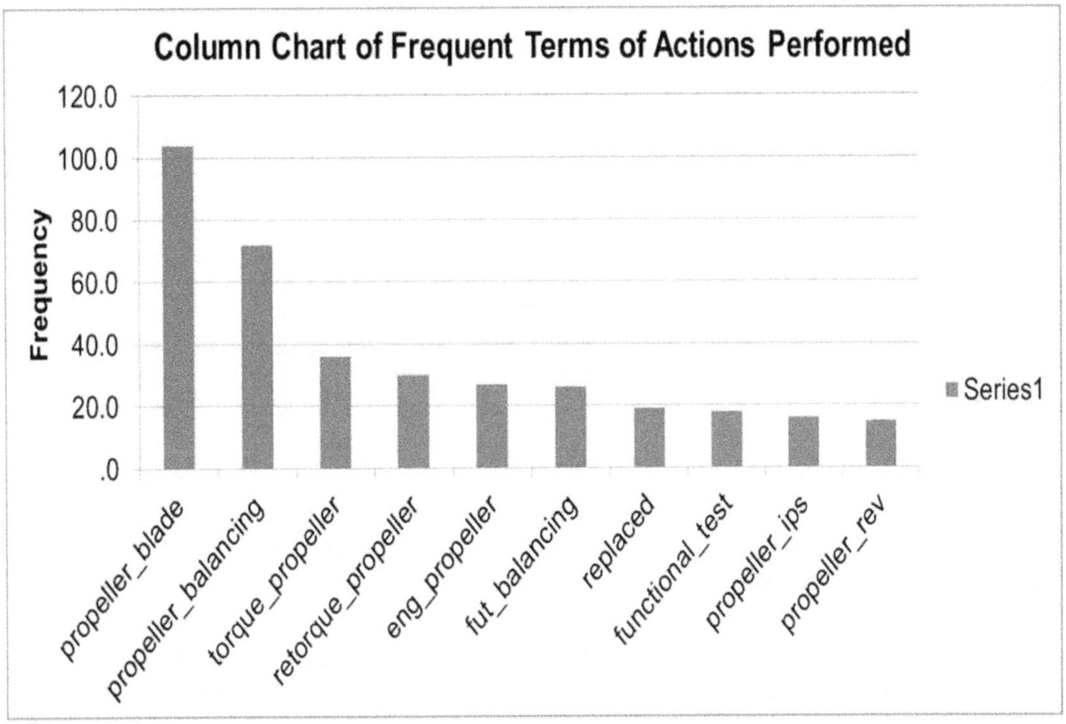

Insights from Association Rule Mining

To complement the results of text mining, association rule mining was also carried out to understand and quantify the relationship between different defective components, based on the 'description' and 'action' attributes. Table 4 shows the association rules derived from frequently reported defective components, of the aircraft propeller (ATA 61) system. Of the derived association rules, the most reported defective component, within the propeller system, are the de-icer, and failure modes including cruise failure. These propeller defects have a high confidence of 1. The propeller blade damage has a highest confidence (0.932). Similarly, the attribute, 'damaged blade' is associated with a high confidence. For the rule 'cruise–>propeller', a further analysis of the defect pointed to a link between the cruise altitude, and a high propeller vibration. The vibration measurement here is closely corelated to the rule; 'ips–>propeller'. As mentioned, IPS represents inches-per-second, which is a unit of measure of the propeller amplitude. Table 5 summarizes association rules for the aircraft propeller.

Based on the above rules, it could be concluded that the vibration of the propeller during the cruise phase necessitated further investigation. Proposals by the maintenance experts include a need for a more detailed investigation of how balancing of the propeller blades was performed during maintenance. This meant a review of the assembly and disassembly procedures of the propeller blade, and where possible, training maintenance crew on optimal maintenance. This insight demonstrated the value of association rules derived from maintenance-related attributes of the aircraft power plant system.

Table 4. Association rules from propeller (ATA61) system description

Rules	Confidence
[blade] ->[propeller]	0.932
[de icer] ->[blade]	1.000
[damaged blade] ->[propeller]	0.938
[cruise] ->[propeller]	1.000
[ips] ->[propeller]	1.000
[torque] ->[propeller]	1.000
[fh] ->[propeller]	0.912
[inspection] ->[propeller]	1.000
[dowty] ->[blade]	0.867

Apart from identifying association rules for defective components, association rules were also extracted for maintenance interventions performed for sub-systems of the aircraft equipment. For instance, the rule shows that pre-dominantly, 'balancing', 'visual inspection', 'torqueing' and 're-torqueing' as the main interventions performed for the propeller blades. Since these actions were recommended to be performed after a certain number of flight hours, the insights from the rule were useful for maintenance planning and allocating maintenance resources more optimally. For example, by knowing the number of reported defects based on history of the aircraft systems, a more optimal planning of maintenance is possible. Moreover, a high recurrence of defects requiring, for example, propeller balancing mean a need to review existing maintenance auditing procedures, especially, verifying that the correct maintenance processes was followed while repairing the propeller blades, and other sub-systems of the aircraft.

Insights from Classification Approaches of the Maintenance Dataset

Table 5 show results derived for different classification algorithms based on the predictive accuracy of the classifier. The accuracy here considers the ratio of correct predictions to overall prediction. For each system of the aircraft equipment, the Support Vector Machine (SVM) algorithm yielded the highest predictive accuracy, based on the labeled data of criticality classification, determined based on the urgency of repair.

Table 6 shows a confusion matrix of the propeller (ATA 61) system, as classified by the SVM algorithm. A classifier performance of 61.76%, corelates to a good prediction of critical defects in much larger datasets containing information of similar defects, and repair lead-times (indicator of urgency to

Table 5. Classification results

	Accuracy			
	ATA 61	**ATA 79**	**ATA 52**	**ATA 53**
Decision Tree	57.56%	69.70%	69.70%	78.43%
Support Vector Machine	61.76%	72.73%	73.13%	79.41%
Naive Bayes	46.64%	66.67%	47.10%	74.51%
Neural Network	44.73%	68.18%	46.42%	74.51%

repair a defect). The SVM classifier was quite effective in classifying urgent repairs, followed a correct prediction of 'non-serious', and 'medium-urgency' defects. Furthermore, to identify which specific defects required urgent maintenance intervention, a decision tree model was used, which allows users to visualize classified defects, which are depicted in a tree-like structure. Example of urgent defects includes, the propeller blade defect and the FUT failure earlier described.

Table 6. Confusion matrix under support vector machine for ATA 61

	true Urgent	**true Not serious**	**true Medium**	**class precision**
pred. Urgent	92.00	33.00	20.00	63.45%
pred. Not serious	9.00	42.00	13.00	65.62%
pred. Medium	9.00	7.00	13.00	44.83%
class recall	83.64%	51.22%	28.26%	

Insights from the classifier were important for optimizing maintenance interventions for the ATA 61 and potentially, other sub-systems of the aircraft equipment. Moreover, as discussed, a predictive approach for maintenance planning is potentially more efficient and effective using classifier models such as used in this study. For instance, the prediction and identification of urgent repairs, allows the maintenance crew to better allocate maintenance resources, audit maintenance processes, and improve the maintenance regime (e.g. through training).

CONCLUSION

This chapter presents a data mining approach for supporting maintenance decisions of aircraft systems. The approach integrates varying data and text mining methods, and embeds a classification algorithm for predicting critical failure events of aircraft equipment based on their repair lead-times. Several systems were evaluated using the proposed approach, including the aircraft propeller (ATA 61) system, for which, important defects such as propeller blade and FUT failures were extracted from maintenance datasets recorded by technicians, pilots and quality assurance staff of a case airline. For the aircraft structure sub-system (ATA 52), several defects were also identified and prioritized for maintenance, such as damaged door seals. Moreover, the proposed methodology was useful for supporting decisions as to allocating maintenance resources, based on the prevalent maintenance activities carried out to repair defects. The approach furthermore supports automating maintenance reporting and e-maintenance activities, since it provides practitioners with a structured approach through which they can use textural and other forms of maintenance-related data to improve repair processes. Furthermore, the proposed approach provides a robust method for improving maintenance auditing processes and identifying violations of audit procedures.

REFERENCES

Ackert, S. P. (2010). *Basics of aircraft maintenance programs for financies: evaluations and insights of commercial aircraft programs*. Accessed from: www.aircraftmonitor.com

Aggarwal, C. C., & Zhai, C. (2012). *Mining text data*. Springer Science & Business Media. doi:10.1007/978-1-4614-3223-4

Berry, M. J. A., & Linoff, G. (1997). *Data Mining Techniques. For Marketing, Sales and Customer Support*. Wiley Computer Publishing.

Borchiellinia, R., Cardub, M., Colella, F., Labagnara, D., Martinetti, A., Patrucco, M., ... Vittorio, V. (2013). A prevention through design approach for the environmental S&H conditions and the ventilation system at an Italian underground quarry. *AIDIC Conference Series, 11*, 51-60.

Chemweno, P., Morag, I., Sheikhalishahi, M., Pintelon, L., Muchiri, P., & Wakiru, J. (2016). Development of a novel methodology for root cause analysis and selection of maintenance strategy for a thermal power plant: A data exploration approach. *Engineering Failure Analysis, 66*, 19–34. doi:10.1016/j.engfailanal.2016.04.001

Chemweno, P., Pintelon, L., Jongers, L., & Muchiri, P. (2016). *i-RCAM: Intelligent expert system for root cause analysis in maintenance decision making*. Paper presented at the 2016 IEEE International Conference on Prognostics and Health Management (ICPHM). 10.1109/ICPHM.2016.7542830

Chemweno, P., Pintelon, L., Muchiri, P. N., & Van Horenbeek, A. (2018). Risk assessment methodologies in maintenance decision making: A review of dependability modelling approaches. *Reliability Engineering & System Safety, 173*, 64–77. doi:10.1016/j.ress.2018.01.011

Chemweno, P., Pintelon, L., Van Horenbeek, A., & Muchiri, P. (2015). Development of a risk assessment selection methodology for asset maintenance decision making: An analytic network process (ANP) approach. *International Journal of Production Economics, 170*, 663–676. doi:10.1016/j.ijpe.2015.03.017

Chen, G., He, W., Leung, L. C., Lan, T., & Han, Y. (2017). Assigning licenced technicians to maintenance tasks at aircraft maintenance base: A bi-objective approach and a Chinese airline application. *International Journal of Production Research, 55*(19), 5550–5563. doi:10.1080/00207543.2017.1296204

Djatna, T., & Alitu, I. (2015). An application of association rule mining in total productive maintenance strategy: An analysis and modelling in wooden door manufacturing industry. *Procedia Manufacturing, 4*, 336–343. doi:10.1016/j.promfg.2015.11.049

Fontaine, O., Martinetti, A., & Mateou, S. (2016). Remote Pilot Aircraft System (RPAS): Just culture, human factors and learnt lessons. *Chemical Engineering Transactions, 53*, 205–210.

García, J. R. R., Martinetti, A., Becker, J. M. J., Singh, S., & van Dongen, L. A. (2019). Towards an Industry 4.0-Based Maintenance Approach in the Manufacturing Processes. In *Handbook of Research on Industrial Advancement in Scientific Knowledge* (pp. 135–159). IGI Global. doi:10.4018/978-1-5225-7152-0.ch008

Gera, M., & Goel, S. (2015). Data mining-techniques, methods and algorithms: A review on tools and their validity. *International Journal of Computers and Applications*, *113*(18).

Gerdes, M. (2019). *Health Monitoring for Aircraft Systems using Decision Trees and Genetic Evolution* (Doctoral dissertation). Luleå, Sweden: Luleå University of Technology, Graphic Production.

King, R., & Curran, K. (2019). Predictive Maintenance for Vibration-Related failures in the Semi-Conductor Industry. *Journal of Computer Engineering & Information Technology*, *8*(1), 1.

Kroes, M. J., Watkins, W. A., Delp, F., & Sterkenburg, R. (2013). *Aircraft Maintenance and Repair* (7th ed.). McGraw-Hill Education.

Liu, T., Abd-Elrahman, A., Morton, J., & Wilhelm, V. L. (2018). Comparing fully convolutional networks, random forest, support vector machine, and patch-based deep convolutional neural networks for object-based wetland mapping using images from small unmanned aircraft system. *GIScience & Remote Sensing*, *55*(2), 243–264. doi:10.1080/15481603.2018.1426091

Mack, D. L., Biswas, G., Koutsoukos, X. D., & Mylaraswamy, D. (2016). Learning bayesian network structures to augment aircraft diagnostic reference models. *IEEE Transactions on Automation Science and Engineering*, *14*(1), 358–369. doi:10.1109/TASE.2016.2542186

Maquee, A., Shojaie, A. A., & Mosaddar, D. (2012). Clustering and association rules in analyzing the efficiency of maintenance system of an urban bus network. *International Journal of System Assurance Engineering and Management*, *3*(3), 175–183. doi:10.100713198-012-0121-x

Martinetti, A., Chatzimichailidou, M. M., Maida, L., & van Dongen, L. (2019). Safety I–II, resilience and antifragility engineering: A debate explained through an accident occurring on a mobile elevating work platform. *International Journal of Occupational Safety and Ergonomics*, *25*(1), 66–75. doi:10.10 80/10803548.2018.1444724 PMID:29473459

Moharana, U., Sarmah, S., & Rathore, P. (2019). Application of data mining for spare parts information in maintenance schedule: A case study. *Journal of Manufacturing Technology Management*, *11*(7), 21. doi:10.1108/JMTM-09-2018-0303

Nimmagadda, S. L., Reiners, T., & Wood, L. (2018). On big data-guided upstream business research and its knowledge management. *Journal of Business Research*, *89*, 143–158. doi:10.1016/j.jbusres.2018.04.029

Olshannikova, E., Ometov, A., Koucheryavy, Y., & Olsson, T. (2015). Visualizing Big Data with augmented and virtual reality: Challenges and research agenda. *Journal of Big Data*, *2*(1), 22. doi:10.118640537-015-0031-2

Rachburee, N., Arunrerk, J., & Punlumjeak, W. (2018). Failure Part Mining Using an Association Rules Mining by FP-Growth and Apriori Algorithms: Case of ATM Maintenance in Thailand. In *IT Convergence and Security 2017* (pp. 19–26). Springer. doi:10.1007/978-981-10-6451-7_3

Reuss, P., Stram, R., Althoff, K.-D., Henkel, W., & Henning, F. (2018). Knowledge engineering for decision support on diagnosis and maintenance in the aircraft domain. In *Synergies Between Knowledge Engineering and Software Engineering* (pp. 173–196). Springer. doi:10.1007/978-3-319-64161-4_9

Rodrigues, R. S., Balestrassi, P. P., Paiva, A. P., Garcia-Diaz, A., & Pontes, F. (2012). Aircraft interior failure pattern recognition utilizing text mining and neural networks. *Journal of Intelligent Information Systems, 38*(3), 741–766. doi:10.100710844-011-0176-1

Sun, T.-L., & Salgado, G. A. M. (2017). *Sustainable Data Collection Framework: Real-Time, Online Data Visualization.* Paper presented at the International Conference on Sustainable Design and Manufacturing. 10.1007/978-3-319-57078-5_6

Thai, V. P., Zhong, W., Pham, T., Alam, S., & Duong, V. 2019, April. Detection, Tracking and Classification of Aircraft and Drones in Digital Towers Using Machine Learning on Motion Patterns. In *2019 Integrated Communications, Navigation and Surveillance Conference (ICNS)* (pp. 1-8). IEEE.

Turnbull, A., Carroll, J., Koukoura, S., & McDonald, A. (2019). Prediction of wind turbine generator bearing failure through analysis of high-frequency vibration data and the application of support vector machine algorithms. *Journal of Engineering (Stevenage, England), 2019*(18), 4965–4969. doi:10.1049/joe.2018.9281

U.S. Department of Transportation. (2003). *Guidance on Aviation Rules and Statutes*. Author.

Wang, Z., & Li, Y. (2017). *Equipment Maintenance Support Decision Method Research Based on Big Data.* Paper presented at the International Conference on Geo-Spatial Knowledge and Intelligence.

Young, T., Fehskens, M., Pujara, P., Burger, M., & Edwards, G. (2010). *Utilizing data mining to influence maintenance actions.* Paper presented at the 2010 IEEE AUTOTESTCON. 10.1109/AUTEST.2010.5613610

Zhang, F., Du, B., Zhang, L., & Xu, M. (2016). Weakly supervised learning based on coupled convolutional neural networks for aircraft detection. *IEEE Transactions on Geoscience and Remote Sensing, 54*(9), 5553–5563. doi:10.1109/TGRS.2016.2569141

Zhang, W., Yang, D., & Wang, H. (2019). Data-Driven Methods for Predictive Maintenance of Industrial Equipment: A Survey. *IEEE Systems Journal, 13*(3), 2213–2227. doi:10.1109/JSYST.2019.2905565

Chapter 12
MRO 4.0:
Mapping Challenges Through the ILS Approach

Henrique Costa Marques
https://orcid.org/0000-0001-9573-8742
Instituto Tecnológico de Aeronáutica, Brazil

Fernando Teixeira Mendes Abrahão
https://orcid.org/0000-0001-8713-0868
Instituto Tecnológico de Aeronáutica, Brazil

Guilherme Conceição Rocha
Instituto Tecnológico de Aeronáutica, Brazil

ABSTRACT

The demand for increased efficiency of production processes, while maintaining quality and safety in the operating environment, are permanent requirements of industrial revolutions. In the information age, data acquisition and its use to affect business strategies are being carried out by sensing production lines, tracking processes, and the product itself throughout its life cycle. Industry 4.0 requires an organizational transformation in terms of culture, process, and technology for the organization to be able to harness the potential of information. This chapter seeks to establish the difficulties and challenges of organizational transformation from the analysis of an aviation MRO company in light of integrated logistics support (ILS). The discussion will lead to the points to be taken into account from all elements of the ILS that will produce a roadmap for decision-makers to follow.

INTRODUCTION

The industrial revolution provides advancement on techniques and processes to create and support products with varying levels of complexity throughout their life cycle, changing the way products are being designed, built, and supported. The fourth revolution is occurring due to the potential use of Informa-

DOI: 10.4018/978-1-7998-3904-0.ch012

tion and Communication Technology (ICT) to create new services and capabilities to help incremental autonomous decision making across all industrial processes. Technologies such as the Internet of Things, wireless sensor networks and virtual environments are bringing new skills so that machines can make inferences in higher-level processes, identifying alternative courses of action and adapting to the new conditions presented at runtime.

In terms of support and design for supportability, such revolution has also brought the ability to track product's parameters during its operation, providing feedback so that future versions of the product can be better designed and improved. Complex systems that allow the observation of product functionalities degradation are being monitored, and maintenance services are being suggested by prognostic and health management (PHM) systems. Taking, for example, commercial aviation systems, the existence of scheduled maintenance plans may be conservative and strict due to the regulatory authorities of the air transportation activity (Gdalevitch, 2000). New aircraft are being monitored by an increasing number of intelligent sensors that allow identification of the remaining useful life of various components and subsystems. For this new information to be helpful and to adapt maintenance plans promptly, it is necessary to ensure that the new alternative predictive maintenance process is as robust as it is the purely preventive one (scheduled).

Aerospace systems maintenance companies, also called Maintenance, Repair and Overhaul (MRO), are even passing through these transformations. MRO processes also evolved from pure mechanisation (MRO 1.0) to the development of assembly lines on maintenance activities (MRO 2.0), to the usage of industrial automation and data processing techniques (MRO 3.0), and the usage of decentralised processes based on autonomous agents and cyber-physical systems, developed over an Internet of Things (IoT) infrastructure (MRO 4.0).

Achieving the highest level of autonomy in maintenance systems implies the ability to prescribe maintenance activities following the item accumulated degradation and possible changes during operations. These operations may or may not follow the exact expected profile, and that is when sensors integrated into a maintenance system can indicate that the system may or may not fly another mission. Aviation future requires this level of awareness and autonomy, and it is called as "smart operations".

Aerospace MRO companies are part of this environment and should be able to increase aircraft availability with safety and profitability (Rodrigues & Lavorato, 2016). Besides the fact that MROs are not in charge of client́ s aircraft fleet management, actually, they are in charge of the aircraft maintenance as asked by the client. During the execution of scheduled maintenance activities, some unexpected maintenance may occur (e.g., cracks or corrosion detection during inspection), generating new unscheduled maintenance tasks, also called non-routine maintenance (Kinnison, 2004). The unscheduled maintenance creates new job tasks, which have to be integrated into the predefined scheduled maintenance packages. Also changes schedules already defined, brings demands in terms of selecting the right staff to handle the unexpected maintenance tasks and keeps tracking of the whole activity, parts and supplies determination and delivery, bringing variability in terms of downtime. The ability to have autonomous decision support systems to prescribe maintenance activities would then represent one essential facet of MRO 4.0 companies.

According to Haroun and Duffuaa, one of the key considerations is to outline a structure that will support maintenance (Haroun & Duffuaa, 2009). Parida and Kumar also addressed that logistic support is one of the vital requirements for maintenance planning (Parida & Kumar, 2009). The MRO 4.0 demands a series of adaptations in infrastructure, processes, and cultural transformation, so that the available capabilities of Industry 4.0 could be implemented in this environment.

The MRO, as an industrial plant, can be considered a complex system. If we add the complexities of the life cycles of the systems that are to be maintained, repaired and overhauled, this would add additional difficulties in terms of planning and scheduling the maintenance activities (Rzevski et al., 2016). In terms of 4.0 paradigm, the MRO requires an integrated and iterative approach to address not only its life cycle peculiarities but also the customer´s assets life cycle characteristics. So, it is crucial to consider the problem scope not only from the life cycle phases of the 4.0 initiative but also the consequences from the life cycle phases of the customer´s assets.

The Integrated Logistic Support (ILS) has been used as a framework to design the logistical support of complex systems, and it is an integrated and iterative approach derived from Systems Engineering for developing the supportability of complex systems throughout their entire life cycles (Abrahão et al., 2019; ASD/AIA, 2018). The ILS approach guides the system engineering process in order to minimise costs and decrease support footprint (demand for support), making the system easier to support and to operate since conception until retirement (Defense Acquisition University, 2011). This work has the hypothesis that if using the same approach to develop the MRO 4.0 framework would apply for the same benefits.

Although initially developed for defence purposes, the ILS and the IPS (Integrated Product Support, a late acronym used by the Defense Acquisition University) it is also widely used in commercial product support or customer service organisations. (Federal Aviation Administration, 2019).

In the case of this work, the MRO 4.0 is the complex system to be supported. This "new" MRO 4.0 system has all life cycle phases ahead and the complexity found on any other complex aerospace system. The MRO 4.0 demands and protocols need proper support in order to achieve customer goals, and the ILS approach considers a comprehensive list of concepts, methodologies, tasks, requirements, and issues (also referred as ILS Elements or IPS Elements) in a time and integrated manner to cover every possible constraint affecting the complex aerospace system logistics support.

This approach is new, although it may have been used unconsciously for the development of other guides since the use and structuring of ILS concepts represent the best practices in terms of coherence and consistency for the development of support solutions (Cegelec, 2020; IT Tech, 2020). The goal is to integrate the MRO 4.0 concepts and characteristics within the logistics support scenario. By doing this, useful conclusions from the MRO 4.0 study regarding constraints and opportunities can become requirements to map the development with all restrictions and opportunities outlined into an Integrated Logistics Support Plan (ILSP) as the path for each adaptation or new implementation to be made.

The design and continuous improvement tasks of complex aerospace systems logistics support, described on ILSP, shall also consider the life cycle perspective and Systems Engineering approach. Some logistics considerations are specific, better suited, or convenient for each phase of the development and support of a system life cycle, even though other factors are sound for the development of the entire system life cycle span. Besides that, according to the Systems Engineering approach, support requirements need to be defined, analysed, and solutions need to be designed, developed, implemented, integrated, tested, and validated to build the system logistics support solution.

Because of this analytical task of evaluating MRO 4.0 process according to ILS Elements, it will be possible to highlight in a synoptic table the following aspects:

- New maintenance demands and requirements;
- The ILS elements that are most affected in this scenario;
- New business opportunities and how they need to be supported;
- Technology gaps and trends; and

- Challenges associated with these new technologies.

In this context, this chapter aims to identify challenges, business opportunities, and technology gaps regarding MRO 4.0 processes, by analysing and scrutinizing each MRO process against ILS Elements framework. The chapter discusses the MRO perspective according to Industry 4.0 novelties related to technologies and methodologies applied on the maintenance and support of complex aerospace systems.

The chapter is divided into six sections. Background section introduces the main concepts, the domain context of MRO 4.0 and establishes the MRO processes and the industry 4.0 capabilities. The section Integrated Logistics Support Framework describes the ILS framework tied to the MRO perspective and presents considerations about supportability constraints. Similarly, the field research section provides an overview of the maintenance challenges faced by flight operations companies and their perspective from the technological opportunities. The discussion section brings reflections about technology gaps, trends, and challenges associated with the available technologies. The future research directions section promotes trends to be developed based on the chapter considerations. Lastly, the conclusion section presents the main contributions of this work.

BACKGROUND

Maintenance may be defined as the set of actions and processes to ensure that a given system could perform its intended functions as established by its reliability and safety characteristics and requirements (Kinnison, 2004). It includes management, supervision, and also administrative actions. Repair and overhaul are levels of maintenance and have the same objective. The term Maintenance, Repair and Overhaul (MRO) may be understood as a complicated process to be developed by the system's operator or by a third party, according to the capabilities and objectives inherent in the operator's business, to keep its assets in working conditions within a required availability level.

Maintenance Approaches

There are two primary forms of maintenance: scheduled (preventive) and unscheduled (corrective) (Kinnison, 2004). While the unscheduled is performed after the occurrence of breakdowns and failures, the scheduled is performed to minimise the likelihood of them occurring. Predictive and prescriptive maintenance are both scheduled/preventive in the way they are conducted in advance of faults and failures.

Predictive maintenance (PdM) includes the capability to identify the future system utilisation to predict its Remaining Useful Life (RUL) given the operational demand. When the RUL forecast of a given component reveals the imminence to achieve a certain threshold, the system has to be maintained by a scheduled maintenance task in the next opportunity.

Prescriptive maintenance is the process to be capable of identifying the maintenance performance at all levels and prescribe a course of action given the evolving operation, being adaptive (Marques & Giacotto, 2019). This means to have situational awareness not only about the operational demand but also in all aspects that involve the maintenance environment and processes. That is why industry 4.0 technologies and processes are suitable for the development of prescriptive maintenance. This capability enables decision support systems to suggest a set of on-demand scheduled maintenance tasks in rapidly changing environments because of the ability to acquire, process and decide over a large amount of data in realtime.

The e-Maintenance research area (Holmberg et al., 2010) was developed from the perception of the need for the evolution of MRO management systems to become even smarter in identifying the potential of ICT to build the process of digitisation of information needed to conduct smart operations within the 4.0 context (Bierer et al., 2016).

MRO Processes

Aviation MRO has activities divided into categories of on-aircraft and off-aircraft maintenance (Kinnison, 2004). The on-aircraft maintenance may be performed inline (at an airport ramp or terminal) or at the hangar and are conducted at the aircraft or in some aircraft's subsystem. When a maintenance team has to repair an aircraft's subsystem out of it (off-aircraft maintenance), such activity will take place in a dedicated workshop. That way, the MRO organisation is structured to support all the involved engineering activities, as seen in Figure 1.

Figure 1. MRO's engineering structure
Source: Adapted from (Kinnison, 2004)

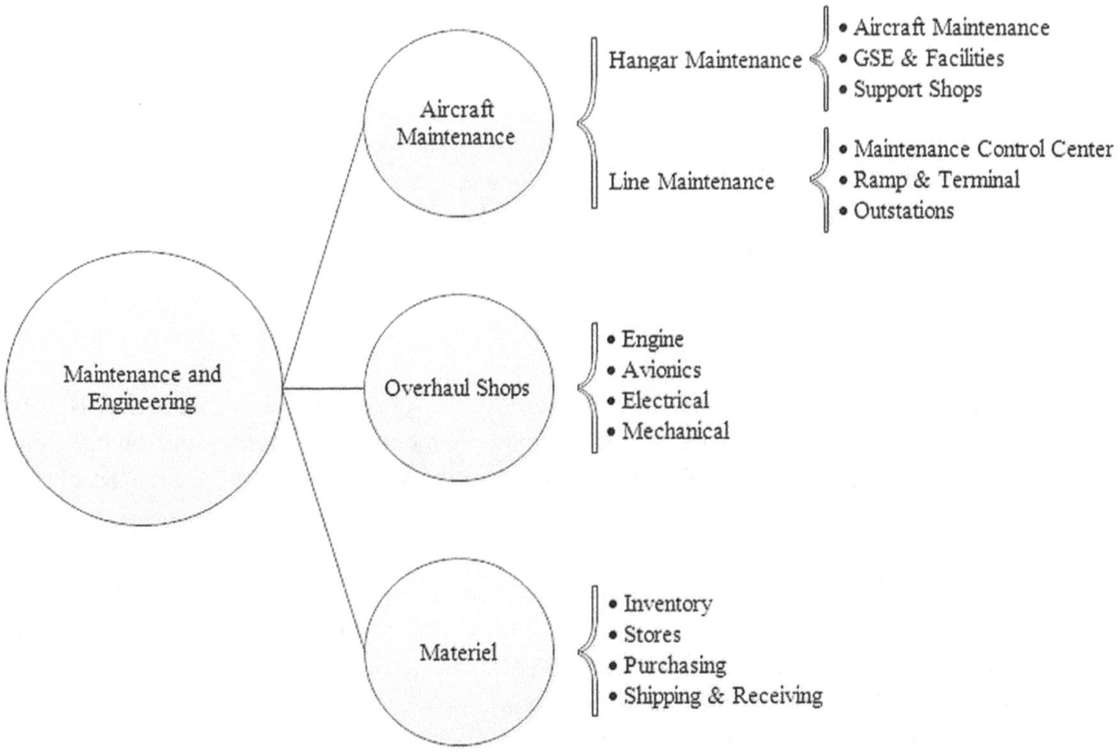

An MRO organisation does not only consist of maintenance engineering activities but also activities related to operations, customer support, ground support equipment (GSE), finance, among others. All the MRO activities may be mapped into processes, as stated in Table 1.

Table 1. MRO's main processes

Engineering, Maintenance Programs and Regulatory	Flight Operations Maintenance	Airframe and Engine Maintenance	Shop, Tool and GSE Maintenance	Supply Chain Maintenance	Customer Relationship Management	Reservations, Airport Operations, Finance and Human Capital
Maintenance Program Management	Maintenance Control	Heavy Maintenance Visit Planning	Shop Maintenance Control	Demand Forecasting	Procurement and Invoice Receipt	Pricing, Marketing and Sales
Configuration and Records Management	Line Station Planning	Heavy Maintenance Production Planning and Control	Shop Production Planning	Optimised Fulfillment and Supply Chain Maintenance	Sales Order and Invoice Generation	Airport Operations
Reliability Analysis	Line Maintenance Execution	Heavy Maintenance Execution	Shop Maintenance Execution	Materiel Receipt	Maintenance Cost Tracking	Flight Operations
Quality Assurance and Engineering Support	Diagnostics and Prognostics	--	Tool Control and GSE	Warehouse Management	Financial Reporting	Finance and Accounting
Technical Document Management	Aircraft Health Management	--	--	--	Financial Analysis	Human Capital Management

Source: (Denis, 2008)

In the present work, each process was translated in terms of tasks and resources that allow its execution and tracking during the activities. From this work, it is then possible to obtain the supportability requirements for each activity. It also identifies the opportunities for using Industry 4.0, enabling technologies in the environment and processes of the MRO organisation.

Industry 4.0 Enabling Technologies for MRO

The MRO core processes are well established and should not vary with new technologies. However, the way to conduct them can be impacted by the tendency to obtain more information about the entire activity, allowing not only greater control, but also an updated perception in a shorter time, accelerating the process management cycle.

Many researchers are conducting experiments with ICT to increase efficiency in maintenance management and execution (Sahay, 2012), as stated before, when considering e-Maintenance as the research line that has introduced this concern since 2000. The 4.0 Industry is based on the knowledge about resources, assets, and processes of the complex system of interest acquired by sensors and human reports, through artificial intelligence advisory for maintenance planning.

That is why technologies like Digital Twin (Esposito et al., 2019; Kraft & Kuntzagk, 2017; Liu et al., 2018; Ríos et al., 2016), Internet of Things (IoT), Big Data Analytics, Augmented Reality and Additive Manufacturing, among others, play an essential role (García et al., 2019; Indra, 2020). These technologies are based on data availability, and intelligence gathered from massive databases. Such technologies are key enablers when considering an MRO organisation as a production line in the sense that a product comes to be disassembled, repaired and reassembled by humans using tools and GSEs, following specific and certified tasks to accomplish the service safely.

The maintenance production line may be sensed in many different ways, using cameras, Radio Frequency Identification (RFID) and IoT devices which generate data from and for mechanics, materiel, tools, and GSE during the service (Edwards et al., 2017; Karakuş et al., 2019; Ucler & Gok, 2015). After that, Big Data Analytics (machine learning and other artificial intelligence method) are used to comprehend and acquire insight about the tools and GSE utilisation and localisation, mechanics efficiency, hangar space distribution, human labour utilisation (Permatasari et al., 2019) for each maintained aircraft and many other applications (Denis, 2008; Dinis et al., 2019; Galar et al., 2012; Lee et al., 2018).

During the maintenance service maintainers have to access manuals, task cards, and the aircraft data from many different databases. To improve that, data fusion and Augmented Reality are being used to give the maintainer the right information at the right time, receiving support from remote experts if needed (Ceruti et al., 2019; Cipresso et al., 2018; Masoni et al., 2017; Utzig et al., 2019).

When spare parts are needed, and the material and blueprint for Additive Manufacturing exist the capability to print the required part reduces the necessity to store and transport, cutting costs and giving the MRO the possibility to plan the part printing at the right time. The cost reduction in parts acquisition and storage in warehouses are the main drivers for this technology research in aviation maintenance (Ceruti et al., 2019; Froes & Boyer, 2019; Zhang & Liang, 2019).

Therefore, in order to implement an MRO 4.0, the organisation needs to go through a conceptual and design phase, in which supportability must be taken into account. For this, the ILS framework is advantageous and consistent, allowing to anticipate problems, demands for process changes, personnel needs, certification demands, among other aspects, in order to allow the correct incorporation of technology.

INTEGRATED LOGISTICS SUPPORT FRAMEWORK

To provide proper reflection and guidance, the ILS framework applied to the MRO 4.0 environment, needs to take into consideration not only the definition of each ILS element but some life cycle phase adjustments as well. In this specific approach (ILS supporting MRO operations) the system been supported is the MRO and its life cycle. Therefore, for each ILS element, a distinction is advisable if the MRO or a subsystem (for example a new shop of the MRO) is on its conceptual/initial production rate, steady-state production rate or its dephasing production rate, since it may lead to slightly different ILS approaches.

The ILS elements

The Integrated Logistics Support elements are twelve: Maintenance; Supply Support; Workforce and Personnel; Training; Computer Resources; Technical Data; Support Equipment; Facilities and Infrastructure; Package, Handling, Storage and Transportation (PHS&T); Design Interface; Sustaining Engineering; and Product Support Manager. The definitions come from two leading publications: The *Integrated Product Support Element Guidebook*, from the Defense Acquisition University (DAU) (Defense Acquisition University, 2011); and the ASD/AIA SX000i *International guide for the use of the S-Series Integrated Logistics Support (ILS) specifications* (ASD/AIA, 2018), from the AeroSpace and Defence Industries Association of Europe / ASD-STAN and Aerospace Industries Association. These are sound references for both commercial and defence ILS applications.

The context of Figure 1 and the content of Table 1 are relevant to understand the ILS approach to MRO and explore the main improvements expected for the MRO 4.0 concept of operations. Perhaps, the most significant new input regarding support aspects from the MRO 3.0 to MRO 4.0 is technology, enabling new ways to develop (based on the updated concept of operations) new, improved, augmented and connected concept of support.

The next discussion focuses on each process described in Table 1 and its relationship within ILS elements under 4.0 schematics and expectations. By default, all ILS elements support all operations; therefore, none can be ignored. However, the objective here is to drive attention for ILS elements that are the most demanded, not just for the support of the process, but also for the expected new demands from the MRO 4.0. In the same vein, Product Support Management is highly used for integration and supervision of the new challenges involved, and Computer Resources as a tool to make it possible.

ILS elements Contributions to MRO Processes

Table 2. Enginering maintenance program and regulatory processes

Engineering, Maintenance Program and Regulatory	Maintenance Program Management	Configuration and Records Management	Reliability analysis	QA and Engineering Support	Technical Document Management

- **Maintenance Program Management:** The ILS Elements frequently involved supporting this process are Maintenance, Computer Resources, and Product Support Management due to the nature of the activity, management, and control. High levels of integration are expected to consider all related issues, and there is an excellent chance to take advantage of MRO 4.0 potential for that.
- **Configuration and Records Management:** The ILS Elements more involved in helping this process are Computer resources, Supply Support, and Technical Data. The data collection would require a coordinated effort to manage data useful for the entire set of analyses and diagnostics needed for MRO 4.0. Not just proper records and control of part numbers, National Stock Numbers, but also data related to the condition of the items (available hours or cycles, remaining useful life – RUL, among others). The intricacies of the MRO tasks in different aircraft, different tail numbers, different remaining useful lives, and various suppliers, highly suggest a strong presence of Computer Resources to help control and track of everything mentioned here. Enough attention is necessary to provide up to date and certified publications and lists of applied parts and service bulletins.
- **Reliability Analysis:** The ILS Elements mainly involved in this process are Sustaining Engineering, Product Support Management, and Computer Resources due to the nature of management and control of the Reliability, Availability, Maintainability, and Safety (RAMS) indicators and parameters. The collection of maintenance data, analysis of RAMS, root cause analysis, development of design changes are typical MRO activities that blend with fleet management activities. With an MRO 4.0 approach, the operator is expected to be able to synchronise this entire set of information with fleet management to maximise fleet availability and scheduled reliability, and minimising maintenance costs by a more efficient usage of maintenance support resources.

- **QA and Engineering Support:** The ILS Elements more involved in this process are Maintenance, Sustaining Engineering, Training, and Product Support Management, mainly because of the nature of the assessments required to provide a proper Maintenance Quality Assurance Program. System performance criteria, internal audits, design reviews, supplier management, safety performance, improvement management, knowledge management, and operator's satisfaction are demanding activities from the investigations and control provided by these ILS Elements' activities.
- **Technical Document Management:** The ILS Element frequently involved in this process is Technical Data. This process is the essence of Technical Data, and the MRO 4.0 would use technology to provide increased security, data compatibility, faster and precise data management, and data interoperability for system integration for the MRO support management and program.

Table 3. Flight Operations Maintenance processes

Flight Operations Maintenance	Maintenance Control	Line Station Planning	Line Maintenance Execution	Diagnostics and Prognostics	Aircraft Health Management

- **Maintenance Control:** The ILS Elements more involved in this process are Maintenance and Product Support Management. Since the 3.0 paradigm, major aircraft components started to collect and provide operation and wear data. Maintenance control in this process accounts for every action taken to ensure the best decisions for inspections, repairs, and overhauls at the MRO. To support this process, a significant amount of information technology and data analysis apply to allow precise control of the MRO activities.
- **Line Station Planning and Line Maintenance Execution**: The ILS Elements mainly involved in these processes are Maintenance and Product Support Management. Maintenance planning and execution activities at the MRO derive from the aircraft's Maintenance Plan. Such actions require scheduling tools integrated with data collection and management to be attainable and effective. Two levels of maintenance are involved with different requirements (station and line), which also demands high levels of integration provided by MRO Support Management.
- **Diagnostics and Prognostics:** The ILS Elements more involved in this process are Maintenance and Sustaining Engineering. Depending on the maintenance philosophy, if pure preventive or predictive, the capability to establish diagnostic and prognostic would be the central core for parts acquisition and separation. The ability to plan the maintenance tasks based on data acquisition from the system utilisation and the understanding of components and subcomponents degradation would give a better forecast capability.
- **Aircraft Health Management:** The ILS Elements more involved in this process are Maintenance, Computer Resources, and Sustaining Engineering. Aircraft Health Management (AHM) requires a set of procedures to be accomplished through data acquired from aircraft sensors, ground support systems that are capable of processing those data to perform diagnosis and to produce prognostics based on aircraft usage from the operational requirements. The technology to accredited this process is still in development and only recently have operators been questioned about their ability to meet standards on how they can replace some maintenance functions with Integrated Aircraft Health Management (IAHM) technologies (Federal Aviation Administration, 2019).

Table 4. Airframe and Engine Maintenance processes

Airframe and Engine Maintenance	Heavy Maintenance Visit Planning	Heavy Maintenance Production Planning and Control	Heavy Maintenance Execution

- **Heavy Maintenance Visit Planning:** The ILS Elements most involved in this process are Maintenance, Facilities and Infrastructure, Support Equipment, Workforce, Supply Support, and Product Support Management. Even though some new generation aircraft rely less on substantial overhauls as a preventive strategy, legacy systems, and individual cases involving significant damages or modifications will continue to make use of heavy maintenance activities. Heavy maintenance requires a very sophisticated process to address the maintenance plan integrated with the entire MRO infrastructure, workforce, support equipment, supplies, and support management to be successful. MRO 4.0 would use sensors, networks, software/hardware, and information technology to help with the accomplishment of Heavy Maintenance tasks in time at competitive and known costs.
- **Heavy Maintenance Production Planning, Control, and Execution:** The ILS Elements most involved in this process are Maintenance and Product Support Management. Maintenance planning and control in this process encompass every decision made to provide the best decisions for planning and control the MRO most demanding activities. To support this process, a significant amount of information technology and data analysis apply to allow precise control of production tasks. Scheduling tools based on new sets of information provided by not just Supply Chain Management (SCM) deliverables, but from the new and improved set of data and decisions from a hangar, in the 4.0 paradigm, expects to improve and augment production planning and controlling significantly.

Table 5. Shop, Tools and GSE Maintenance processes

Shop, Tool and GSE Maintenance	Shop Maintenance Control	Shop Production Planning	Shop Maintenance Execution	Tool Control and GSE

- **Shop Maintenance Control/Execution and Production Planning:** These processes are similar to what happens with Airframe and Engine Maintenance. The ILS Elements most involved in these processes are Maintenance, Support Equipment, and Product Support Management. Maintenance planning and control in this process address every decision made to provide the best decisions for planning and control of the MRO activities. To support these processes, a significant amount of information technology and data analysis applications is necessary, allowing precise control of production tasks and general programming activities. Scheduling tools based on new sets of information provided under the 4.0 premises should lead to innovation and improvements on customer's satisfaction and reduced risks. The difficulty is based on the non-routine maintenance that happens during scheduled maintenance and a repair that makes use of the shops were not previously identified. To make use of previous knowledge could help in terms of the schedule of possible maintenance tasks in advance of the identification of the corrective maintenance to be done.

- **Tool Control and GSE:** The ILS Elements more involved in this process are Support Equipment, Computer Resources, and Product Support Management. GSE management is vital to assure that the MRO activity is resilient and consistent supporting activities and the business case. GSE items may be simple, but it may be a quite complex piece of equipment. In some cases, supporting a specific GSE may require a dedicated support system. Tool control is part of the GSE management and is a natural candidate for improvements expected from the 4.0 paradigm. Sensors will be part of the tools, facilities, and infrastructure, letting to accuracy in terms of configuration management, location, calibration management, and usage.

Table 6. Supply chain maintenance processes

Supply Chain Maintenance	Demand Forecasting	Optimised Fulfillment and Supply Chain Maintenance	Materiel Receipt	Warehouse Management

- **Demand Forecasting:** The ILS elements more involved in this process are Supply Support, Sustaining Engineering, Computer Resources, and Product Support Management. Supplies would respond to a big part of the costs involved in the MRO business. Not just consumable items but repairable items are needed in stock to attend the demands from the MRO activities. There is no crystal ball, but some predictive analytics are handy to provide information for the MRO management on what, when, and how much to procure to satisfy the maintenance lines. Some integration with Sustaining Engineering is more prone to occur with new technologies allowing traceability of RAMS performances throughout systems been supported by the MRO. This shall lead maintainers to keep RAMS tendencies under control and act just in time to update levels, procurement, and management of supplies.
- **Optimised Fulfillment and Supply Chain Maintenance:** The ILS Elements more involved in this process are Supply Support, Computer Resources, and Product Support Management. MRO 3.0 to MRO 4.0 is under a new paradigm also in terms of the supply chain approach. The new speed of the information chain permits to go further than "reduction in costs" and "improved service levels", but to add value to the chain, increasing flexibility. For the MRO, the new computer resources framework (4.0) will enable the system to operate this way via the introduction of faster processors and addressing all layers of the maintenance decision process, free of integration gaps (human-made system integrations). By the way, it is quite common that systems integration happens manually at MROs not adherent to the 4.0 paradigm, which makes the process slow and susceptible to human errors.
- **Materiel Receipt:** The ILS Elements more involved in this process are Supply Support and PHS&T. Processing material receipt information is core to the MRO. The activity is in charge of the Supply Support Element and demands a considerable amount of attention, traceability, and control.

- **Warehouse Management:** The ILS Elements more involved in this process are PHS&T and Product Support Management. This process is a core activity for the PHS&T element, which directly affects key performance indicators such as RAMS. Storage, packing, and handling issues influence the overall MRO performance causing errors, delays, and unexpected costs. The 4.0 paradigm could help integrate the system's demands for PHS&T with its features and controls to better contribute to the MRO goals.

Table 7. Customer relationship management processes

Customer Relationship Management	Procurement and Invoice Receipt	Sales Order and Invoice Generation	Maintenance Cost Tracking	Financial Reporting	Financial Analysis

- **Procurement and Invoice Receipt and Sales Order and Invoice Generation:** The ILS elements most involved in these processes are Computer Resources and Product Support Management, as they are deliverables from sales and procurement management. The entire set of billing / sales-related documents is the result of business management actions.
- **Maintenance Cost Tracking:** The ILS Elements more involved in this process are Computer Resources, Maintenance, and Product Support Management. Cost tracking is crucial to keep the business case balanced. Therefore the total approach to the maintenance and management of all MRO assets provided by the ILS strategy permits top management to stay aware of the cost performances of all MRO systems and subsystems.
- **Financial Reporting:** The ILS Elements more involved in this process are Computer Resources and Product Support Management. Computer Resources element as a function of the intensive use of data to control and manage cost and finance information within the 4.0 paradigm. The Product Support Management element is in charge of the overall finance control involved in all MRO support activities.
- **Financial Analysis:** The ILS Elements most demanded in this process are Computer Resources and Product Support Management. Since such analysis is a function of the intensive use of data to control and manage financial information within the 4.0 paradigm, Product Support Management would require intensive use of Computer Resources to help on the development of financial deliverables for the business case as a whole.

Table 8. Reservation, airport operations, finance and human capital processes

Reservations, Airport Operations, Finance and Human Capital	Pricing, Marketing and Sales	Airport Operations	Flight Operations	Finance and Accounting	Human Capital Management

- **Pricing, Marketing, and Sales:** The ILS Elements more involved in this process are Computer Resources and Product Support Management. The Product Support Management element is in charge of the overall costs' control involved in all MRO support activities. Therefore, it delivers crucial information for this process with all the information related to the MRO support capacity to attend the market's demands and price information.

- **Airport Operations:** The ILS Elements more involved in this process are Facilities and Infrastructure, Support Equipment, and Product Support Management. Most MROs are located at airports since the airport infrastructure is necessary for the arrival and departure of aeroplanes visiting the MRO. In this case, the entire MRO ramp activity is MRO support operations that take place at the airport resources or, at least, at the airport interface. The MRO Support Management should integrate these activities with the airport management in the best possible way to provide the MRO with the capability to operate its infrastructure at the ramp.
- **Flight Operations:** Maintenance management interfaces with operational management unequivocally for the sake of performance indicators. One depends on the other. So, all maintenance activities need alignment with the flight operations to be developed. The ILS Elements more involved in this process are Maintenance and Product Support Management. In this specific case, the support effort aligns with the traditional air fleet operation support, since the expected support is devoted to plan and execute preflight, inter flight and overnight inspections, as they, for a receipt, tests, and delivery flight activities.
- **Finance and Accounting:** The ILS Elements more involved in this process are Computer Resources and Product Support Management. Computer Resources element as a function of the intensive use of data to control and manage cost information within the 4.0 paradigm. The Product Support Management element is in charge of the overall costs' control involved in all MRO support activities.
- **Human Capital Management:** The ILS Elements more involved in this process are the Workforce and Personnel, Computer Resources, Training and Product Support Management. Human Capital Management is the essence of the Workforce and Personnel Element. Due to the evolution from 3.0 to 4.0, such activities would need support from Computer Resources in order to be effective. The entire set of tasks, subtasks, decisions, forecasts, control, and monitoring of the workforce suggests a high number of computer resources involved. Not just the typical workforce allocation and schedules are included, but a lot of human factor interactions are expected. These analyses and decisions/outputs will require a high level of integration with Product Support Management and Training to accomplish the MRO's objectives.

After the understanding of the involved ILS elements in each MRO's process, the key elements to understand the gaps and challenges are now easier to follow.

FIELD RESEARCH

Field research was conducted with 3 (three) MRO from international airlines with complex operations. For each MRO, several interviews have been conducted with maintenance personnel involved both on management and daily operational activities, aiming to investigate main demands, challenges and technologies for MRO 4.0 deployment, considering the combination of ILS element with MRO process. A RAMS specialist conducted the interviews. The results are compiled and summarised on the following synoptic Table 9.

Table 9. MRO's demands, challenges, and technologies through ILS elements

MRO Processes / ILS elements	1- Maintenance	2- Supply Support	3- Workforce and Personnel	4- Training
A- Engineering, Maintenance Programs and Regulatory	Demands: Compliance with certification requirements related to maintenance procedures Technologies: Virtual Reality Simulation Challenges: Simulation representativeness	Demands: Compliance with certification requirements related to Supply Support Technologies: Shelf-life test bench Challenges: Test bench representativeness	Demands: Compliance with certification requirements related to Personnel Technologies: Digital signature control and blockchain Challenges: Digital signature authenticity guarantee	Demands: Compliance with certification requirements related to Training Technologies: Virtual Reality Simulation Challenges: Training effectiveness verification and accreditation
B- Flight Operations Maintenance	Demands: Maintenance records management Technologies: Computerized Maintenance Management System (CMMS) and blockchain Challenges: Maintenance records authenticity guarantee	Demands: Optimize Supply Support Management Technologies: Stock and availability simulation Challenges: Simulation representativeness	Demands: Personnel authorisation management Technologies: Human resources management systems Challenges: Alerts messages when authorisation is about to expire	Demands: Training effort management Technologies: Human resources management systems Challenges: Training effectiveness verification
C- Airframe and Engine Maintenance	Demands: Maintenance records management Technologies: CMMS, Enterprise Resource Planning (ERP), and blockchain Challenges: Maintenance records authenticity guarantee	Demands: Optimize Supply Support Management Technologies: Stock and availability simulation Challenges: Simulation representativeness	Demands: Personnel authorisation management Technologies: Human resources management systems Challenges: Alerts messages when authorisation is about to expire	Demands: Training effort management Technologies: Human resources management systems Challenges: Training effectiveness verification
D- Shop, Tool and GSE Maintenance	Demands: Tools provision for the maintenance team Technologies: RFID Challenges: Tools management at maintenance facilities	Demands: N/A Technologies: N/A Challenges: N/A	Demands: Provide individual control of tools usage Technologies: Individual toolboxes and RFID Challenges: Guarantee that there is no lost tool	Demands: Assure that mechanics know how to use tools correctly Technologies: VR (Virtual Reality) and AR (Augmented Reality) simulations Challenges: Simulation representativeness
E- Supply Chain Maintenance	Demands: Suppliers maintenance records management Technologies: CMMS and machine learning algorithms (Prescriptive Maintenance) Challenges: Rogue unit identification and assistance on No-Fault Found issues investigation	Demands: Stock optimisation Technologies: Stock and availability simulation and PBL (Performance-Based Logistics) contract Challenges: Simulation representativeness and PBL contract fairness	Demands: Personnel authorisation management Technologies: Human resources management systems Challenges: Alerts messages when authorisation is about to expire	Demands: Training effort management Technologies: Human resources management systems Challenges: Training effectiveness verification
F- Customer Relationship Management	Demands: Maintenance cost tracking Technologies: CMMS, ERP, and blockchain Challenges: Maintenance records authenticity guarantee	Demands: Procurement and Invoice Receipt Technologies: E-procurement systems Challenges: Procurement costs minimisation	Demands: Technicians evaluation Technologies: Human resources management systems Challenges: Monitor technicians' performance	Demands: Training scheduling Technologies: Human resources management systems Challenges: Monitor technicians' performance
G- Reservations, Airport Operations, Finance and Human Capital	Demands: Maintenance investment forecast Technologies: Forecasting systems Challenges: Forecasting representativeness	Demands: Supply Support investment forecast Technologies: Forecasting systems Challenges: Forecasting representativeness	Demands: Human Resources capacity planning Technologies: Human resources management systems and Forecasting systems Challenges: Forecasting representativeness	Demands: Training demand forecasting Technologies: Human resources management systems and Forecasting systems Challenges: Forecasting representativeness

MRO Processes / ILS elements	5- Computer Resources	6- Technical Data	7- Support Equipment	8- Facilities and Infrastructure
A- Engineering, Maintenance Programs and Regulatory	Demands: Computer resources design to show compliance with certification requirements Technologies: Modeling and Simulation Tools Challenges: Simulation representativeness	Demands: MRO technical data specification and design to show compliance with certification requirements Technologies: Document editors and Augmented Reality Systems Challenges: Documents and augmented reality integration	Demands: Support Equipment specification and design to show compliance with certification requirements Technologies: Lessons Learned Database Challenges: Keeping the database updated	Demands: Facilities and Infrastructure specification and design to show compliance with certification requirements Technologies: Lessons Learned Database Challenges: Keeping the database updated
B- Flight Operations Maintenance	Demands: Maintenance records management Technologies: CMMS and blockchain Challenges: Maintenance records authenticity guarantee	Demands: Maintenance records guard Technologies: Cloud datacenter and blockchain Challenges: Data security assurance	Demands: Support Equipment availability assurance Technologies: CMMS and ERP Challenges: Monitor support equipment physical location and calibration status	Demands: Facilities and Infrastructure maintenance Technologies: N/A Challenges: N/A
C- Airframe and Engine Maintenance	Demands: Maintenance records management Technologies: CMMS and blockchain Challenges: Maintenance records authenticity guarantee	Demands: Maintenance records guard Technologies: Cloud datacenter and blockchain Challenges: Data security assurance	Demands: Support Equipment availability assurance Technologies: CMMS and ERP Challenges: Monitor support equipment physical location and calibration status	Demands: Facilities and Infrastructure maintenance Technologies: N/A Challenges: N/A
D- Shop, Tool and GSE Maintenance	Demands: Computer resources availability assurance Technologies: CMMS and ERP Challenges: Computer resources performance degradation should be monitored	Demands: Tool and GSE maintenance records guard Technologies: Cloud datacenter and blockchain Challenges: Data security assurance	Demands: Support Equipment availability assurance Technologies: CMMS and ERP Challenges: Monitor support equipment physical location and calibration status	Demands: Tool and GSE adequate storage facilities provisioning Technologies: N/A Challenges: N/A

continued on following page

Table 9. Continued

MRO Processes / ILS elements	1- Maintenance	2- Supply Support	3- Workforce and Personnel	4- Training
E- Supply Chain Maintenance	Demands: Suppliers' computer resources availability assurance Technologies: CMMS and ERP Challenges: Monitor computer resources performance degradation	Demands: Suppliers' maintenance records guard Technologies: Cloud datacenter and blockchain Challenges: Data security assurance	Demands: Suppliers' Support Equipment availability assurance Technologies: CMMS and ERP Challenges: Monitor support equipment physical location and calibration status	Demands: Suppliers' adequate storage facilities provisioning Technologies: N/A Challenges: N/A
F- Customer Relationship Management	Demands: Computer resources specification for Procurement and Invoice Receipt, Sales Order and Invoice Generation, Maintenance Cost Tracking, Financial Reporting and Financial Analysis Technologies: N/A Challenges: N/A	Demands: Procurement and Invoice Receipt records guard Technologies: Cloud datacenter and blockchain Challenges: Data security assurance	Demands: N/A Technologies: N/A Challenges: N/A	Demands: N/A Technologies: N/A Challenges: N/A
G- Reservations, Airport Operations, Finance and Human Capital	Demands: Computer resources investment forecast Technologies: Forecasting systems Challenges: Forecasting representativeness	Demands: Technical Data record and guard systems investment forecast Technologies: Forecasting systems Challenges: Forecasting representativeness	Demands: Support Equipment investment forecast Technologies: Forecasting systems Challenges: Forecasting representativeness	Demands: Facilities and Infrastructure investment forecast Technologies: Forecasting systems Challenges: Forecasting representativeness
MRO Processes / ILS elements	**9- Packaging, Handling, Storage and Transportation**	**10- Design Interface**	**11- Sustaining Engineering**	**12- Product Support Management**
A- Engineering, Maintenance Programs and Regulatory	Demands: Packaging, Handling, Storage and Transportation design to show compliance with certification requirements Technologies: Modeling and Simulation Tools Challenges: Simulation representativeness	Demands: Design Interface specification to show compliance with certification requirements Technologies: Modeling and Simulation Tools Challenges: Simulation representativeness	Demands: Sustaining Engineering specification to show compliance with certification requirements Technologies: Lessons Learned Database Challenges: Keeping the database updated	Demands: Product Support Management specification to show compliance with certification requirements Technologies: Lessons Learned Database Challenges: Keeping the database updated
B- Flight Operations Maintenance	Demands: Monitor Packaging, Handling, Storage and Transportation operations during maintenance Technologies: External cameras Challenges: Assure that cameras capture the technicians' movements correctly, and measure associated elapsed time	Demands: System design for high maintainability Technologies: CAD Tools and Virtual Reality Simulation Challenges: Simulation representativeness	Demands: Continuous monitoring of Sustaining Engineering Demands Technologies: Failure Reporting Analysis and Corrective Action Systems (FRACAS) Challenges: FRACAS database completeness assurance	Demands: Change control management Technologies: Change control management and Configuration Control Board (CCB) tools Challenges: Tradeoff analysis execution
C- Airframe and Engine Maintenance	Demands: Monitor Packaging, Handling, Storage and Transportation operations during maintenance Technologies: Hangar cameras Challenges: Assure that cameras capture the technicians' movements correctly, and measure associated elapsed time	Demands: System design for high maintainability Technologies: CAD Tools and Virtual Reality Simulation Challenges: Simulation representativeness	Demands: Continuous monitoring of Sustaining Engineering Demands Technologies: FRACAS Challenges: FRACAS database completeness assurance	Demands: Change control management Technologies: Change control management and CCB tools Challenges: Tradeoff analysis execution
D- Shop, Tool and GSE Maintenance	Demands: Monitor Packaging, Handling, Storage and Transportation operations related to Tools and GSEs Technologies: Hangar and external cameras Challenges: Assure that cameras capture the technicians' movements correctly, and measure associated elapsed time	Demands: System design for the smooth operation of Tools and GSEs Technologies: CAD Tools and Virtual Reality Simulation Challenges: Simulation representativeness	Demands: Continuous monitoring of Sustaining Engineering Demands related to Tools and GSEs Technologies: FRACAS Challenges: FRACAS database completeness assurance	Demands: Tools and GSEs Change control management Technologies: Change control management and CCB tools Challenges: Tradeoff analysis execution
E- Supply Chain Maintenance	Demands: Monitor Suppliers' Packaging, Handling, Storage and Transportation operations Technologies: Hangar and external cameras Challenges: Assure that cameras capture the technicians' movements correctly, and measure associated elapsed time	Demands: System design for high maintainability at Suppliers' facilities Technologies: CAD Tools and Virtual Reality Simulation Challenges: Simulation representativeness	Demands: Continuous monitoring of Sustaining Engineering demands related to Supplier's operation Technologies: FRACAS Challenges: FRACAS database completeness assurance	Demands: Change control management related to Supplier's operation Technologies: Change control management and CCB tools Challenges: Tradeoff analysis execution
F- Customer Relationship Management	Demands: Packaging, Handling, Storage and Transportation operations procurement Technologies: E-procurement systems Challenges: Procurement costs minimisation	Demands: Customer feedback gathering during the design phase Technologies: Virtual Reality Simulation Challenges: Simulation representativeness	Demands: Customer dashboards visibilities related to Sustaining Engineering demands status Technologies: FRACAS Challenges: FRACAS database completeness assurance	Demands: Customer dashboards visibilities related to Change Controls Management Technologies: Change control management and CCB tools Challenges: Tradeoff analysis execution
G- Reservations, Airport Operations, Finance and Human Capital	Demands: Packaging, Handling, Storage and Transportation investment forecast Technologies: Forecasting systems Challenges: Forecasting representativeness	Demands: Engineering effort for Design Influence investment forecast Technologies: Forecasting systems Challenges: Forecasting representativeness	Demands: Sustaining Engineering investment forecast Technologies: Forecasting systems Challenges: Forecasting representativeness	Demands: Product Support Management investment forecast Technologies: Forecasting systems Challenges: Forecasting representativeness

DISCUSSION

The most significant difference proposed by the approach suggested in this chapter lies in the fact that the support structures previously developed for MROs 1, 2 and 3 were never systemic/integrated and did not even need to be as systemic as the structures for MRO 4.0. In many cases, simple workshops ended up evolving into larger structures and with more tools without, however, contemplating all the integration necessary for the correct management of the supply chain and operations. There was no systematic application of a methodology that integrated not only all elements of the Integrated Logistic Support, but also integrated the operations of the MRO with the elements of the ILS of the aircraft and operators that make use of the MROs. The 4.0 paradigm does this mainly through information technology, but with each improvement resulting from the information supported by a compatible and integrated logistics structure.

Referring to Table 9, ILS elements are displaced in columns numbered from 1 to 12 and the MRO processes in rows A through G. Each combination of ILS element with the MRO process composes a cell. For instance, when we refer to the table's cell 5-A, we will be mentioning the Computer Resources element in the Engineering, Maintenance Program and Regulatory process.

By analysing deeply Table 9, one can notice that there are several challenges to be overcome and that the key to making the transition to MRO 4.0 is the combination of correct usage and integration between processes and technologies.

For instance, Virtual Reality Simulation is a kind of technology that can help MRO to overcome several demands associated with different combinations of MRO Process and ILS Elements, as listed below:

- Compliance with certification requirements related to maintenance procedures (cell 1-A)
- Compliance with certification requirements related to Training (cell 4-A)
- Assure that mechanics know how to use tools correctly (cell 4-D)
- System design for high maintainability (cells 10-B, 10-C)
- System design for the smooth operation of Tools and GSEs (cell 10-D)
- System design for high maintainability at Suppliers' facilities (cell 10-E)
- Customer feedback gathering during the design phase (cell 10-F)

It also can be noticed that most of the technologies that enable MRO 4.0 implementation aim to automatise processes or to improve efficiency. On the other hand, a significant part of the challenges is associated with model or simulation representativeness.

Another demand that 4.0 technologies could help is in the digitisation for running paperless processes requested by the digital thread approach (Backes et al., 2017; Gharbi et al., 2018; Linnimaa, 2016):

- Maintenance records and digital signature authenticity guarantee (cells 1-B, 1-C, 1-F and 3-A)
- Human resources allocation and task support (cells 3-B, 3-C, 3-D, 3-E and 3-F)
- Documents and Augmented Reality integration and data security assurance (cells 6-A, 6-B, 6-C, 6-D, 6-E and 6-F)
- Databases updating during operations (cells 8-A, 11-A, 11-B, 11-C, 11-D, 11-E, 11-F and 12-A)

FUTURE RESEARCH DIRECTIONS

Based on the assumption that current MRO organisation primary difficulty is to know in advance the aircraft to be maintained since the operator asks for a maintenance slot not being specific in which condition the aircraft will arrive, and also the model and type of aircraft to be maintained, to manage job cards in a digitised environment would be the main driver to establish the MRO 4.0 organisation properly.

Trends

The following research directions are pointed out from the previous discussions.

Digitisation of MRO Shop Floor

To support all forecasts demand from parts acquisition, job cards, and workforce allocation is necessary to acquire data and make it available to a series of digital processes. An MRO digital twin would be only available if all resources are mapped and have its digital counterpart to make simulations available and give decisionmakers the ability to foresee the many situations depending on the aircraft to be received next (cells: 1-B, 1-C, 1-D, 1-G, 2-E, 2-G, 3-D, 3-G, 4-G, 5-A, 5-C, 5-G, 6-A, 6-C, 6-D, -G, 7-E, 7-G, 8-G, 9-A, 9-G, 10-A, 10-F, 10-G, 11-G, and 12-G).

Prescriptive maintenance (also called Cognitive Maintenance) will be the next step in making job cards available in a dynamic environment characterised by the short-term planning sectors that have to define the resources to be planned. Such maintenance approach depends on the available data of previous maintenance performed and recorded. Depending on factors such as the plane model an type, the maintenance team that performed similar activities, the check to be performed, aircraft age, spares availability, and Service Bulletins to be included the Artificial Intelligence can suggest different courses of actions with assessment in the revenues and outcomes.

The trends in research are database interoperability (pattern definition), machine learning to support business intelligence of maintenance activities, optimisation for dynamic allocation of shared resources, and digitisation of all necessary MRO's logistics information, including technical data (digital thread) (*Modernizing MRO*, 2018).

Guarantee of Maintenance Transactions

Another trend is the use of blockchain technologies to guarantee the transactions of the activities that support the maintenance, either of confirmation of the tasks performed or of the exchange of components (cells: 1-B, 1-C, 1-F, 3-A, 5-B, 5-C, 6-B, 6-C, 6-D, and 6-E) or the acquisition of replaceable items (cell: 6-F) (Kearney, 2019).

Virtual and Augmented Reality

Virtual Reality could improve aircraft maintainability that could reduce downtime during maintenance activities. An MRO company could use the technology to support training for maintenance teams (cells: 4-A and 4-D) and also during the task planning (cells: 1-A, 10-B, 10-C, 10-D, 10E, and 10-F). The Augmented Reality technology could enhance the mechanic's efficiency during a service bulletin execution or some activities that are not usual in daily job cards (cell 6-A). The usability of VR and AR in the industry is still in its infancy. Much effort has to be done until a massive utilisation can be expected (Martinetti et al., 2019).

CONCLUSION

The chapter provided an overview of Maintenance, Repair, and Overhaul (MRO) processes that could be enhanced by technologies from the Industry 4.0 paradigm like Machine Learning, IoT, and Big Data Analytics. Also, the Integrated Logistics Support (ILS) framework was briefly described, and the relationship between ILS elements and an overall MRO process was pointed out. From the mapping of these relationships were pointed the demands, problems, and difficulties to be faced for an organisation that intends to make the transition to the 4.0 paradigm or to keep enhancing its capabilities.

The primary concern of an MRO continues to be to know in advance the job cards to be planned since the operator did not give this information as soon as the MRO would like to have. Since this would not be different in the days to come, the main contribution to this scenario is to have the capability to be fast enough to replan and to reduce the variability of its processes. Given that, Prescriptive Maintenance would be the primary trend to be pursued in the e-Maintenance research area for companies that would like to transit to the MRO 4.0 level.

To achieve such level would have to change the way data is gathered from the company´s processes, and digitisation of logistics information is mandatory. Because of that, human resources has to be capable of using the digital environment to keep tracking of the evolving situation and to receive data in its context.

Wherever feasible, such companies are likely to have a large volume of aircraft to be received for one year and have already digitised technical data from many supported aircraft, which will promote smoother integration into their maintenance management systems. Organisations will need a cultural shift from paper to full digital registration. These are premisses to be competitive in a market that is growing in size but requires the same level of safety and efficiency, two things that are hard to achieve at the same time.

REFERENCES

Abrahão, F. T. M., Filho, J. N. M., Neto, L. P. D., & de Mesquita, A. C. P. (2019). Development of the AeroLogLabTOOL®. *SPOLM2019*, 14. https://www.marinha.mil.br/spolm/sites/www.marinha.mil.br.spolm/files/DEVELOPMENT%20OF%20THE%20AEROLOGLABTOOL.pdf

ASD/AIA. (2018, July). *ASD/AIA SX000i – International guide for the use of the S-Series Integrated Logistics Support (ILS) specifications*. https://www.sx000i.org/

Backes, M., Miwa, F., Okajima, C., Souza, A., Jr., & Tkacz, D. (2017). *Digital records: Improving efficiency in commercial aviation*. https://repositorio.itl.org.br/jspui/handle/123456789/11

Bierer, A., Götze, U., Köhler, S., & Lindner, R. (2016). Control and Evaluation Concept for Smart MRO Approaches. *Procedia CIRP*, *40*, 699–704. doi:10.1016/j.procir.2016.01.157

Cegelec. (2020). *Integrated Logistic Support – Maintaining systems in Operating and Security Conditions*. Cegelec Defense. https://www.cegelec-defense.com/en/infrastructures-networks/moc-msc/

Ceruti, A., Marzocca, P., Liverani, A., & Bil, C. (2019). Maintenance in Aeronautics in an Industry 4.0 Context: The Role of Augmented Reality and Additive Manufacturing. *Journal of Computational Design and Engineering, S2288430018302781*(4), 516–526. doi:10.1016/j.jcde.2019.02.001

Cipresso, P., Giglioli, I. A. C., Raya, M. A., & Riva, G. (2018). The Past, Present, and Future of Virtual and Augmented Reality Research: A Network and Cluster Analysis of the Literature. *Frontiers in Psychology, 9*, 2086. doi:10.3389/fpsyg.2018.02086 PMID:30459681

Defense Acquisition University. (2011, December 1). *Integrated Product Support (IPS) Element Guidebook.* https://www.dau.edu/tools/t/Integrated-Product-Support-(IPS)-Element-Guidebook-

Denis, M. Wm. (2008, September 8). *MRO Technologies & Services: Point of View.* https://www.slideshare.net/guesta9496c4/aviation-mro-it-emergence-of-saas-and-convergence-of-bpo

Dinis, D., Barbosa-Póvoa, A., & Teixeira, Â. P. (2019). Valuing data in aircraft maintenance through big data analytics: A probabilistic approach for capacity planning using Bayesian networks. *Computers & Industrial Engineering, 128*, 920–936. doi:10.1016/j.cie.2018.10.015

Edwards, T., Bayoumi, A., & Lester Eisner, M. (2017). Internet of Things – A Complete Solution for Aviation's Predictive Maintenance. In Y. Bahei-El-Din & M. Hassan (Eds.), *Advanced Technologies for Sustainable Systems: Selected Contributions from the International Conference on Sustainable Vital Technologies in Engineering and Informatics, BUE ACE1 2016, 7-9 November 2016, Cairo, Egypt* (pp. 167–177). Springer International Publishing. 10.1007/978-3-319-48725-0_16

Esposito, M., Lazoi, M., Margarito, A., & Quarta, L. (2019). Innovating the Maintenance Repair and Overhaul Phase through Digitalization. *Aerospace, 6*(5), 53. doi:10.3390/aerospace6050053

Federal Aviation Administration. (2019). *Flight Standards – Draft Advisory Circulars (ACs) Open for Comment.* https://www.faa.gov/aircraft/draft_docs/media/afx/AC_43-218_Coord_Copy.pdf

Froes, F. H., & Boyer, R. (2019). *Additive manufacturing for the aerospace industry.*

Galar, D., Gustafson, A., Tormos, B., & Berges, L. (2012). Maintenance Decision Making based on different types of data fusion. *Ekspolatacja i Niezawodnosc - Maintenance and Reliability, 14*(2), 135–144.

García, J. R. R., Martinetti, A., Becker, J. M. J., Singh, S., & van Dongen, L. A. M. (2019). Towards an Industry 4.0-Based Maintenance Approach in the Manufacturing Processes. In V. González-Prida Diaz & J. Pedro Zamora Bonilla (Eds.), *Handbook of Research on Industrial Advancement in Scientific Knowledge* (pp. 135–159). IGI Global. doi:10.4018/978-1-5225-7152-0.ch008

Gdalevitch, M. (2000, November 1). *MSG-3, The Intelligent Maintenance.* Aviation Pros. https://www.aviationpros.com/engines-components/article/10388498/msg3-the-intelligent-maintenance

Gharbi, A., Briceno, S. I., & Mavris, D. N. (2018). *Enabling the Digital Thread in Commercial Aircraft Companies. In 2018 Aviation Technology, Integration, and Operations Conference.* American Institute of Aeronautics and Astronautics. doi:10.2514/6.2018-4008

Haroun, A. E., & Duffuaa, S. O. (2009). Maintenance Organization. In M. Ben-Daya, S. O. Duffuaa, A. Raouf, J. Knezevic, & D. Ait-Kadi (Eds.), *Handbook of maintenance management and engineering* (pp. 4–5). Springer. doi:10.1007/978-1-84882-472-0_1

Holmberg, K., Adgar, A., Arnaiz, A., Jantunen, E., Mascolo, J., & Mekid, S. (Eds.). (2010). *E-maintenance*. Springer London. doi:10.1007/978-1-84996-205-6

Indra. (2020). *Logistics & Maintenance*. Indra. https://www.indracompany.com/en/logistics-maintenance

Karakuş, G., Karşıgil, E., & Polat, L. (2019). The Role of IoT on Production of Services: A Research on Aviation Industry. In N. M. Durakbasa, & M. G. Gencyilmaz (Eds.), *Proceedings of the International Symposium for Production Research 2018* (pp. 503–511). Springer International Publishing. 10.1007/978-3-319-92267-6_43

Kearney, K. (2019). Four Ways Blockchain will Change Aircraft Maintenance [Web Blog]. Honeywell Aerospace. https://aerospace.honeywell.com/content/aero/en/us/home/learn/about-us/blogs/2019/08/four-ways-blockchain-will-change-aircraft-maintenance.html

Kinnison, H. A. (2004). *Aviation maintenance management*. McGraw-Hill.

Kraft, J., & Kuntzagk, S. (2017). *Engine Fleet-Management: The Use of Digital Twins From a MRO Perspective*. V001T01A007-V001T01A007. doi:10.1115/GT2017-63336

Lee, C.-H., Shin, H.-S., Tsourdos, A., & Skaf, Z. (2018). New Application of Data Analysis Using Aircraft Fault Record Data. *Journal of Aerospace Information Systems*, *15*(5), 297–306. doi:10.2514/1.I010577

Linnimaa, S. (2016). *Information management in aircraft maintenance*. https://lutpub.lut.fi/handle/10024/121873

Liu, Z., Meyendorf, N., & Mrad, N. (2018). The role of data fusion in predictive maintenance using digital twin. *AIP Conference Proceedings*, *1949*(1), 020023. doi:10.1063/1.5031520

Marques, H. C., & Giacotto, A. (2019). *Prescriptive Maintenance: Building Alternative Plans for Smart Operations*. doi:10.3384/ecp19162027

Martinetti, A., Marques, H. C., Singh, S., & van Dongen, L. (2019). Reflections on the Limited Pervasiveness of Augmented Reality in Industrial Sectors. *Applied Sciences (Basel, Switzerland)*, *9*(16), 3382. doi:10.3390/app9163382

Masoni, R., Ferrise, F., Bordegoni, M., Gattullo, M., Uva, A. E., Fiorentino, M., ... Di Donato, M. (2017). Supporting Remote Maintenance in Industry 4.0 through Augmented Reality. *Procedia Manufacturing*, *11*, 1296–1302. doi:10.1016/j.promfg.2017.07.257

Modernizing MRO. (2018, January 2). *Aerospace America*. https://aerospaceamerica.aiaa.org/departments/modernizing-mro/

Parida, A., & Kumar, U. (2009). Maintenance Productivity and Performance Measurement. In M. Ben-Daya, S. O. Duffuaa, A. Raouf, J. Knezevic, & D. Ait-Kadi (Eds.), *Handbook of maintenance management and engineering* (pp. 17–37). Springer. doi:10.1007/978-1-84882-472-0_2

Permatasari, C. I., Yuniaristanto, Sutopo, W., & Hisjam, M. (2019). Aircraft maintenance manpower shift planning with multiple aircraft maintenance licenced. *IOP Conference Series. Materials Science and Engineering*, *495*, 012023. doi:10.1088/1757-899X/495/1/012023

Ríos, J., Morate, F. M., Oliva, M., & Hernández, J. C. (2016). Framework to support the aircraft digital counterpart concept with an industrial design view. *International Journal of Agile Systems and Management, 9*(3), 212–231. doi:10.1504/IJASM.2016.079934

Rodrigues, D., & Lavorato, P. (2016). Maintenance, Repair and Overhaul (MRO) Fundamentals and Strategies: An Aeronautical Industry Overview. *International Journal of Computers and Applications, 135*(12), 21–29. doi:10.5120/ijca2016908563

Rzevski, G., Knezevic, J., Skobelev, P., Borgest, N., & Lakhin, O. (2016). Managing aircraft lifecycle complexity. *International Journal of Design & Nature and Ecodynamics, 11*(2), 77–87. doi:10.2495/DNE-V11-N2-77-87

Sahay, A. (2012). *Leveraging Information Technology for Optimal Aircraft Maintenance, Repair and Overhaul (MRO)*. Elsevier. doi:10.1533/9780857091437

Tech, I. T. (2020). *Integrated Logistics – IT Tech Direct*. https://www.ittechdirect.net/integrated-logistics/

Ucler, C., & Gok, O. (2015). Innovating General Aviation MRO's through IT: The Sky Aircraft Management System - SAMS. *Procedia: Social and Behavioral Sciences, 195*, 1503–1513. doi:10.1016/j.sbspro.2015.06.452

Utzig, S., Kaps, R., Azeem, S. M., & Gerndt, A. (2019). Augmented Reality for Remote Collaboration in Aircraft Maintenance Tasks. *2019 IEEE Aerospace Conference*, 1–10. 10.1109/AERO.2019.8742228

Zhang, X., & Liang, E. (2019). Metal additive manufacturing in aircraft: Current application, opportunities and challenges. *IOP Conference Series. Materials Science and Engineering, 493*, 012032. doi:10.1088/1757-899X/493/1/012032

ADDITIONAL READING

Ben-Daya, M., Duffuaa, S. O., Raouf, A., Knezevic, J., & Ait-Kadi, D. (Eds.). (2009). *Handbook of Maintenance Management and Engineering*. Springer London; doi:10.1007/978-1-84882-472-0

Blanchard, B. S., & Fabrycky, W. J. (2011). *Systems engineering and analysis* (5th ed.). Boston: Prentice-Hall.

Dinis, D., Barbosa-Póvoa, A., & Teixeira, Â. P. (2019). A supporting framework for maintenance capacity planning and scheduling: Development and application in the aircraft MRO industry. *International Journal of Production Economics, 218*, 1–15. doi:10.1016/j.ijpe.2019.04.029

López-Ramos, L. A., Cortés-Robles, G., Roldán-Reyes, E., Alor-Hernández, G., & Sánchez-Ramírez, C. (2019). The Knowledge-Based Maintenance: An Approach for Reusing Experiences in Industrial Systems. In J. L. García Alcaraz, L. Rivera Cadavid, R. G. González-Ramírez, G. Leal Jamil, & M. G. Chong Chong (Eds.), *Best Practices in Manufacturing Processes* (pp. 505–523)., doi:10.1007/978-3-319-99190-0_23

Palmarini, R., Erkoyuncu, J. A., Roy, R., & Torabmostaedi, H. (2018). A systematic review of augmented reality applications in maintenance. *Robotics and Computer-integrated Manufacturing*, *49*, 215–228. doi:10.1016/j.rcim.2017.06.002

Uhlmann, E., Stark, R., Rethmeier, M., Baumgarten, J., Bilz, M., & Geisert, C. … Reinkober, S. (2015). Maintenance, Repair and Overhaul in Through-Life Engineering Services. In L. Redding & R. Roy (Eds.), Through-life Engineering Services (pp. 129–156). doi:10.1007/978-3-319-12111-6_9

Wibowo, A., Tjahjono, B., & Tomiyama, T. (2017). Designing Contracts for Aero-engine MRO Service Providers: Models and Simulation. *Procedia CIRP*, *59*, 246–251. doi:10.1016/j.procir.2016.10.124

KEY TERMS AND DEFINITIONS

Augmented Reality: Is the technology that allows the visualisation of virtual objects or information in front of the real-world objects of interest through a data fusion process and data projection on the person´s field of view.

E-Maintenance: Is the research area that identifies the possible usage of Information Technology and Telecommunications to enhance the maintenance management and execution for complex systems.

Failure Reporting Analysis and Corrective Action Systems (FRACAS): Is a system (often run using software) that provides a process for reporting, classifying, analysing failures and planning corrective actions in response to identified failures.

Integrated Logistics Support Plan (ILSP): It is a document containing how the support concept will be implemented, prescribes actions for each ILS element and tasks that will be required for the system/equipment supportability according to each of its life cycle phase.

Maintenance, Repair, and Overhaul (MRO): In aviation are mostly all maintenance activities that contribute to ensuring the safety and airworthiness of all aircraft following the guidelines of the international aviation authorities.

Remaining Useful Life: The remaining lifetime of a component or item in terms of hours of use or cycles. The measure aims to inform the time remaining to take any maintenance action before the failure of the component or item.

Situational Awareness: The capability to perceive the elements in the environment in time and space and the comprehension of their meaning and the status projection in the near future.

Chapter 13
Role of Additive Manufacturing in Industry 4.0 for Maintenance Engineering

Arun Kumar
Indian Institute of Technology, Delhi, India

Gurminder Singh
SIMAP Lab, Université Grenoble Alpes, France

Ravinder Pal Singh
Indian Institute of Technology, Delhi, India

Pulak Mohan Pandey
Indian Institute of Technology, Delhi, India

ABSTRACT

The chapter describes the role of additive manufacturing (AM) in Industry 4.0 (I4.0) for maintenance engineering. A brief introduction of the fourth industrial revolution and related technologies has been included. The different AM processes with significant contributions in the relevant industry sectors have been discussed along with suitable examples. Difference between the manufacturing capabilities of conventional and AM technologies has also been presented. Owing to its high degree of design freedom, AM helps to reduce the spare parts inventory cost, component assembly cost, and can replace the discontinued parts easily. A case study presenting these key distinctive features of AM, which make it an indispensable technology for I4.0, are also discussed. Furthermore, the barriers to the adoption of AM technology by manufacturers and possible remedial actions are also discussed in brief. The knowledge gaps in terms of materials and design tools for AM have been identified and a probable road ahead has been discussed.

DOI: 10.4018/978-1-7998-3904-0.ch013

INTRODUCTION

Mechanical production plants of the late 18th century, based on water and steam power, marked the era of the first industrial revolution, now known as Industry 1.0. The advent of the 20th century was marked by ever more present electricity and factory electrification. This paved the way for the second industrial revolution, i.e. Industry 2.0, characterized by mass labour production based on electrical energy. The third industrial revolution, also known as the Digital revolution occurred in the late 20th century. This phase of the industrial revolution, Industry 3.0, was characterized by automatic production based on electronics and internet technology (Galati & Bigliardi, 2019). In 2011 at Hannover Messe Industry Fair, Germany, a new term "*Industry 4.0*" was coined which is assumed as the fourth industrial revolution. Industry 4.0 focuses on adopting a higher level of automatization, connecting physically to the virtual world (Alcácer & Cruz-Machado, 2019).

Figure 1 presents the timeline of four different phases of the industrial revolution. Industry 4.0 is the convergence of industrial production and communication and information technologies (Hermann, Pentek, & Otto, 2016). Different researchers have given the definition of Industry 4.0 in their adaptive meanings. Shafiq et al. (Shafiq, Sanin, Toro, & Szczerbicki, 2015) identified Industry 4.0 as a "connected and continuously available resource handling scheme" by incorporating cyber-physical systems into the conventional manufacturing setups to achieve the objectives of intelligent factory characterized by resource efficiency and adaptability. (Lu, 2017) described Industry 4.0 as the "cyber-physical systems (CPS) based on heterogeneous data and knowledge integration". (Leyh, Martin, & Schaffer, 2017) described Industry 4.0 from a production perspective as the setup in which intelligent workpiece can independently coordinate their paths through the factory and machines communicate in real-time with the respective warehouse.

Figure 1. Revolution of industry
Source: Hermann, Pentek, & Otto, 2016

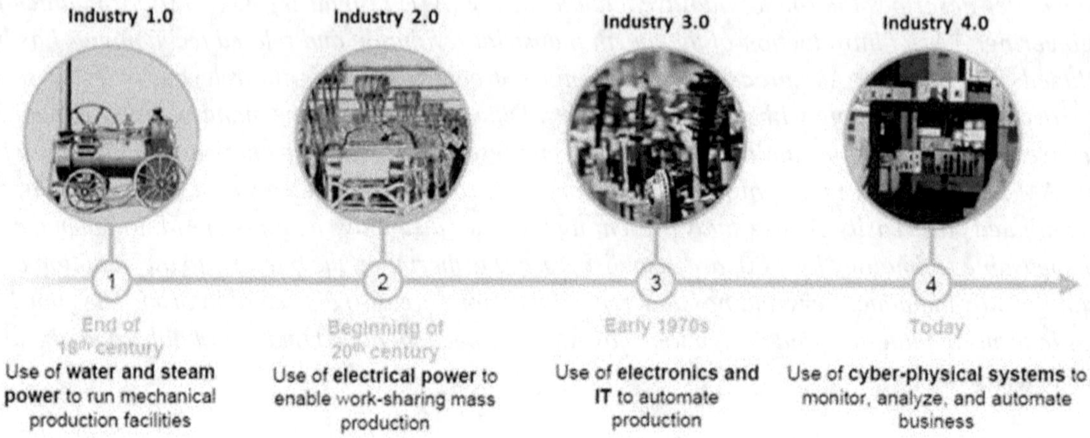

Industry 4.0 has resulted in a paradigm shift from centrally controlled to the decentralized production process. Due to constant communication of resources and machines throughout the complete production cycle, the product is not only aware of its current and target state but also can navigate within the smart

setup while instructing engines to perform the required processes (Hermann et al., 2016). (Pereira & Romero, 2017) categorised the impact and influence of Industry 4.0 into six main areas:

1. Industry,
2. Products and services,
3. Business models and market,
4. Economy,
5. Work environment and
6. Skills development.

Unlike previous industrial revolutions, Industry 4.0 goes a step further and along with improving productivity also influences the whole supply chain from product development and engineering to logistics. Different new generation tools such as internet of things, cloud computing, data analysis, cyber-physical systems, etc., and manufacturing fields such as additive manufacturing, micro-machining, etc., has developed the industry scenario in recent times.

Additive manufacturing (AM), a layer-by-layer process has become an important part of the present industry. It has been used for different applications of Industry 4.0 and has marked its self as a 'game-changer' for the maintenance aspects of the present industry. Many industries have started using different techniques of the AM to create prototypes for the design proposes. However, they have identified AM application more essential for the efficient spare part management system. Example, if a company wants to buy a replacement part and identified that they have to buy the spare part only in bulk, then use of AM for the fabrication of the custom part will be more cost-effective. AM has acted as a lifesaver to many of the companies having unique old machinery. It can be used to fabricate the obsolete parts, whose fabrication has been stopped by the equipment manufacturer.

Technologies Interrelated to INDUSTRY 4.0

Different authors (Alcácer & Cruz-Machado, 2019)(A. Gilchrist, 2016)(Saucedo-Martínez, Pérez-Lara, Marmolejo-Saucedo, Salais-Fierro, & Vasant, 2018) have identified nine major-related technologies or backbone of Industry 4.0 as shown in Figure 2.

Big data analytics

(Megahed & Jones-Farmer, 2013) described big data as a large, diverse and complex dataset collected from several sensors, instruments and computer-based interactions. (Chen, Mao, & Liu, 2014) identified three characteristic features (3*Vs*) to describe big data namely *variety*, *velocity* and *volume*. Volume signifies the amount or size of data whereas variety and velocity relate to the nature and speed of data accrual (Gölzer & Fritzsche, 2017). Big data analytics provide better decision-making insights to the organization. After an extensive literature review, (Cui, Kara, & Chan, 2020) identified 17 major applications (ref. Figure 3) of big data in the manufacturing domain. It is worth noticing that the prime focus of big data analytics is on monitoring and prediction. In a manufacturing environment, big data find its applications in production management, machine maintenance, and quality management.

Figure 2. Technologies related to industry 4.0 (Dalmarco, Ramalho, Barros, & Soares, 2019)

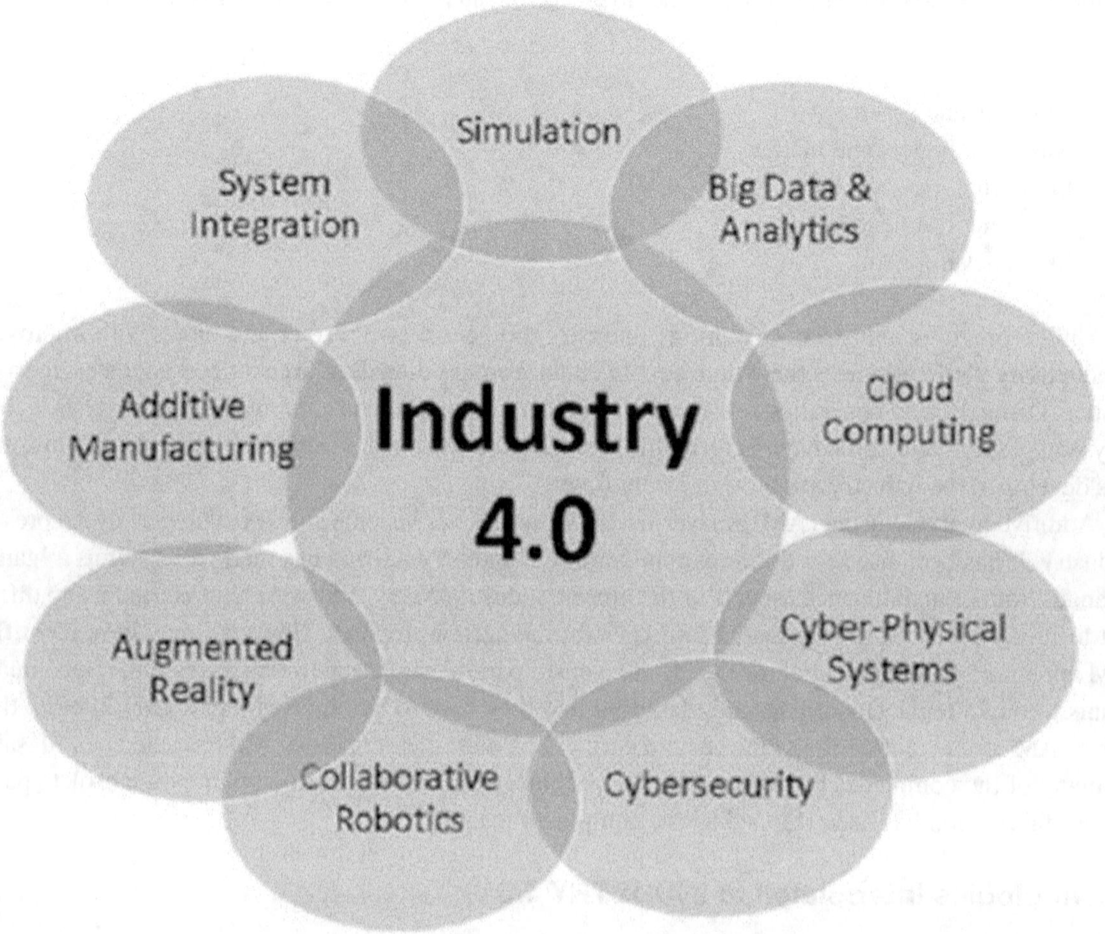

Internet of Things (IoT)

The term Internet of things, famously known as *IoT,* is composed of two words, "internet" and "things". Internet refers to the system of interconnected computer networks using standard internet protocol suite to serve users globally. The second word "things" in the IoT refers to the real objects in the physical or the material world. These "things" are not necessarily only non-living objects (such as parts and machinery) but very well can be living things also such as person, plants, and animals (Alam, Vats, & Kashyap, 2018). Thus, IoT can be described as a comprehensive network of objects interacting and sharing information. It allows things-things, human-things, and human-human communication. Figure 4 depicts a typical Industrial IoT layout.

IoT makes use of various technologies such as Radio Frequency Identification (RFID), Wireless Sensing Networks (WSN), Wireless Fidelity (Wi-Fi), Near Field Communication (NFC), Bluetooth, Barcode, Internet Protocol (IP) and Electronic Product Code (EPC). IoT technology is not only revolutionizing industries, but is significantly affecting various other fields such as medicine and health-care, education, and governance.

Figure 3 Big data applications (Cui et al., 2020)

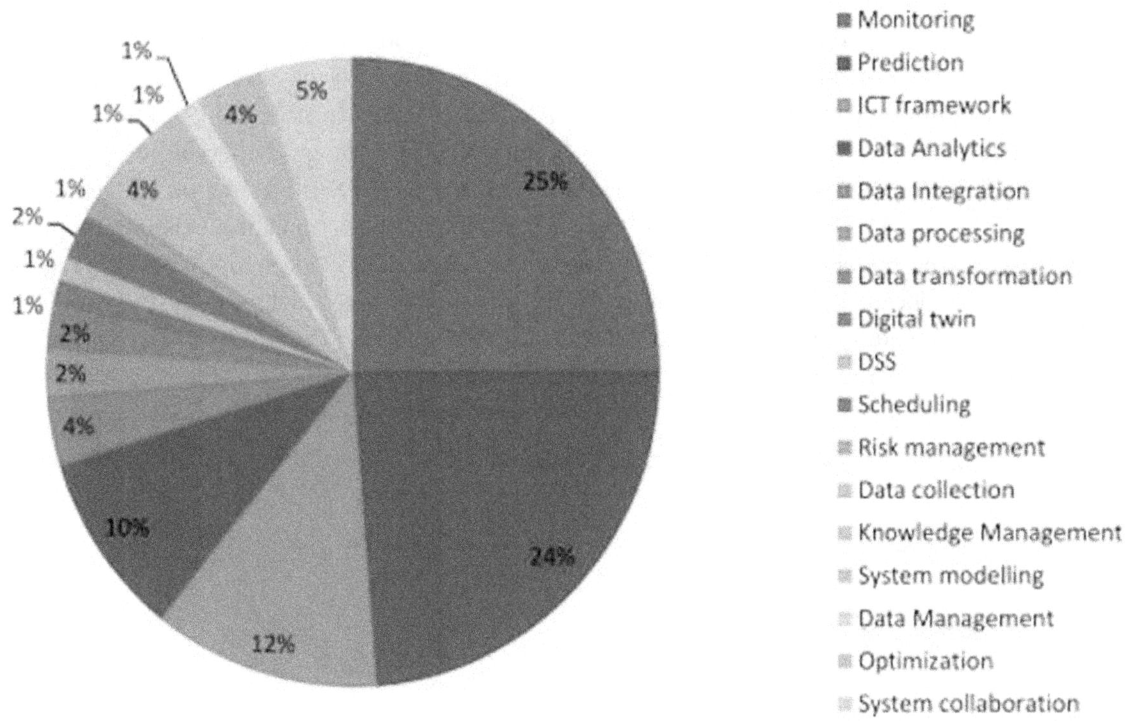

Figure 4. Typical industrial-IoT (Alcácer & Cruz-Machado, 2019)

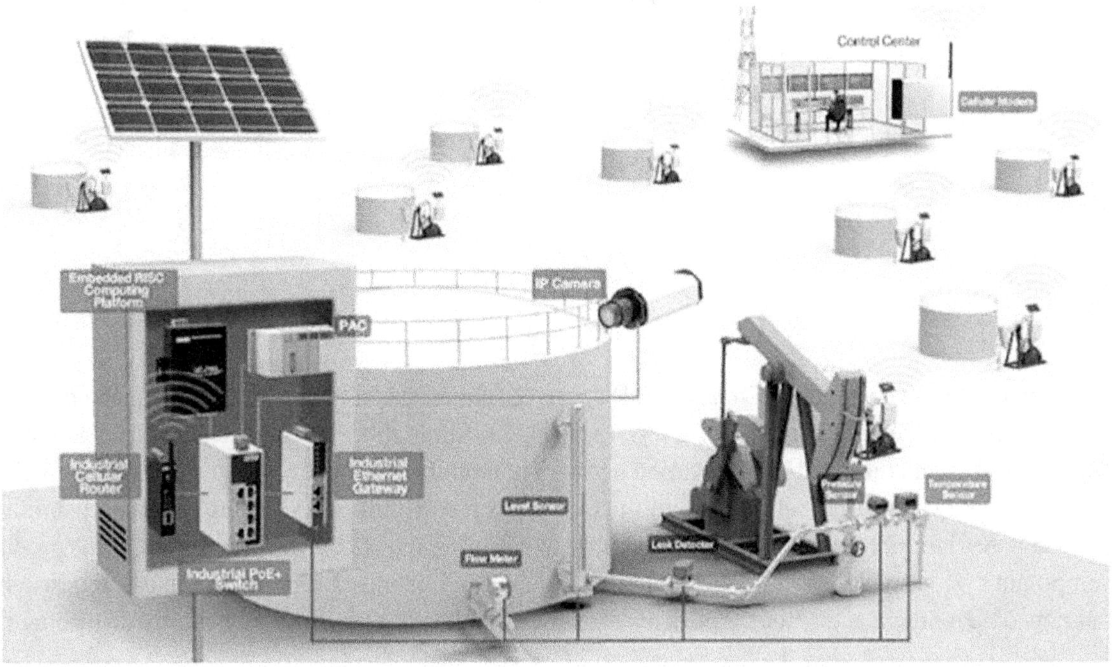

Cyber-Physical Systems

Cyber-Physical Systems (CPS) is coined from the merger of two words, *cyber* and *physical*. Here, *cyber* refers to electronic systems and *physical* refers to real objects or things. The physical component interacts with the physical world by creating its virtual copy(Alcácer & Cruz-Machado, 2019). Lee et al.(Lee, Bagheri, & Kao, 2015) described cyber-physical systems as a technology managing interconnected system between its physical assets and computational capabilities. CPS is an integral component for Industry 4.0. Integrating CPS with production, logistics and services can transform existing setups into Industry 4.0 factories. Sophia Keil (Keil, 2017) presented a schematic CPS system for a manufacturing setup in which an embedded system in the sense of CPS is integrated with machines (ref. Figure 5). This embedded system includes sensors to collect relevant data for analysis. The results of the data further govern or decide the interaction between connected components.

Figure 5 Schematic structure of manufacturing Cyber-Physical System(Keil, 2017)

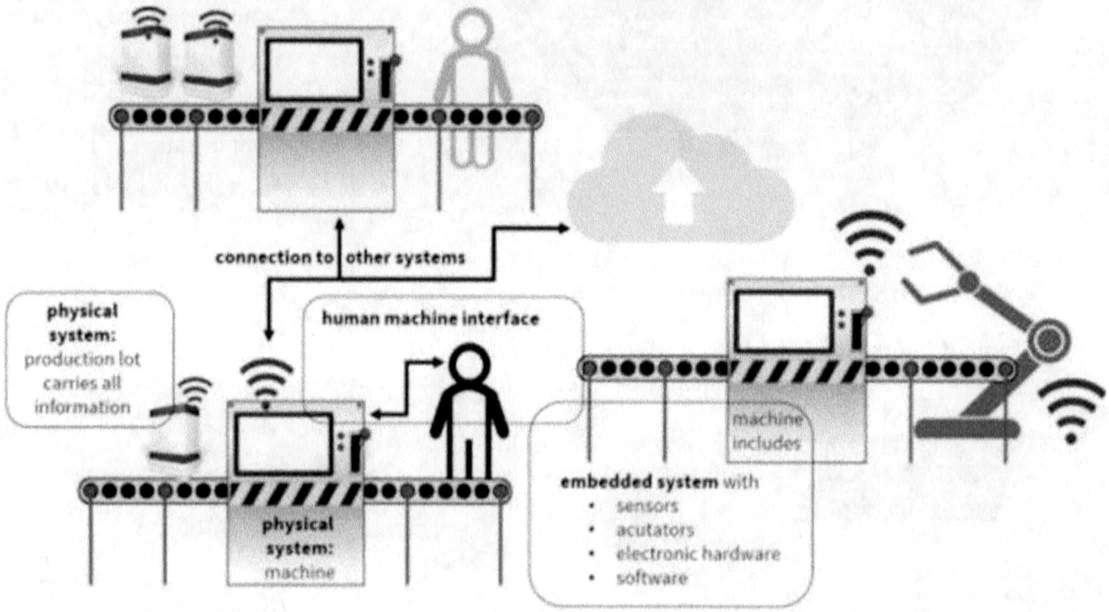

Simulation

Simulation modelling handbook (Chung, 2004) describes simulation as a process of creating and experimenting with a mathematical model of a physical system. Simulation is a powerful tool for analyzing complex systems and thus find its application not only in manufacturing but also in varied fields such as healthcare, supply chain, marketing, and military(Negahban & Smith, 2014). need for mass customization, product variation, complexity and intricacy are ever-increasing. Thus, the complexity of the product development system is also increasing. Simulation and modelling play a crucial role in gaining insights about new development and resource policies for such complex systems without actually disturbing the system(Mourtzis, Doukas, & Bernidaki, 2014). Simulation also enables the experiments for the validation of products, process or system configuration.

Augmented Reality

Augmented Reality (AR) can be described as a variation of Virtual Reality (VR) (Billinghurst, Clark, & Lee, 2014). VR technology completely engages a user in a virtual environment. In this user is not able to see the real environment surrounding him/her. On the other hand, in AR user can see his real surroundings and the virtual environment is superimposed on the real world.

In manufacturing domain AR find its application in assembly, maintenance and repair. Instructions are more effectively and efficiently understood if received in the form a physical-virtual superimposed experience instead in the form of texts and pictures. Figure 6 shows an example of AR in which a virtual lamp is placed over a real table and the planned path of the robotic arm shown by virtual red lines.

Figure 6. a) A virtual lamp superimposed on a real table b) path of robotic arm shown by virtual lines (Billinghurst et al., 2014)

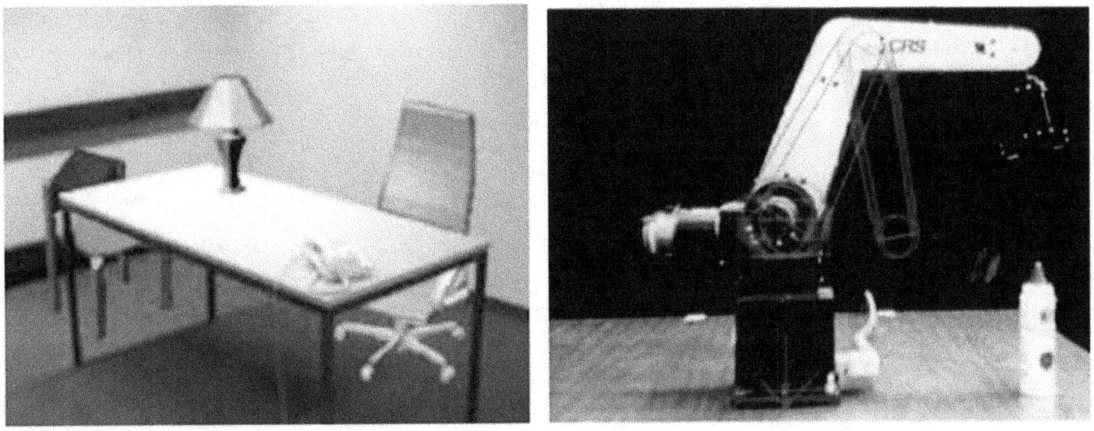

Additive Manufacturing

Additive Manufacturing (AM) is a layer-by-layer manufacturing process to fabricate 3D realistic prototypes from a 3D designed model. Generally, models are designed using CAD software or reverse-engineering scanned model or model generated from medical scanning method CT scan or cloud computing data (S. Singh, Ramakrishna, & Singh, 2017). AM techniques give the advantage to fabricate complex shape geometries at a reasonable cost and without the use of any additional machining or tooling. The complex shapes such as rocket propellers, foams, heat exchangers, etc. can be easily fabricated within a few hours (Pandey & Singh, 2019; G. Singh & Pandey, 2019e, 2019b, 2020). AM is rapidly advancing in the field of modern industry to minimize waste and to improve the productivity of the work (Bose, Vahabzadeh, & Bandyopadhyay, 2013; Jacobs, Grunert, Mohr, & Falk, 2008; Khas, Pandey, & Ray, 2015). There are different types of AM techniques available in the market based on different types of manufacturing principles. Several types of materials can be used in different AM techniques. Many authors (Chong, Ramakrishna, & Singh, 2018; Farid et al., 2015; Ngo, Kashani, Imbalzano, Nguyen, & Hui, 2018; Patra & Young, 2016; Poomathi et al., 2018; Prakash, Singh, Pabla, Sidhu, & Uddin, 2018; S. Singh, Prakash, & Ramakrishna, 2019; S. Singh, Singh, Gupta, Prakash, & Singh, 2019; S. Singh & Singh, 2016; Tack, Victor, Gemmel, & Annemans, 2016; Wang, Jiang, Zhou, Gou, & Hui, 2017; Wong & Hernandez, 2012) have reviewed the different types of AM techniques with their merits and demerits in literature. Apart from the present additive manufacturing techniques (having high capital cost), different rapid tooling or indirect additive manufacturing methods are also reported in the literature which required low capital and manufacturing cost (Gill & Kaplas, 2009; Mun, Yun, Ju, & Chang, 2015; G. Singh & Pandey, 2018, 2019d, 2019d, 2019a, 2019c; J. P. Singh & Pandey, 2018; S. Singh & Singh, 2015). Generally used AM techniques (in the present industry) based on different materials are shown in Figure 7.

Figure 7. Classification of AM techniques based on materials used in the present industry

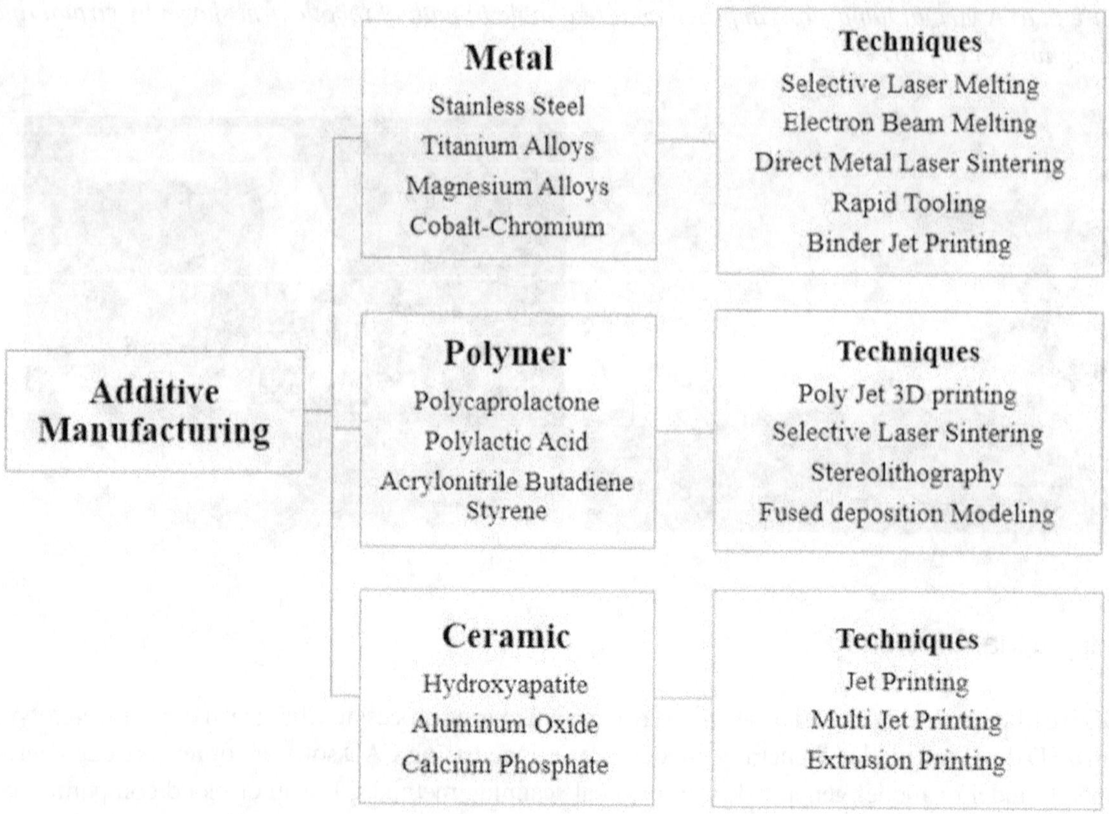

Fused deposition modelling (FDM) is one of the cheapest techniques of the AM family. It is based on the thermal deposition of the thermoplastic material based thin wire or filament such as Acrylonitrile Butadiene Styrene (ABS), Poly-lactic acid (PLA), etc. The plastic gets solidifies layer-by-layer after extrusion from the hot extruder and a solid model can be obtained. SLA (stereolithography) is based on the laser or light curing of a photo sensible liquid to obtain a 3D shape. The method is also efficient for thermoset plastic material such PMMA, rubber, etc. There are many other AM techniques such as SLS (selective laser sintering), Polyjet 3D printing, etc. which can be used for the fabrication of the polymer materials. These techniques are highly recommended for the fabrication of non-structural parts.

However, to obtain solid model of high strength material, melting of metallic powder is necessary. There are number of AM equipment available using melting of metallic powder for the fabrication of metal 3D objects using CAD design. SLM (selective laser melting) is one of them using high power laser source for the melting of powder layer-by-layer. EBM (electron beam melting) is the non-reflecting AM technique and recommended for the complex shape fabrication of high thermal conductive materials such as copper, brass, aluminum, etc. The fabricated solid parts obtained from these techniques have similar structural properties as per the machined or cast metals. There are other new-emerging metal AM techniques such as direct metal laser deposition, laser emerging net shaping, etc. The design freedom in AM become more interesting from the Industry 4.0 point of view. The complex shape based on the thin truss structures and repeated in terms of unit cell (known as lattice) can be easily fabricated by AM for high structural efficiency. Most of the earlier 3D printers used technology that required them to operate at high temperatures. The printers were bulky as well. However, the latest 3D printers from leading manufacturers like Stratasys, Markforge, Makerboat, EOS, etc. have bought about significant changes in the way printers work. The new emerging AM printers has become the important tool of the present industry scenario.

DISTINCTIVE FEATURES OF AM

As discussed in the previous section, AM makes use of CAD data for realizing the product and thus is inherently capable of making complex shapes which are not possible to fabricate with the conventional machining process. Along with this design freedom, there are other distinctive features of AM which makes it an essential component of the new phase of industrial revolution, i.e. Industry 4.0.

Design Freedom

AM is different from other manufacturing process and thus offers unprecedented design freedom and opportunities. The traditional approach is Design for Manufacturing (DFM) to minimise manufacturing difficulties (Rosen, 2014). A new approach namely Design for Additive Manufacturing (DfAM) is emerging as a supplementary tool for selecting optimal process parameters for AM (Dilberoglu, Gharehpapagh, Yaman, & Dolen, 2017). Rosen, 2014 identified the following capabilities of AM which provide it with an extra degree of design freedom for customization and improving performance and functionality:

1. **Shape Complexity:** AM makes it possible to build virtually any shape.
2. **Material Complexity:** It also enables to manufacture of parts with complex material compositions.
3. **Hierarchical Complexity:** this refers to the multi-scale of features, sub-features, etc. that are possible with AM.
4. **Functional Complexity:** Functional devices can be fabricated directly in some AM machines, by embedding components and kinematic joints while parts are being built.

Customization for Mass Production and Small Volume Production Both

Plastics processing companies are currently facing two main challenges. One is increasing the complexity of the parts and the other is the increasing variations within the product lots, even down to one-off pieces. He proposed that these different challenges can be effectively addressed by combining injection moulding, AM and I 4.0 technologies (Gaub, 2016). Similarly, Siemens, Phonak, Widex, and the other hearing aid manufacturers which are utilizing powder bed fusion (PBF) based techniques, such as selective laser sintering, to manufacture hearing aid shells. Boeing uses PBF based technologies to produce ducts for F-18 jets. Also, many European companies are fabricating hip implants using PBF methods. It clearly shows the capability of AM to enable one-off, custom manufacturing of small and large volume production (Rosen, 2014).

Part Consolidation

AM allows developing a design for an assembly in such a manner that an overall number of components can be reduced by consolidating them while maintaining the same functionality as of the original construction. Part consolidation inevitably minimises the number of elements, but also leads to an increase in complexity. Such complex parts will be challenging to produce by conventional manufacturing methods but owing its high degree of design freedom and free-form fabrication, AM enables part consolidation. It also helps in reducing part fabrication and tooling time. (Kamal & Rizza, 2019) provided part consolidation examples for the aerospace industry in their study. Figure 8 shows an example of part consolidation for a jet engine cowl latch.

Figure 8. AM component consolidation of a jet engine cowl latch (A) initial latch handle assembly made up of 5 components, (B) AM redesigned handle with components consolidated into a single component, (C) additive manufactured jet engine cowl latch (Kamal & Rizza, 2019).

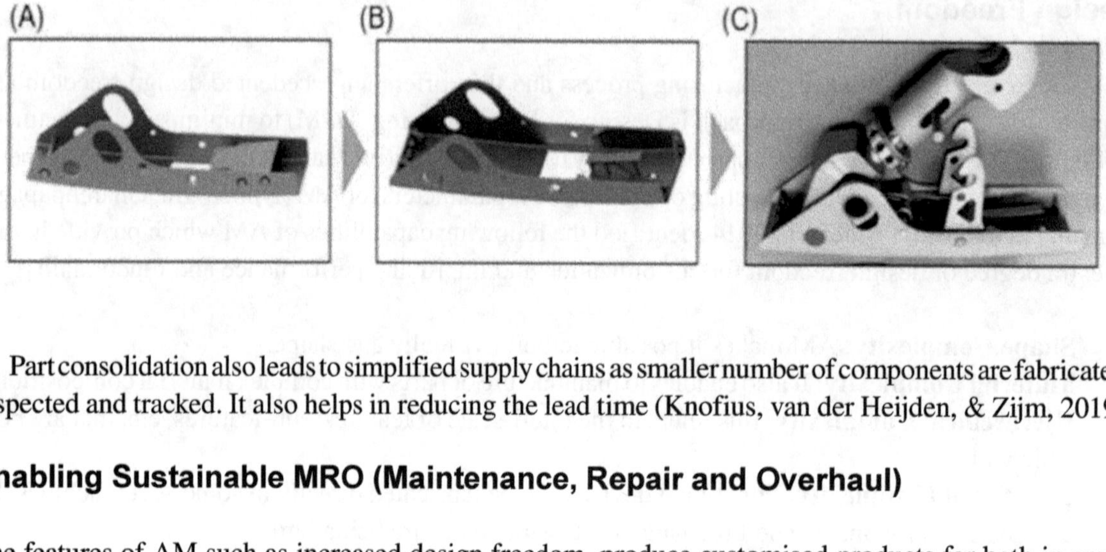

Part consolidation also leads to simplified supply chains as smaller number of components are fabricated, inspected and tracked. It also helps in reducing the lead time (Knofius, van der Heijden, & Zijm, 2019).

Enabling Sustainable MRO (Maintenance, Repair and Overhaul)

The features of AM such as increased design freedom, produce customised products for both in small and large quantities, option for topology optimisation, and part consolidation make it a major disruptive technology for the supply chain of spare parts. It can significantly reduce the time-to-market and cost of manufacturing. End-users can adapt and manufacture complex and customized spare parts themselves, thus reducing their dependence on suppliers and spare part manufactures. A simple process flow describing the same is shown in Figure 9 (Wits, García, & Becker, 2016).

Figure 9. Process flow for AM to enable end-user

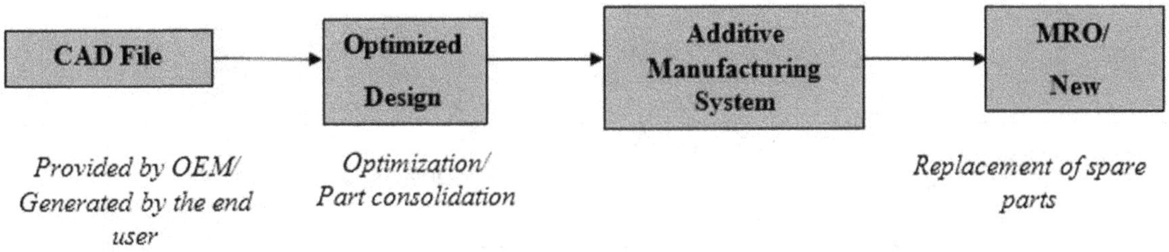

AM FOR MAINTENANCE AND SAFETY IN CONTEXT OF INDUSTRY 4.0: A CASE STUDY

In a study of maintenance in aeronautics in the paradigm of I 4.0, AM and AR are discussed as the technologies which can produce spare parts and support maintenance respectively (Ceruti, Marzocca, Liverani, & Bil, 2019). Authors described aeronautics as a complex and demanding field in both the design and maintenance terms. Providing spare parts in a shorter duration of time is another challenge faced by local operators in the civil aviation sector. Authors have proposed that I 4.0 and related technologies can be implemented in the aeronautics field to support design, maintenance, in-flight structural health and others. AM and AR are the two technologies which can support most of the on-ground maintenance operations while other technologies related to I 4.0 can support in-flight operations.

Operations like inspection, replacement of damaged parts, refilling of lubricants and gases (in hydraulic accumulators and damping cylinders), and fixing of coatings falls under the category of MRO (Maintenance, Repair, and Overhaul) activities. MRO activities are critical for safety in the aeronautics industry. Thus, these are strictly regulated by national aeronautical authorities such as FAA (Federal Aviation Authority) and EASA (European Aviation Safety Authority). All the commercial operators are required to prepare a Continuous Airworthiness Maintenance Program (CAMP) which is included in their Operation Specifications (OpSpecs). Airworthiness Review Certificate (ARC) is a document attesting the maintenance tasks performed. Different types of maintenance checks are presented in Table 1.

In this study, AM has been proposed as a disruptive technology which has the potential to solve the logistics related problems of metal spare parts. Traditional manufacturing and AM route for fabrication have been compared in Figure 10.

Provided acceptance of AM fabricated metals from due authorities, it is possible not only to shorten the supply chain of spare parts, AM also enables to topologically optimize the product. It can easily fabricate complex geometries without changing the setup with minimum cost implications. Several companies have adopted AM build parts in commercial aircraft, but their applications are now limited to non-structural components. Boeing B787 aircraft has almost 30 AM fabricated parts. Airbus A350XWB

Table 1. Maintenance checks for commercial/ civil aircraft

S.No.	Check type	Operations	Carried after	Duration
1	Check-A (Light Check)	Checking and inspection of passenger cabin, internal and external structure, engine pylons, control surfaces, engines	200 -300 cycles	1-2 days
2	Check-B (Light Check)	Checks conducted in A along with a deep inspection of engines, structural elements, all movable parts, wings, composite materials (looking for crack or delamination)	2000 flight hours	1-4 days
3	Check-C (Heavy Maintenance)	Apart from checks conducted in A and B, several components and groups are disassembled and carefully inspected (engines and pylons)	3500 flight hours	8-15 days
4	Check-D (Heaviest Maintenance/ Overhaul)	Aircraft is completely disassembled and both the external and internal structure are inspected in each detail	18,000-26,000 flight hours	60 days

Figure 10. Advantage of Parts build via AM route over traditional manufacturing route

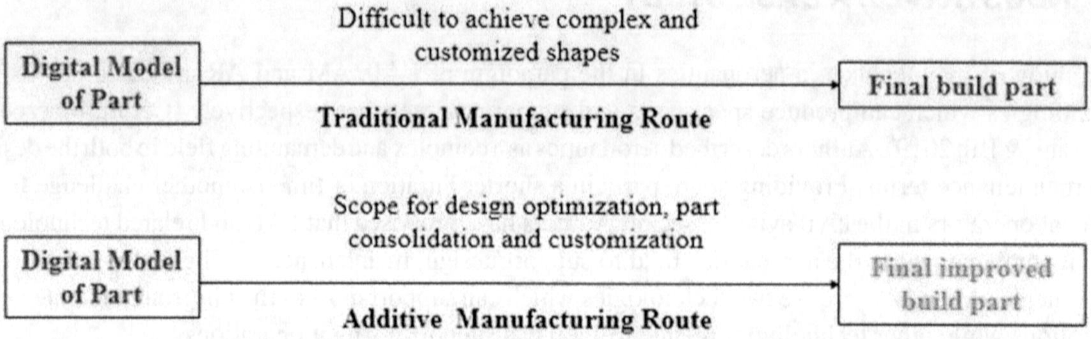

uses almost 1000 ULTEM 9085 (Fused Deposition Modelling) parts. But the application of AM for structural components need extensive study.

Case Study: SLAT Extension Mechanism Bracket

In this case study, the SLAT extension mechanism bracket having overall dimensions as shown in Figure 11 is fabricated using the AM process. This bracket is used in the leading edge of an aircraft wing.

This bracket can be easily manufactured for the above-given dimensions with the traditional process. But fabricating a lattice-based model with the same functionality is not possible with conventional methods of manufacturing. AM has the advantage of manufacturing complex parts and thus in this study more efficient design strategy, based on lattice structures (shown in Figure 12), is adopted for designing this bracket.

FEM (Finite Element Method) analysis for both the original dense part and lattice-based parts was carried out in this study. Authors have considered two merit indices for comparing traditional dense part

Figure 11. SLAT extension mechanism with dimensions(Ceruti et al., 2019)

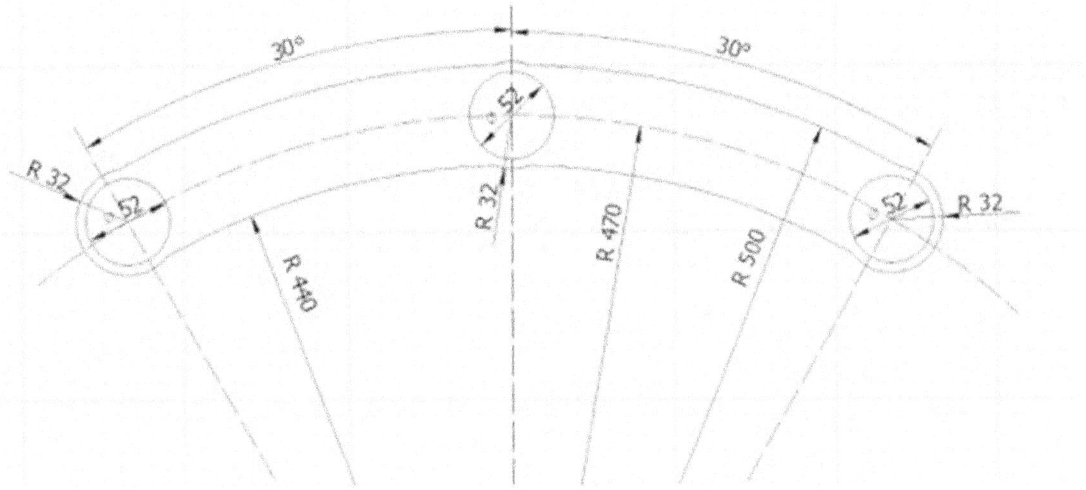

Figure 12. CAD models of the original part (top right), and lattice-based design of SLAT extension mechanism bracket

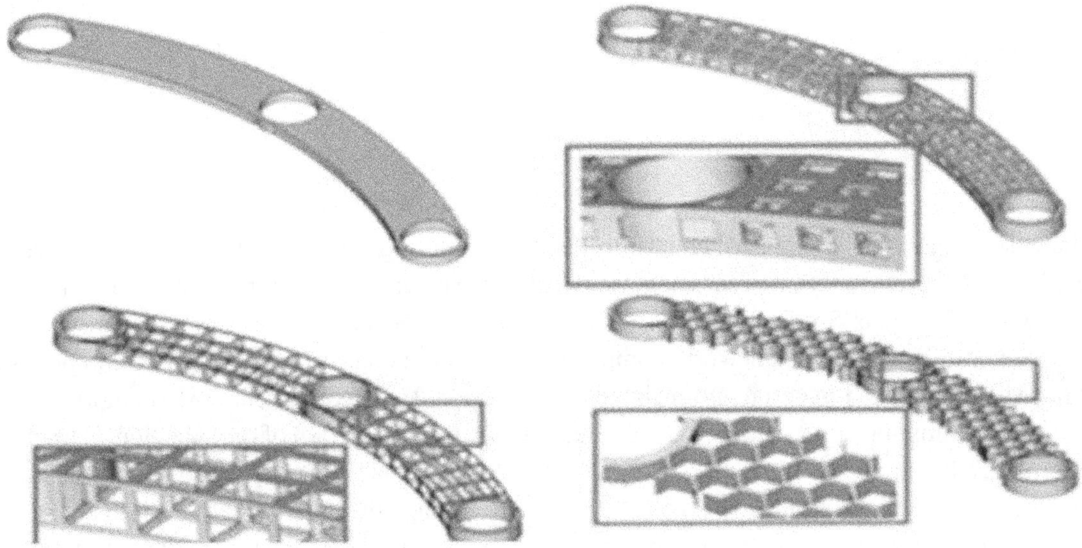

and AM build lattice-based parts. The first index was the ratio of the mass of AM lattice-based part to the mass of traditional dense part for equal displacement. While the other index was the ratio of maximum displacement of AM builds part to the maximum displacement of traditional dense part for the same mass. Results indicated that increase in stiffness can be achieved by adopting the lattice-based part.

In the case study presented in the section above, it is described that AM build parts are used in non-structural components. Use of AM build structural parts is not in practice. One of the reasons is the absence of regulations for AM build. The trend will change but requires extensive study for the AM builds parts for structural applications.

BARRIERS TO INTEGRATION OF AM IN MANUFACTURING SYSTEMS IN INDUSTRY 4.0

Despite several advantages of AM, companies are hesitant to integrate AM because of four major barriers shown in Figure 13 (Yi, Gläßner, & Aurich, 2019).

Figure 13. Barriers to integration of AM technologies

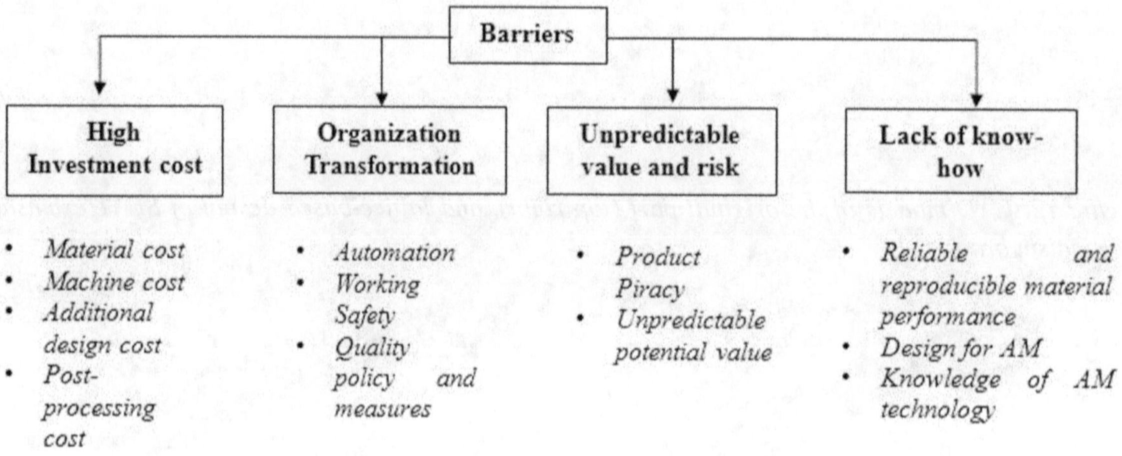

The cost of introducing the AM setup into the manufacturing system and the additional cost associated with material and post-processing are the major barriers to the adoption of AM technology. A framework of Product-Service System (PSS) is a proposed business pattern to overcome this cost barrier. Within this business framework, companies enter into a contract with AM machine providers for the AM setup, related material and maintenance services. Thus, equipment cost is amortized over longer period. Adoption of AM technology also leads to the transformation of organizational structures. Furthermore, adopting AM leads to reorganization of the supply chain and requires new safety measures. These changes lead to an additional cost which in turn discourage the adoption of AM into the system. Some companies prefer to stay with the conventional manufacturing system because they are not able to predict and define the benefits of adopting the AM approach. Companies have also stated product piracy and intellectual property protection as some of the barriers towards the adoption of AM technologies. Lack of knowledge related to AM is also another major barrier. Companies need to build up their technical know-how in two major domains, namely, design for AM and knowledge of material performance.

ROAD AHEAD

In the current market scenario of high competition, it is not the bigger fish that eats the smaller fish, but it is the faster fish that eats the slower fish. Companies target at achieving production of complicated shapes at higher speed and lower cost (Mehrpouya, Dehghanghadikolaei, & Fotovvati, 2019). AM has the potential to reduce the time required from the conception of a product to its design and development, and finally to market by 10-50%. Figure 14 shows the major benefits of adopting AM technology.

Figure 14. Key capabilities of AM in I4.0 setup

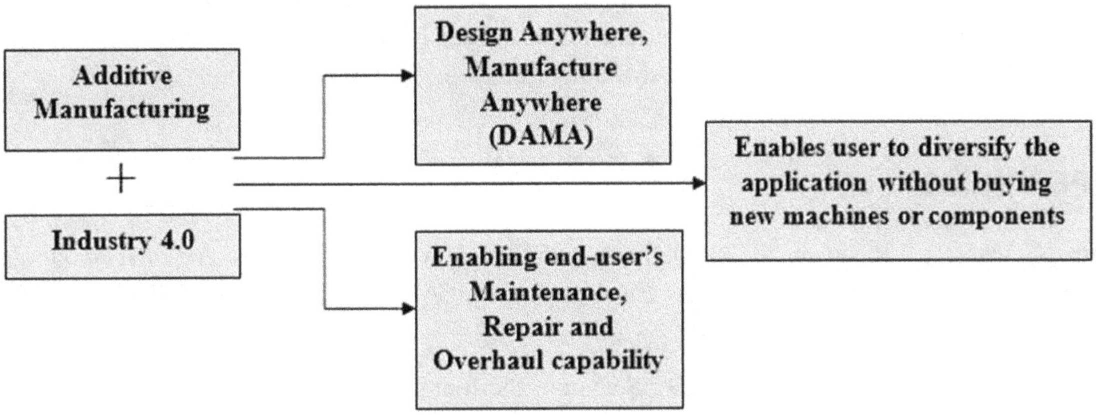

But the AM technology is still not mature and need further research in the field of materials and design tools (software) for AM to provide a cost-effective and faster method for production in smart manufacturing (I4.0) setup. In the case study presented in the section above, it is described that AM build parts are used only in non-structural components for aeronautics applications. Use of AM build structural parts is still not in practice. One of the reasons is the absence of regulations for AM build parts. In future, this trend will change but requires extensive study for the AM build parts for structural applications. Also, dedicated design tools, that can effectively utilize the design freedom offered by AM, need to be developed. The limited number of available AM standards poses a challenge for companies to enhance their knowledge base in this field. At present, there are two organizations that have started work in this direction. These organisations have proposed to publish standard series for addressing issues related to AM such as basic principle, terms, process categories, and, material tests (Yi et al., 2019). In future, these AM standards will work as guidelines and assist in AM applications.

CONCLUSION

The present industry is at the threshold of its new revolution called industry 4.0. This new industrial revolution (I 4.0) has marked the paradigm shift from a centrally controlled production process to a decentralized production process. Technologies related to I 4.0 has been briefly discussed in the present chapter. The distinctive features (shape complexity, material and functional complexity) and key capabilities (increased design freedom, part consolidation and customization for both small and large production volumes) of Additive Manufacturing in the framework of I 4.0 have also been discussed. AM has been identified as one of the key disruptive technologies in the I 4.0 framework for the supply chain of spare parts. Further, a case study has also been included in the chapter to discuss the advantages of AM over the traditional manufacturing process. The case study clearly showed that the highly complex shapes (in this case a lattice-based shape) can be easily achieved through AM which are otherwise impossible or extremely difficult to realize through traditional manufacturing process. Despite being a 'game changer' for maintenance and spares industry, there are several barriers to the adoption of AM in modern manufacturing systems, which have been presented in this chapter. The two major barriers that impede the integration of AM to manufacturing systems are high investment cost and lack of technical know-how. A framework (Product-Service-System) to overcome the high investment barrier has also been included in this chapter. In the end, a road map has been presented for the future research work in the field of material and design tools for AM in the framework of I 4.0.

REFERENCES

Alam, N., Vats, P., & Kashyap, N. (2018). Internet of Things: A literature review. *2017 Recent Developments in Control. Automation and Power Engineering, RDCAPE, 2017*(May), 192–197. doi:10.1109/RDCAPE.2017.8358265

Alcácer, V., & Cruz-Machado, V. (2019). Scanning the Industry 4.0: A Literature Review on Technologies for Manufacturing Systems. *Engineering Science and Technology, an International Journal.* doi:10.1016/j.jestch.2019.01.006

Billinghurst, M., Clark, A., & Lee, G. (2014). A survey of augmented reality. *Foundations and Trends in Human-Computer Interaction, 8*(2–3), 73–272. doi:10.1561/1100000049

Bose, S., Vahabzadeh, S., & Bandyopadhyay, A. (2013). Bone tissue engineering using 3D printing. *Biochemical Pharmacology, 16*(12), 496–504. doi:10.1016/j.mattod.2013.11.017

Ceruti, A., Marzocca, P., Liverani, A., & Bil, C. (2019). Maintenance in Aeronautics in an Industry 4.0 Context: The Role of Augmented Reality and Additive Manufacturing. *Journal of Computational Design and Engineering*, 1–11. doi:10.1016/j.jcde.2019.02.001

Chen, M., Mao, S., & Liu, Y. (2014). Big data: A survey. *Mobile Networks and Applications, 19*(2), 171–209. doi:10.100711036-013-0489-0

Chong, L., Ramakrishna, S., & Singh, S. (2018). A review of digital manufacturing-based hybrid additive manufacturing processes. *International Journal of Advanced Manufacturing Technology, 95*(5-8), 2281–2300. doi:10.100700170-017-1345-3

Chung, C. A. (2004). *Simulation Modelling Handbook: A practical approach.* Academic Press.

Cui, Y., Kara, S., & Chan, K. C. (2020). Manufacturing big data ecosystem: A systematic literature review. *Robotics and Computer-integrated Manufacturing, 62*, 101861. doi:10.1016/j.rcim.2019.101861

Dalmarco, G., Ramalho, F. R., Barros, A. C., & Soares, A. L. (2019). Providing industry 4.0 technologies: The case of a production technology cluster. *The Journal of High Technology Management Research, 30*(2), 100355. doi:10.1016/j.hitech.2019.100355

Dilberoglu, U. M., Gharehpapagh, B., Yaman, U., & Dolen, M. (2017). The Role of Additive Manufacturing in the Era of Industry 4.0. *Procedia Manufacturing, 11*(June), 545–554. doi:10.1016/j.promfg.2017.07.148

Farid, S., Shirazi, S., Gharehkhani, S., Mehrali, M., Yarmand, H., Simon, H., ... Osman, A. (2015). A review on powder-based additive manufacturing for tissue engineering : Selective laser sintering and inkjet 3D printing. *Science and Technology of Advanced Materials, 16*, 1–20. doi:10.1088/1468-6996/16/3/033502 PMID:27877783

Galati, F., & Bigliardi, B. (2019). Industry 4.0: Emerging themes and future research avenues using a text mining approach. *Computers in Industry, 109*, 100–113. doi:10.1016/j.compind.2019.04.018

Gaub, H. (2016). *Customization of mass-produced parts by combining injection molding and additive manufacturing with Industry 4.0 technologies*. Reinforced Plastics. doi:10.1016/j.repl.2015.09.004

Gilchrist, A. (2016). *Introducing Industry 4.0. In Industry 4.0* (pp. 195–215). Berkeley, CA: Apress. doi:10.1007/978-1-4842-2047-4

Gill, S. S., & Kaplas, M. (2009). Comparative study of 3D printing technologies for rapid casting of aluminium alloy. *Materials and Manufacturing Processes, 24*(12), 1405–1411. doi:10.1080/10426910902997571

Gölzer, P., & Fritzsche, A. (2017). Data-driven operations management: Organisational implications of the digital transformation in industrial practice. *Production Planning and Control, 28*(16), 1332–1343. doi:10.1080/09537287.2017.1375148

Hermann, M., Pentek, T., & Otto, B. (2016). Design principles for industrie 4.0 scenarios. *Proceedings of the Annual Hawaii International Conference on System Sciences*, 3928–3937. 10.1109/HICSS.2016.488

Jacobs, S., Grunert, R., Mohr, F. W., & Falk, V. (2008). 3D-Imaging of cardiac structures using 3D heart models for planning in heart surgery : A preliminary study. *Interactive Cardiovascular and Thoracic Surgery, 7*(1), 6–9. doi:10.1510/icvts.2007.156588 PMID:17925319

Kamal, M., & Rizza, G. (2019). *Design for metal additive manufacturing for aerospace applications. Additive Manufacturing for the Aerospace Industry*. Elsevier Inc. doi:10.1016/b978-0-12-814062-8.00005-4

Keil, S. (2017). Design of a cyber-physical production system for semiconductor manufacturing design. *Proceedings of the Hamburg International Conference of Logistics (HICL)*, 318–340.

Khas, K. S., Pandey, P. M., & Ray, A. R. (2015). Design and development of a device to measure the deformities of clubfoot. *Proceedings of the Institution of Mechanical Engineers. Part H, Journal of Engineering in Medicine, 229*(3), 194–204. doi:10.1177/0954411915574758 PMID:25833995

Knofius, N., van der Heijden, M. C., & Zijm, W. H. M. (2019). Consolidating spare parts for asset maintenance with additive manufacturing. *International Journal of Production Economics, 208*, 269–280. doi:10.1016/j.ijpe.2018.11.007

Lee, J., Bagheri, B., & Kao, H. A. (2015). A Cyber-Physical Systems architecture for Industry 4.0-based manufacturing systems. *Manufacturing Letters, 3*, 18–23. doi:10.1016/j.mfglet.2014.12.001

Leyh, C., Martin, S., & Schaffer, T. (2017). Industry 4.0 and Lean Production-A matching relationship? An analysis of selected Industry 4.0 models. *Proceedings of the 2017 Federated Conference on Computer Science and Information Systems, FedCSIS 2017, 11*, 989–993. 10.15439/2017F365

Lu, Y. (2017). Industry 4.0: A survey on technologies, applications and open research issues. *Journal of Industrial Information Integration, 6*, 1–10. doi:10.1016/j.jii.2017.04.005

Megahed, F. M., & Jones-Farmer, L. A. (2013). A Statistical Process Monitoring Perspective on "Big Data". *Frontiers in Statistical Quality Control*, 1–21.

Mehrpouya, M., Dehghanghadikolaei, A., Fotovvati, B., Vosooghnia, A., Emamian, S. S., & Gisario, A. (2019). The Potential of Additive Manufacturing in the Smart Factory Industrial 4.0: A Review. *Applied Sciences (Basel, Switzerland), 9*(18), 3865. doi:10.3390/app9183865

Mourtzis, D., Doukas, M., & Bernidaki, D. (2014). Simulation in manufacturing: Review and challenges. *Procedia CIRP, 25*, 213–229. doi:10.1016/j.procir.2014.10.032

Mun, J., Yun, B. G., Ju, J., & Chang, B. M. (2015). Indirect additive manufacturing based casting of a periodic 3D cellular metal - Flow simulation of molten aluminum alloy. *Journal of Manufacturing Processes, 17*, 28–40. doi:10.1016/j.jmapro.2014.11.001

Negahban, A., & Smith, J. S. (2014). Simulation for manufacturing system design and operation: Literature review and analysis. *Journal of Manufacturing Systems, 33*(2), 241–261. doi:10.1016/j.jmsy.2013.12.007

Ngo, T. D., Kashani, A., Imbalzano, G., Nguyen, K. T. Q., & Hui, D. (2018). Additive manufacturing (3D printing): A review of materials, methods, applications and challenges. *Composites Part B, 143*, 172–196. doi:10.1016/j.compositesb.2018.02.012

Pandey, P. M., & Singh, G. (2019). Effect of unit cell shape and strut size on flexural properties of ordered Copper foam. *International Journal of Engineering Science, 12*(3), 73–79.

Patra, S., & Young, V. (2016). A Review of 3D Printing Techniques and the Future in Biofabrication of Bioprinted Tissue. *Cell Biochemistry and Biophysics, 74*(2), 93–98. doi:10.100712013-016-0730-0 PMID:27193609

Pereira, A. C., & Romero, F. (2017). A review of the meanings and the implications of the Industry 4.0 concept. *Procedia Manufacturing, 13*, 1206–1214. doi:10.1016/j.promfg.2017.09.032

Poomathi, N., Singh, S., Prakash, C., Patil, R. V., Perumal, P. T., & Barathi, V. A. … Maheshwari, N. U. (2018). Bioprinting in ophthalmology: Current advances and future pathways. *Rapid Prototyping Journal*. doi:10.1108/RPJ-06-2018-0144

Prakash, C., Singh, S., Pabla, B. S., Sidhu, S. S., & Uddin, M. S. (2018). Bio-inspired low elastic biodegradable Mg-Zn-Mn-Si-HA alloy fabricated by spark plasma sintering. *Materials and Manufacturing Processes, 00*(00), 1–12. doi:10.1080/10426914.2018.1512117

Rosen, D. W. (2014). Research supporting principles for design for additive manufacturing: This paper provides a comprehensive review on current design principles and strategies for AM. *Virtual and Physical Prototyping*, *9*(4), 225–232. doi:10.1080/17452759.2014.951530

Saucedo-Martínez, J. A., Pérez-Lara, M., Marmolejo-Saucedo, J. A., Salais-Fierro, T. E., & Vasant, P. (2018). Industry 4.0 framework for management and operations: A review. *Journal of Ambient Intelligence and Humanized Computing*, *9*(3), 789–801. doi:10.100712652-017-0533-1

Shafiq, S. I., Sanin, C., Toro, C., & Szczerbicki, E. (2015). Virtual engineering object (VEO): Toward experience-based design and manufacturing for industry 4.0. *Cybernetics and Systems*, *46*(1-2), 35–50. doi:10.1080/01969722.2015.1007734

Singh, G., & Pandey, P. M. (2018). Design and Analysis of Long-Stepped Horn for Ultrasonic Assisted Sintering. *21st International Conference on Advances in Materials and Processing Technology (AMPT).*

Singh, G., & Pandey, P. M. (2019a). Experimental investigations into mechanical and thermal properties of rapid manufactured copper parts. *Proceedings of the Institution of Mechanical Engineers. Part C, Journal of Mechanical Engineering Science*, *0*(0), 1–14. doi:10.1177/0954406219875483

Singh, G., & Pandey, P. M. (2019b). Rapid manufacturing of copper components using 3D printing and ultrasonic assisted pressureless sintering: Experimental investigations and process optimization. *Journal of Manufacturing Processes*, *43*, 253–269. doi:10.1016/j.jmapro.2019.05.010

Singh, G., & Pandey, P. M. (2019c). Topological ordered copper graphene composite foam: Fabrication and compression properties study. *Materials Letters*, *257*, 126712. doi:10.1016/j.matlet.2019.126712

Singh, G., & Pandey, P. M. (2019d). Ultrasonic Assisted Pressureless Sintering for rapid manufacturing of complex copper components. *Materials Letters*, *236*, 276–280. doi:10.1016/j.matlet.2018.10.123

Singh, G., & Pandey, P. M. (2019e). Uniform and graded copper open cell ordered foams fabricated by rapid manufacturing: Surface morphology, mechanical properties and energy absorption capacity. *Materials Science and Engineering A*, *761*(June), 138035. doi:10.1016/j.msea.2019.138035

Singh, G., & Pandey, P. M. (2020). Neck growth kinetics during ultrasonic-assisted sintering of copper powder. *Proceedings of the Institution of Mechanical Engineers. Part C, Journal of Mechanical Engineering Science*, *0*(0), 1–11. doi:10.1177/0954406220904108

Singh, J. P., & Pandey, P. M. (2018). Fabrication and assessment of mechanical properties of open cell porous regular interconnected metallic structure through rapid manufacturing route. *Rapid Prototyping Journal*, *24*(1), 138–149. doi:10.1108/RPJ-04-2015-0043

Singh, S., Prakash, C., & Ramakrishna, S. (2019). 3D printing of polyether-ether-ketone for biomedical applications. *European Polymer Journal*, *114*(February), 234–248. doi:10.1016/j.eurpolymj.2019.02.035

Singh, S., Ramakrishna, S., & Singh, R. (2017). Material issues in additive manufacturing: A review. *Journal of Manufacturing Processes*, *25*, 185–200. doi:10.1016/j.jmapro.2016.11.006

Singh, S., Singh, N., Gupta, M., Prakash, C., & Singh, R. (2019). Mechanical feasibility of ABS/HIPS-based multi-material structures primed by low-cost polymer printer. *Rapid Prototyping Journal*, *25*(1), 152–161. doi:10.1108/RPJ-01-2018-0028

Singh, S., & Singh, R. (2015). Wear modelling of Al-Al2O3functionally graded material prepared by FDM assisted investment castings using dimensionless analysis. *Journal of Manufacturing Processes*, *20*, 507–514. doi:10.1016/j.jmapro.2015.01.007

Singh, S., & Singh, R. (2016). Fused deposition modelling based rapid patterns for investment casting applications : A review. *Rapid Prototyping Journal*, *22*(2), 123–143. doi:10.1108/RPJ-02-2014-0017

Tack, P., Victor, J., Gemmel, P., & Annemans, L. (2016). 3D - printing techniques in a medical setting : A systematic literature review. *Biomedical Engineering Online*, *15*(1), 1–21. doi:10.118612938-016-0236-4 PMID:27769304

Wang, X., Jiang, M., Zhou, Z., Gou, J., & Hui, D. (2017). 3D printing of polymer matrix composites : A review and prospective. *Composites. Part B, Engineering*, *110*, 442–458. doi:10.1016/j.compositesb.2016.11.034

Wits, W. W., García, J. R. R., & Becker, J. M. J. (2016). How Additive Manufacturing Enables more Sustainable End-user Maintenance, Repair and Overhaul (MRO) Strategies. *Procedia CIRP*, *40*, 693–698. doi:10.1016/j.procir.2016.01.156

Wong, K. V., & Hernandez, A. (2012). A Review of Additive Manufacturing. *International Scholarly Research Network*, *2012*, 1–10. doi:10.5402/2012/208760

Yi, L., Gläßner, C., & Aurich, J. C. (2019). How to integrate additive manufacturing technologies into manufacturing systems successfully: A perspective from the commercial vehicle industry. *Journal of Manufacturing Systems*, *53*, 195–211. doi:10.1016/j.jmsy.2019.09.007

Chapter 14
Monetised Risk Values and Cost–Benefit Evaluation of Maintenance Options for Aging Equipment

Maria Chiara Leva
Technical University of Dublin, Ireland

Micaela Demichela
https://orcid.org/0000-0001-5247-7634
Politecnico di Torino, Italy

Gabriele Baldissone
https://orcid.org/0000-0001-7015-8995
Politecnico di Torino, Italy

ABSTRACT

In this chapter, the authors present an overview of methods that can be used to evaluate risks and opportunities for deferred maintenance interventions on aging equipment, and underline the importance to include monetised risk considerations and timeline considerations, to evaluate different scenarios connected with the possible options. Asset managers are compelled to continue operating aging assets while deferring maintenance and investment due to the constant pressure to reduce maintenance costs as well as short-term budget constraints in a changing market environment. Monetised risk values offer the opportunity to support risk-based decision-making using the data collected from the field. The chapter presents examples of two different methods and their practical applicability in two case studies in the energy sector for a company managing power stations. The use of the existing and the new proposed solutions are discussed on the basis of their applicability to the concrete examples.

DOI: 10.4018/978-1-7998-3904-0.ch014

BACKGROUND

Asset managers are often required to maintain operational continuity with aging assets while deferring maintenance and investment. The consequences of such decisions are rarely immediate, deferring maintenance and investment can result in cost reduction in the short term, however it also required on the other end to set up an "intelligent prognostics" system, which can measure, control, and alert the operating personnel, detecting unavoidable risk degradation (Mc Nett 2016).

Overall the situation calls for better monitoring and control on ageing equipment, to quantify the impact of operating modes on system reliability, to accurately estimate their residual life and to adapt the maintenance strategy, while respecting safety, regulation and operational performance.

Any power plant is required to supply the amount of energy demanded by the market and to comply with the regulatory requirements defined by government laws. To attain the objective, one of the most important aspects is to guarantee technical availability. This feature is not always easily achieved: during operation, the equipment that are used the most are gradually deteriorating, until they reach a deterioration failure, or other types of failures, such as fatigue or corrosion, induced by the specific operating conditions of the equipment itself.

New opportunities are given by monitored systems in modern process plants, whose data have to be integrated in DCS (distributed control systems) and PLC (programmable logic controller) to prevent potential dangerous outcomes. Data gained through the automated monitoring and control systems, but also through inspections, are fundamental and can bee used to support risk- based decision making, and ultimately the risk management of ageing equipment. To understand, identify, and manage critical states in aging or deterioration, it is necessary to develop mathematical models that represent the aging process to show the deterioration of power equipment, and determine the cause of aging. A review of the most recurrent causes of trips in a power generation company in The republic of Ireland showed that 43% of all the trips are attributed to equipment aging as root cause and in those 43% more than 65% explicitly mention equipment aging as the primary causes. Although aging and deterioration effects are unavoidable, it is desirable to find a way to slow down the deterioration rate, and to extend equipment's service life and this could be obtained by reducing exposure to the operating, environmental or transient conditions that cause or exacerbate deterioration.

This chapter presents a risk-based assessment for decision related to different maintenance intervention options as applied to the context of power generation (Darabnia & Demichela, 2013a & b).

Asset Management and Risk Analysis

The standard ISO 5500 (ISO 55001, 2014 & ISO 55002, 2014) provides guidelines and industry bets practices for asset management that can apply to organisations of any size, in any sector. The ISO 55000 family stem from PAS 55, a publicly available specification introduced by the British Standards Institution (BSI). In the section of ISO 55001 that deals with planning to achieve asset management objectives there is a clause (Clause 6.2.2) that states the following:

"When planning how to achieve its asset management objectives, the organization shall determine and document….actions to address risks and opportunities associated with managing the assets, taking into account how these risks and opportunities can change with time, by establishing processes for:

- Identification of risks and opportunities;
- Assessment of risks and opportunities;
- Determining the significance of assets in achieving asset management objectives;
- Implementation of the appropriate treatment and monitoring of risks and opportunities."

Appropriate asset management strategies allows companies to achieve risk reduction, opportunity identification or process improvement, which can be identified early in the implementation, and can be exploited to demonstrate returns and gain stronger stakeholder support. An asset management system can help in gaining an understanding of assets, their performance, the risks associated with managing assets, it supports a long-term and sustainable approach to decision making and the organization's risk-based decision making processes can become more effective by addressing asset and financial risks together, and by balancing performance, costs and risks.

A company asset management policy should also take into consideration the evolution of PACS (protection, automation and control system) as they should be developed keeping in mind the possibility of future extension, modification and functional upgrading until complete system refurbishment.

Evaluating Risks and Opportunities for Deferred Maintenance on Aging Equipment

The ability to perform a risk-based decision making based on big data collected from the field and, whenever possible, integrated with the DCS data requires new capabilities for the design of operations. This will reduce human error in safety critical industries, demonstrating how an appropriate model Risk Assessment can significantly improve training market share and business and safety performance (Demichela, Pirani, Leva, 2014).

Managing power plants to operate... till the end of the paragraph ...uses minimum fuel at night" there were two references missing i had to redraft the paragraph to add them. please change the entire paragraph for" In the energy sectors Managing power plants to operate at low cost and without excessive damage is the goal of every plant owner. However the variability of power sources (E.g. Wind generation) coupled with the deregulation of the electricity sector required that older base-load units were either shut down or operated at part-load levels more often (Troy et al. 2010). This cycling operation has demanding effects on the components leading to increased outages and substantial costs. For those units the forecast aging outcomes are difficult to estimate. The challenge expands progressively because "cycling operation" is a broad concept that encompasses load-following, low load, hot starts, warm starts, and cold start of a plant with different lengths of time between operations. Is it more financially convenient to shut down a particular unit at night to save fuel and incur the multitude of cycling costs or stay online at minimum load? Some studies have reported that damage manifests itself in terms of known past and future maintenance and or replacements costs, forced outages and deratings from cycling. Typically the damage mechanism is fatigue and corrosion of the boiler tubes (Lefton et al 2006).

Evaluating the likelihood of occurrence of equipment failures and their consequences for the power plant operation, through a quantitative analysis, integrating probability and consequence, are process well obtained by the Risk Analysis Method. It attempts to focus on the following objectives:

1. To develop a data structure supporting consistent hazard identification and risk rating across different sites;
2. To develop equivalent severity and frequency scales for different loss types and for application across different business units, such as operations, maintenance, finance, HR etc, in order to guarantee a uniform risk analysis;
3. To embed the risk analysis within a risk management process and share good practices across the company.

When optimizing a process scheme or planning equipment maintenance, it's common to use methods, such as risk assessment, which can show the cost impact of the proposal solution. This number could be used to prioritize a series of items that have been risk assessed. Those methods require a great deal of data both for the assessment of probabilities and assessment of consequences. After the scope definition it is necessary to quantify and then evaluate the risk, in order to obtain a proper solution, to which is related a proper acceptability/tolerance (Leva et al. 2016).

A decision generally deals with three elements: alternatives, consequences, and preferences. The alternatives are the possible choices for consideration. Now it's clear that the amount of decision making procedures is considerable, but each method shows the alternatives in a different way, going to remark what it might be the eventual solution. The goal is to choose the best alternative with the proper consideration of uncertainty. Decision analysis provides methods for quantifying preferences trade-offs for performance along multiple decision attributes while taking into account risk objectives.

In the following part we present the use of a proposed customised methods to identify valuable maintenance/repair overhaul options for ageing equipment through a risk based evaluation of different options.

Data Availability in Industry 4.0 and Risk Assessment Prospective

Industry 4.0 aims to revolutionize industrial realities through the use of new approaches and new technologies. Maintenance management may also be affected by this revolution.

One of the first applications of industry 4.0 in maintenance management concerns the analysis and management of Big Data. In this sense Lee et al. (2015) proposes its application in the maintenance of Cyber infrastructure, instead Wan et al. (2017) compares the results of traditional methodologies for the definition of maintenance with Big Data based approaches, and Kezunovic et al. (2013) proposes the use of Big Data for the management of Power System maintenance.

Several companies begin to have large databases on the operating status of the equipment and on maintenance, also the adoption of smart equipment will further increase the size of these databases. But already under current conditions the dimension of these databases make them difficult to analyse with traditional methodologies. In addition, the traditional methodologies foresee to analyse the variables individually but in this way the combined effects of several variables remain hidden or difficult to observe. But with the innovations connected to Big Data, the analysis of these databases will become easier and multidimensional, in order to obtain reliable and detailed results that facilitate planning and risk assessment of maintenance activities. This type of approach can be particularly attractive for aging plant as they may have an already large dataset, even if in many cases not always complete and consistent, whose knowledge can guide maintenance choices to extend the operating life of the system and or improve its performance.

In addition, the use of equipment developed within industry 4.0 (smart equipment) can encourage the adoption of approaches related to prognostic maintenance (Bagheri et al, 2015, Chiu et al., 2017, Ferreiro et al. 2016). This is possible because the equipment is matched with different sensors that can identify the warning signals of the faults in order to carry out a punctual maintenance plan. Furthermore, in this way it will be possible to foresee a dynamic risk assessment linked to the real conditions of the system.

With the adoption of these approaches, Industry 4.0 aims to reduce maintenance costs and risks as optimal choices can be made on maintenance strategies.

In this context the estimation of likelihood of failures for individual equipment and the plant overall can benefit from the use the data to gain more realistic figures for risk exposure and likelihood of failure than the value we were currently able to use for the industrial case studies we had to work with were the estimation of likelihood of failure still heavily rely on technical experience.

Current Methods and Proposed Solutions in Concrete Case Studies

The case studies aims to compare some technical risk assessments, to underline what needs to be improved, but mostly to describe the percentage of uncertainty and inaccuracy, which is derived from a calculation and a processional choice most often based on technical experience, which would require a greater theoretical basis. The following method is applied to the analysis of two case studies in company operating in the energy sector across multiple locations in Ireland. The organisation highlighted the need to advance the identification, analysis and management of risks across the technical assets of the business, and to facilitate comparison and tracking of options in relation to corrective actions. A project team was therefore assembled, with representatives from different stations and specialism, to create a risk analysis process and template to meet the business' needs. The overall approach for risk evaluation of possible interventions in the company is reported in Figure 1.

The approach has been applied to two examples. The first example is about a Generator Rotor hub in a dam. The outer rim of the rotor had a series of stacked steel laminations onto which the rotor poles were attached. The defects that were identified appeared to be largely related to poor welding technique/procedure during construction. So, the risk was around how to best address the potential defect, to avoid structural failures of rotor hub in service, an event that would have a noticeable impact such as damaging the generator poles/stator and possibly injuring personnel. To prevent this phenomena the company needs to decide when to replace the rotor and if do this replacement using a newly designed component. These questions are answered with a technical risk assessment achieved by focusing on technical/operational considerations, together with commercial ones around the possible applicable solutions and how they could be realised, identifying all the critical aspects through the support of a risk matrix and corresponding monetary and likelihood intervals to support the relevant scenarios severity and probability estimations.

The second example concerns the improvement that could be obtained by using a more specific and detailed method. This is demonstrated in a case study to evaluate options for revamping of a vortex finder of a Circulating Fluidised Bed Combustion Boiler (CFBC) in a fossil fuel power generating station. Hot cyclones separate the circulating bed material from the flue gas stream leaving the furnace at high temperature (nromally between 800-900 DegC). The solid particles separated from the flue gas are recirculated back into the furnace through the loop seal. A vortex finder is used to enhance the separation efficiency of the hot cyclone. It is also used alongside the outlet Flue to allow the flue gas containing the fine ash particles to exit the cyclones (Das & Bhattacharya 1990).

The options available to address the problem are reported in Table 1.

Figure 1. Overall approach for risk evaluation of possible interventions

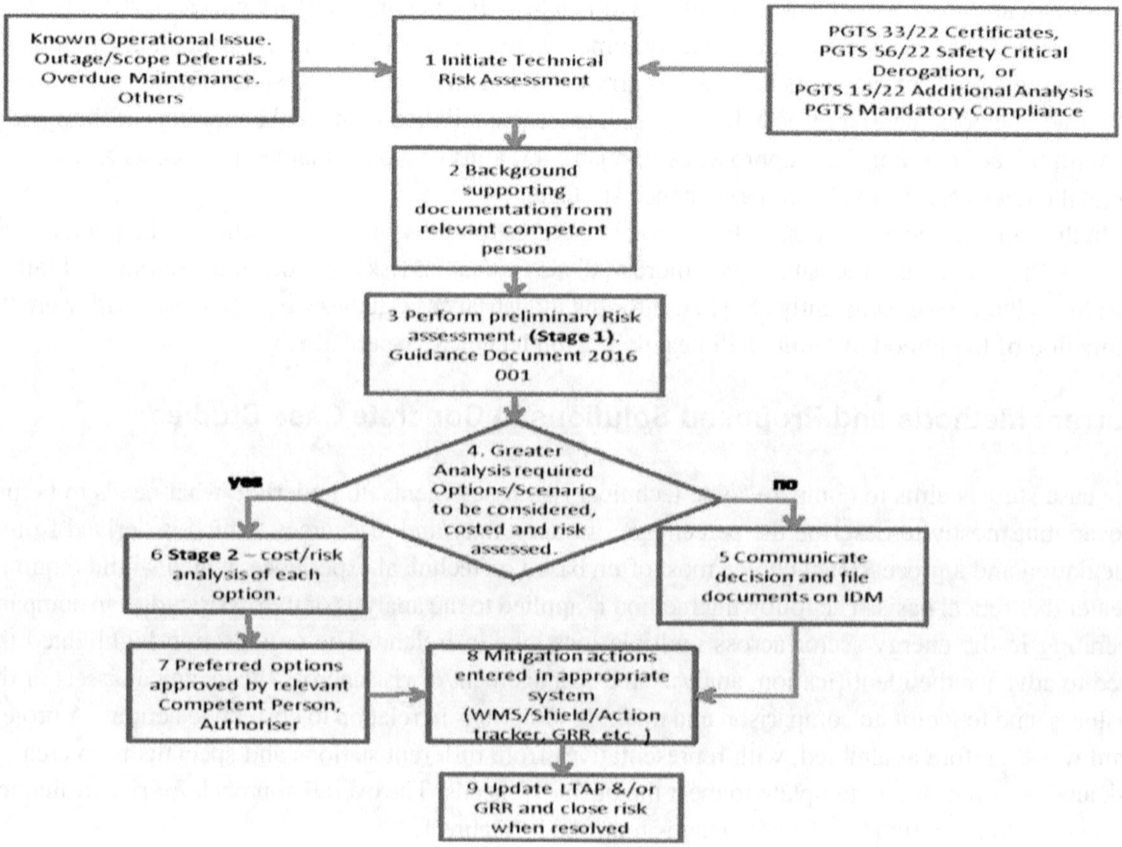

Table 1.Summary of options for intervention

Option	Description	Downtime
0	Do nothing to 2022 (Risk Baseline)	
1	Replace Vortex Finder & Panels 2021; repair/inspect in meantime	2 weeks outage 2018, 2019, 2020 4 weeks in 2021
2	Replace Vortex Finder 2022; repair/inspect in meantime	2 weeks outage 2018, 2019, 2020 3 weeks in 2021; 4 weeks 2022
3	Replace Vortex Finder & Panels 2021; repair/inspect in meantime	2 weeks outage 2018, 2019, 2020 5 weeks in 2021
4	Replace Vortex Finder & Panels 2022; repair/inspect in meantime	2 weeks outage 2018, 2019, 2020 3 weeks in 2021; 5 weeks in 2022
5	Partial panel replacement until 2021; repair/inspect in meantime VF replacement in 2021	3 weeks outage 2018, 2019, 2020 5 weeks in 2021
6	Partial panel replacement until 2022; repair/inspect in meantime VF replacement in 2022	3 weeks outage 2018, 2019, 2020, 2021 5 weeks in 2022

The analysis started by assessing the risk of the current situation if no intervention occurs (baseline case example in Figure 2).

Figure 2. Part of the template used for risk evaluation of current situation

Return Rate 8.40%										
	Year (for NPV)			0	1	2	3	4	5	6
	Impact Category	Rating	Year	2018	2019	2020	2021	2022	2023	2024
	Technical Performance	3	Estimate if available	€20,000	€20,000	€20,000	€20,000	€20,000	€20,000	€20,000
	Financial	1	Estimate if available	€50,000	€50,000	€50,000	€50,000	€50,000	€50,000	€50,000
	Safety	1	Estimate if available	€10,000	€10,000	€10,000	€10,000	€10,000	€10,000	€10,000
	Environmental impact	1	Estimate if available	€10,000	€10,000	€10,000	€10,000	€10,000	€10,000	€10,000
Risk profile Do nothing to 2022	Reputation	1	Estimate if available	€10,000	€10,000	€10,000	€10,000	€10,000	€10,000	€10,000
	Likelihood/Prob			0.00001	0.00055	0.0055	0.055	0.55	0.55	0.55
	Max impact			3	3	3	3	3	3	3
	Prob			1	2	3	4	5	5	5
			Risk (base case)	3	6	9	12	15	15	15
			€Monetised Risk (Risk profil	€ 1	€ 55	€ 550	€ 5,500	€ 55,000	€ 55,000	€ 55,000

After the risk assessment for the current situation (see Figure 2), each option was broke down into the actual costs of the planned interventions plus the monetized values of the risk connected to the intervention itself; the residual risk left after the intervention was also then evaluated. A sum of risk rated costs was then computed by considering for each options the loss of revenue directly proportional to the length of time needed to perform each intervention plan, the actual repair costs (evaluating all the components involved such as materials, equipment, labor etc.), the mitigation costs that refer to the options as for instance a replacement with the same design, could be obtained considering continuous maintenance regime. The evaluation also needs to take into account the performance penalties that could be connected with the options (such as inability to run at full speed power etc.)

A risk rated cost was also assessed in relation to health and safety incidents and/or the risk of process safety events connected with the intervention (such as working at height etc..). No simulation or iterative calculation is supporting this evaluation, as it simply requires expert judgment from the asset management team with an appropriate anchor point to justify the chosen estimates. Last but not least the method required to evaluate a monetized risk value of the residual risk left after the intervention through which a value of the risk rated benefits can be subsequently added to.

The option evaluation requested from the power generation company has to analyse both the investment and the monetised risk exposures that the company has to afford throughout the years. First of all, the risk profile has to be analysed in order to better understand where the risky situation has a higher severity between the five categories of impact: Technical, Financial, Safety, Environmental and Reputational. This kind of analysis has to be done for all the estimated remaining life of the plant (e.g. 7 years or more) and then it will be possible to evaluate that the risk rating will increase progressively during the years if there are no interventions and therefore investments to reduce the risk. Figure 3 reports the risk index profile over the years for baseline case (option 0). Figure 4 on the other hand reports the risk index profiles evaluated over the years for the other options considered.

Figure 3. Risk index profile over the years for baseline case

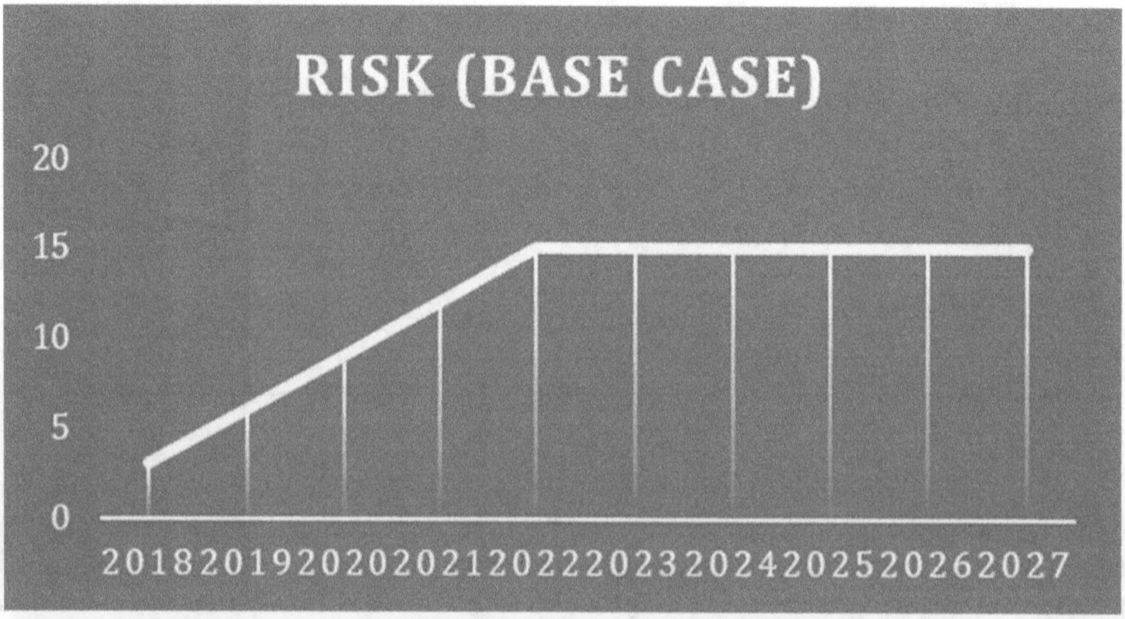

The risk rated benefit used for prioritizing the options is simply obtained by comparing the benefit given by the current risk exposure minus the residual risk exposure against the costs which is obtained by the evaluation of risk rated costs of the various mitigation plans as explained above. The higher the benefits the better the option.

The comparing of risk index profile is in itself not sufficient the option evaluation needs to compare also the different option of investment that is possible to afford in the plant consider the remaining life of the plant itself. In this context let's consider two different types of investments:

Downtime cost;
CAPEX

It is also possible to add other costs as NDT (No Destructive Testing)

Downtime cost is referred to the period during which a piece of equipment is not functional or cannot work. It can be due to technical failure, machine adjustment, maintenance, or non-availability. So, the average downtime cost is usually built into the price of goods produced, to recover its cost from the sales revenue. CAPEX (Capital Expenditure) is the money a company spends to acquire or upgrade productive assets in order to increase the capacity or efficiency for more than one accounting period. The evaluation also needs to take into account the performance penalties that the company could incur into in connection with the options (such as inability to run at full power etc.)

A risk rated cost also needs to consider the potential health and safety/ process safety and or other risk exposure connected with the intervention itself (e.g. operator injury during refurbishment due to the hazard of working at height etc.). No simulation or iterative calculation is supporting this evaluation, as it simply requires expert judgment from the asset management team with an appropriate anchor point to justify the chosen estimates. Last but not least the method required to evaluate a monetized risk

Figure 4. Reports the risk index profiles evaluated over the years for the other options considered

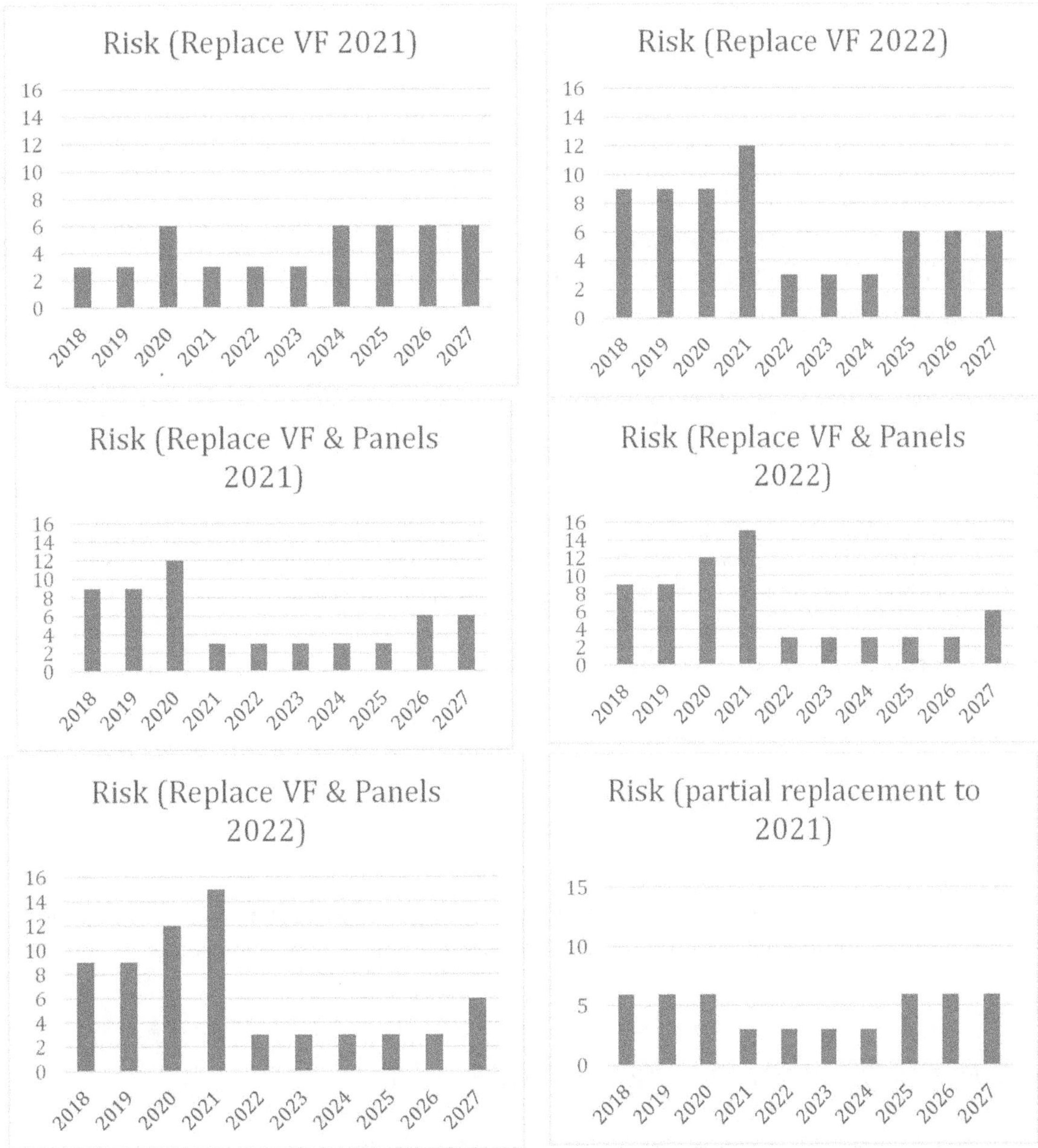

value of the residual risk left after the intervention through which a value of the risk rated benefits can be subsequently added to in each year.

The Net Present Value (NPV) is the value of all future cash flows (positive and negative) over the entire life of an investment discounted to the present. NPV analysis is a form of intrinsic valuation and is used extensively across finance and accounting for determining the value of capital projects that involves cash flows. It is used as a measure of the investment's return. The formula for calculating the NPV is reported below in Eq. (1).

$$NPV(i,N) = \sum_{t=0}^{N} \frac{R_t}{(1+i)^t} \tag{1}$$

Where:

R_t is the monetised amount at the time *t (which can be an actual cash flow in terms of costs or a financial benefit)*

i is the return rate, that is the profit earned from the investment (for the case study of the power generation company a return rate of 8,4% is use)

t is the time of the cash flow.

In summary the evaluation of the best investment option has to compare the benefit resulting from the reduced monetised risk exposure, adding potential other benefit (e.g. reselling the used asset when the new one has been bought), considering the Net Present Value at the time of the decision to be taken, divided by the Net Present values of the costs and monetised risk exposure resulting for each options

Therefore, the comparison between the NPV of the total reduction of monetised risk exposure plus other benefits derived from an option over the years of the remaining life of the plant divided by the NPV of the costs and potential monetised risk exposure associated to that option can give an evaluation index that can highlight which option should be the best one. The option with the higher Evaluation Index should be the best compromise between monetised risk exposure and investment and to be indicative of a good option it should be above 1.

$$Evaluation\ index = \frac{\sum_{t=0}^{N} NPV(reduced\ monetised\ risk\ \exp osure + other\ benefits)}{\sum_{t=0}^{N} NPV(\cos ts\ associated\ to\ option)} \tag{2}$$

Table 2 below shows the option evaluation index actually obtained for the six different options outlined above over an investment for a period of 7 years since all the options were below 1 none of them was viable.

Table 2. Example of table to calculate the Option evaluation index

Profile/Option	NPV of cost	NPV of monetised risk exposure and other benefits	Evaluation Index
Risk Profile	-	€ 202,413	-
Option 1	€1,813,577	€ 201,879	**0.11**
Option 2	€2,061,988	€ 196,119	**0.10**
Option 3	€2,077,412	€ 196,254	**0.09**
Option 4	€2.450.226	€ 153,104	**0.06**
Option 5	€2.813.686	€ 201,809	**0.08**

Figure 5 represents then the evaluation index represented for comparison across the various options. The benefit of this approach is that the evaluation index can be compared not only between options within an individual project but also across projects at portfolio levels if needed. Thus this provides a common evaluation index across options and projects easy to understand and act upon for decision-making. None of the options available for this project seems to be worth doing unfortunately as the costs outnumber the benefits. For an option to be viable the index should be >1.

Figure 5. Example of comparison of evaluation index calculated for the various options of the case study

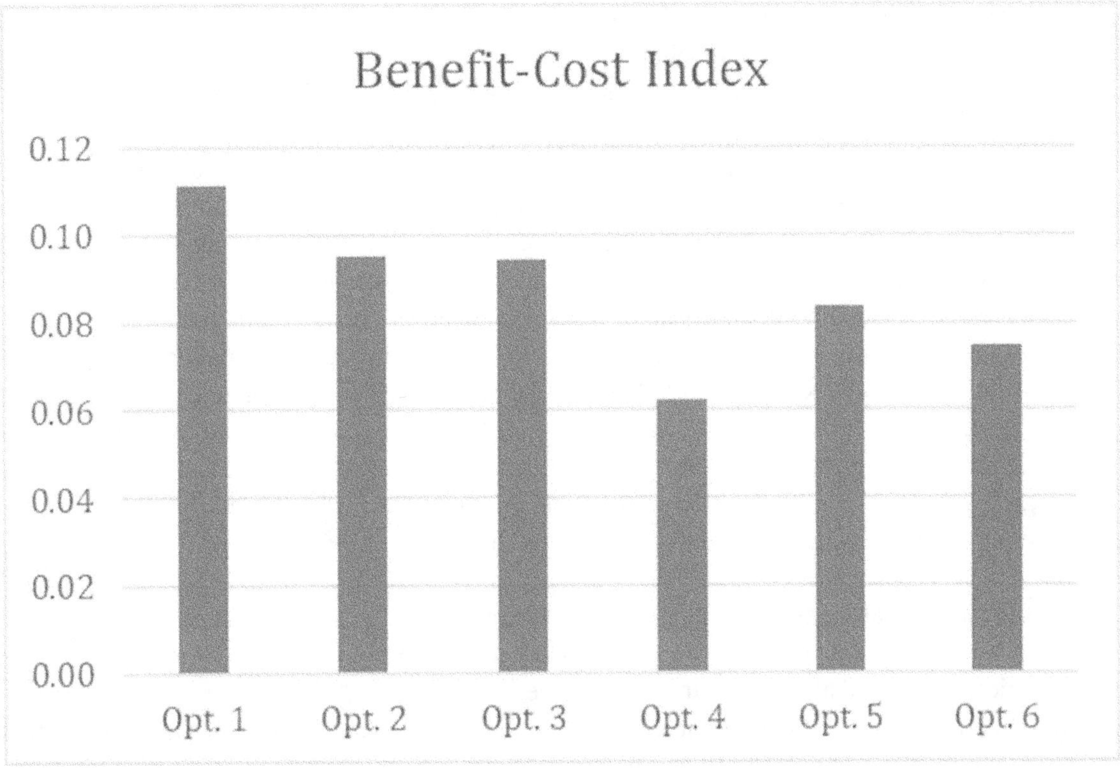

A New Prospective for the Risk Assessment: The Dynamic Decision Analysis

A decision tree can be a very feasible approach for this type of risk assessment. The method should be applicable to aging and new equipment, with a different value of the risk. Market demand often pushes companies to work till end of life ageing plant equipment, making a proper risk assessment even more essential.

A good starting point for the development of the logical-probabilistic model can be a functional analysis of the system, exploiting also, where available, the information contained in the analysis already carried on with traditional methodologies. The last case study was about a common procedure for LP rotor gas turbines risk assessment in which the decision was divided:

- **Option 0:** LP Module cover lift for inspection, without new blade stored,

- **Option A:** LP Module cover lift for inspection, with a stock of new blades stored (7 blades are usually stored, based on the previous company experience)
- **Option B:** LP Module cover lift and replacement of all the blades.
- **Option C:** LP inner block replacement (rotors & carriers)

Each option has a different work scope and cost, associated to a risk rating based on the risk matrix, gained through a possible value of the impact and the likelihood of the solution adopted. In this study a dynamic event tree method was used to calculate the probability of the gas turbine blade rupture. It was based on a structural reliability analysis in order to quantify the behaviour of several critical components of structures subject to uncertain loadings, boundary and geometrical conditions and material parameters.(Millwater & Wu 1993). Turbine failure modes are generally described by frequency, stress corrosion and erosion, creep fatigue; the computation of the probability of failure requires the selection of a particular stress condition, function of time.

The Integrated Dynamic Decision Analysis (IDDA) - described by Remo Galvagni [Clementel & Galvagni 1984; Galvagni & Clementel 1989) allows to carry on the risk analysis in a dynamic way, taking into account process time dependant occurrences, optimisation of procedures for LPG tanks maintenance and testing (Gerbec, Baldissone, & Demichela, 2016). The IDDA method is based on a logical-probabilistic modelling of the system, integrated with its phenomenological modelling. The IDDA analysis was applied in Baldissone et al. (2017) to formaldehyde plant production, in Baldissone et al. (2014 & 2016) to a VOCs treatment plant and Demichela (2014) to the risk-based design on an allyl-chloride production plant. The interaction between logical–probabilistic and phenomenological model could be easily shown in Figure 6.

The logical-probabilistic model, based on the general logic theory, is built according to its own syntactic system to shape an enhanced event tree structure, through:

1. The functional analysis of the system and the construction of a list of levels, with questions and affirmations on the functionality of each element; each level represents the elementary matter of the logical model and also a node in an event tree,
2. The construction of a 'reticulum' indicating the addresses (subsequent level) to be visited depending on the response in each level, and a comment string that allows the user to read the logical development of a sequence;
3. The association to each level of a probability value, which represents the expected degree of occurrence of a failure or an unwanted event and of an uncertainty ratio, which represents the distribution of the probability.
4. The definition of the logical and probabilistic constraints, which allow modifying the run time of the model, to fit it to the current knowledge status.

The elaboration of the logical – probabilistic model, described in the input file through the IDDA software, returns all the possible sequences of events that the system could undergo, depending on the knowledge disclosed in the input model, together with their probabilities of occurrence.

In the logical modelling of the different options, it has been considered how the maintenance procedures are carried on depending on the number of damaged blades.

The Event Tree representation for the option A is reported in Figure 7 (Millwater & Wu 1993).

Figure 6. The interaction between logical–probabilistic and phenomenological model

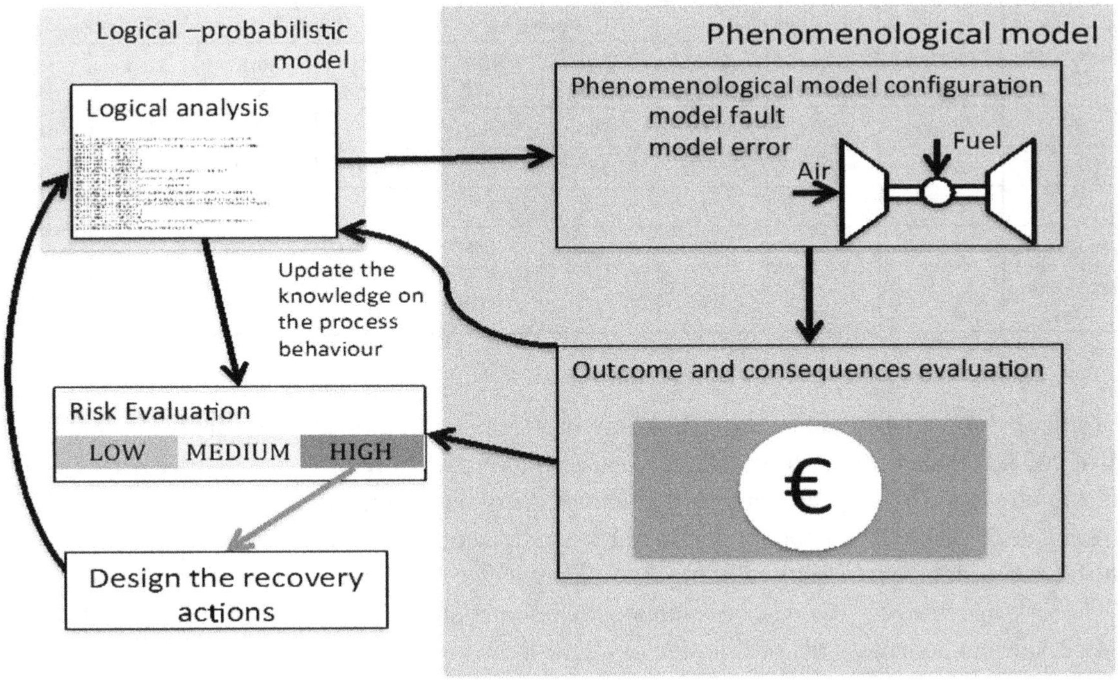

Figure 7. Event Tree developed for Option A

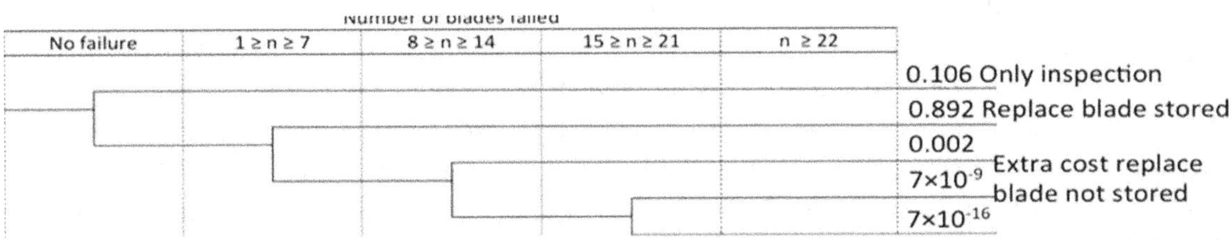

A phenomenological model, together with the logical modelling, must be prepared in order to describe the physical behaviour of the system. The phenomenological model could influence the updating of the logical model generating a better description of the real behaviour of the system, i.e. indicating if, after the failure of a piece of equipment, the other components are able to compensate its dearth and complete the operation, or if cumulative effects can appear and diverge the system from its normal behaviour.

The phenomenological model can provide a direct estimation of the consequences for each single sequence in order to obtain a risk estimation, the evaluation of the overall risk of the system and the expected value of the consequence. The latter is calculated as a weighted average of the consequences, according to their probability.

The phenomenological model allows calculating the costs for the different maintenance options for each scenario described by the logical model. The basic costs have been supplied by the plant management. Table 3 reports the maintenance cost and the risk value for the differ options.

Table 3. Risk evaluation for the different options

Blades fault	Probability	Option 0		Option A		Option B		Option C	
		Cost [k€]	Risk [k€]	Cost [k€]	Risk [k€]	Cost [k€]	Risk [k€]	Cost [k€]	Risk [k€]
0	0.106	1800	190.3	1800	190.8	3400	360.4	5800	614.8
1≤blades≤7	0.892	2050	1829.2	1850	1650.2	3400	3032.8	5800	5173.6
>7	0.002	2050	4.1	2050	4.1	3400	6.8	5800	11.6
Total Risk [€]	-	-	2023	-	1845	-	3400	-	5800
Yearly Risk [€/y]	-	-	506	-	461	-	425	-	725

Through this model the following results have been obtained: Option 0 (no spare parts stored) has a risk of 2023k€, that decreases to 1845k€ for Option A (7 blades stored), since the spare parts are available and no plant trip is required longer than the planned one. Option A modelling also confirms the company decision on the number of blades to be stored, since the failure of 7 blades or less is the one that shows the higher probability of occurrence (about 99%).

The risk increases for the other two options: Option B shows a risk of 3399k€ and Option C of 5799k€, since in both cases the substitution of all the blades (Option B) and also of the rotors and carriers (Option C) involve higher costs.

On the other hand, these two last options have as a consequence an extension of the maintenance period: 8 years for the B and C Options against 4 years for 0 and A Options. Thus actualising the risk values to the yearly maintenance risk, the following figures are obtained:

Option B with a risk value of 425k€/y appears to be the most convenient, against the 461 k€/y for Option A, 506k€/y for Option 0 and 725 k€/y for Option C.

Option C appears to be in both cases the less convenient, but it should be considered that the complete renovation of the inner parts of the turbine should bring also to an improvement of the plant productivity, that should compensate the higher investment costs. Unfortunately the productivity data were not still available when this paper was extended, thus the model does not take into account at the moment this aspect.

Multivariable Fuzzy Logic Approach

Another possible approach to the problem could be multivariable approach. The use of this approach can be explained by the fact that different maintenance strategies can have different pros and cons. In these cases, a multivariable approach can integrate the different peculiarities into a single index. The difficulty of these approaches is to compare different aspects and in giving the correct weight to the different aspects.

A possible approach of this type applied to the decision between multiple maintenance alternatives can use Fuzzy logic (Baldissone et al. 2018). In order to apply this approach, 6 input variables were identified:

- Cost (C); comparison between the economical cost conveniences
- Time (T); comparison of the time required before the next planned maintenance

- Performance (P); comparison between the performances of the equipment
- Duration (D); comparison between the equipment stops due to the maintenance activity
- Risk (R); comparison between the possible unwanted events due to the maintenance activity
- Safety (S); comparison between the risks related to the occupational safety for the operators carrying on the maintenance

To generalise the methodology, relative variables can be used, referring to a reference maintenance alternative. For example the variable Cost (C) is obtained according to the following equation:

$$C = \frac{c_0}{c_1} \quad (3)$$

Where c_0 is the cost of the reference maintenance activity and c_1 is the cost of the analyzed one.

The variables have been divided into 3 trapezoidal membership functions, by way of example the Figure 8 shows the cost variable.

Figure 8. Cost variable membership function

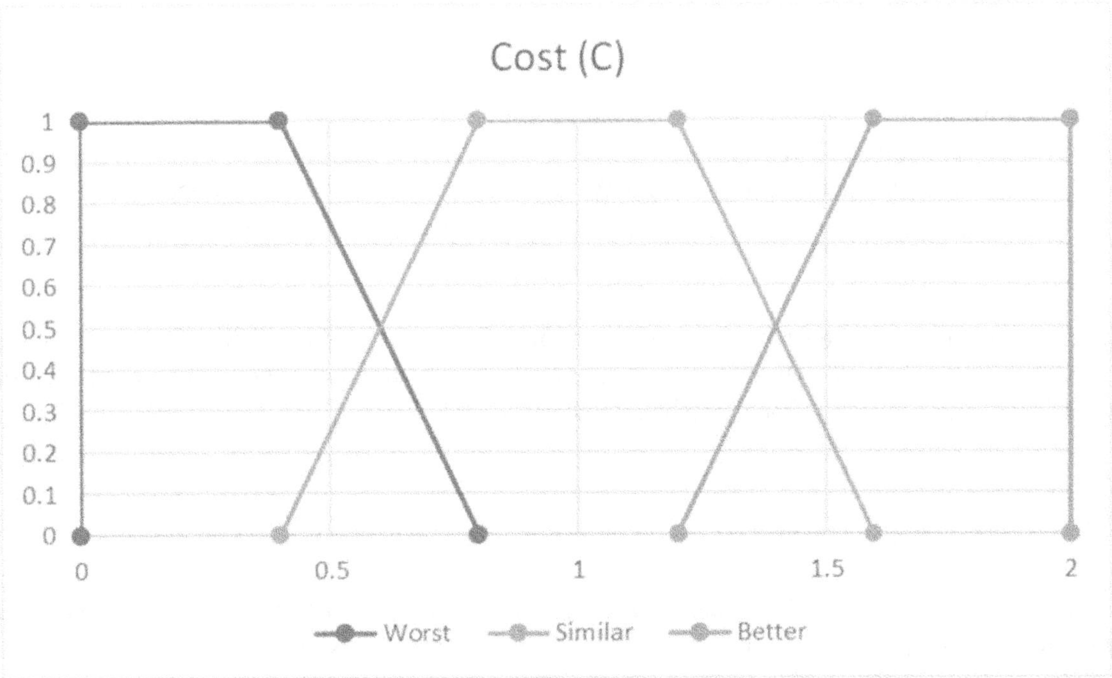

The output variable represents a global judgement on the opportunity of adopting the analysed maintenance strategy with respect to the reference one. The output variable is divided in 3 membership functions (Figure 9):

Figure 9. Output variable membership function

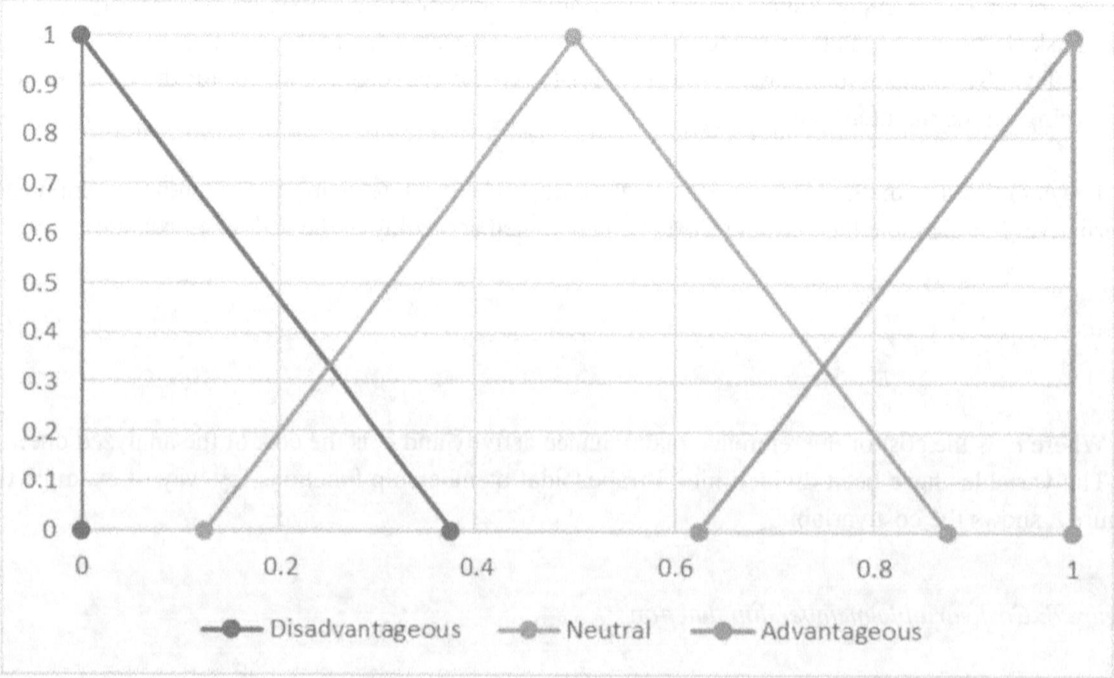

- **Disadvantageous:** The analysed maintenance strategy has a worse global performance with respect to the reference one;
- **Neutral:** The analysed maintenance strategy and the reference one are equivalent;
- **Advantageous:** Globally, the analysed strategy has a better performance with respect to the reference one.

To link the input variables with the output variables, 729 "IF .. THAN ..." type rules are required, representing all the possible permutations of the membership functions of the input variables. In this case the following equation is applied to obtain the rules.

$$\sum_i w_i \cdot W_{ij} = W_o \tag{4}$$

Where w_i is the weight of the variable i, in this case w_i is 1 for all the input variables, because are considered for all input variables the some degree of importance. W_{ij} is the weight of the membership function j in the variable i, in this case are assigned -1 for the worst membership function, 0 for similar and +1 for better. W_0 is used to define the output membership function. If W_0 is lower than 0, the "disadvantageous" membership function is assigned, representing the situation of more declining conditions than increasing ones. If W_0 is equal to 0, the membership function "neutral" is assigned, representing equivalent pros and cons. The membership function "Advantageous" is assigned if W_0 is higher than 0, because in this case the better performance of the analysed maintenance strategy with respect to the standard one are prevailing on the lower performance.

This methodology is tested of the case study presented previously relating to the maintenance of the Low Pressure Turbine (LP) of a Power plant Company.

The standard maintenance strategy usually adopted in the Company is: the LP turbine is opened, the blades are tested and, in case of failed blades, they are replaced. In order to minimise maintenance stops, a small amount (7) of new blades is stored. Two other maintenance strategies have been proposed:

- **Strategy 1:** Lifting of the LP module cover lifting, and replacement of all the blades, with a decrease of the maintenance time since the test of the blades is made after the turbine has been refurbished and restarted;
- **Strategy 2:** LP inner block replacement (rotor and carriers), with a power increase.

The data relating to the various maintenance alternatives are shown in Table 4.

Table 4. Case study data

Maintenance strategy	Cost [M€]	Interval between next maintenance [y]	Power (MW)	Maintenance duration [d]	Monetary risk [k€]	Safety (risk index)
Standard	0.5	4	260	35	1845	6
1	1.02	8	260	60	3399	7
2	2.42	8	263.5	38	5799	10

The result of the Strategy 1 analysis is neutral (0.453): in fact, the disadvantages are balanced by the advantages with respect to the reference maintenance, the Cost (C), Duration (D) and Economical Risk (R) related to the Strategy one are slightly worse, the Performance (P) and Safety (S) are similar and the Time (T) is better.

The result of the Strategy 2 analysis is Disadvantageous (0.146), because the disadvantages are not balanced by advantages. For Strategy 2, Cost (C) and Economical Risk (R) are worse, Safety (S) is similar – worse, Duration (D) and Performance (P) are similar and Time (T) is better.

Following the results of the methodology, the plant managers were able to compare the strategies based on different variables and decided to keep the original maintenance strategy developed on the basis of operational experience.

CONCLUSION AND WAY FORWARD

As discussed above, most of industrial equipment is nowadays used beyond the useful life foreseen at the design stage. To maintain the requested productivity level and the operational safety the equipment thus need a more frequent maintenance. The extra-maintenance requires extra-costs that need to be optimised, in terms of frequency and effectiveness, against the chance of renewing the equipment itself.

This paper has described three case studies where this issue has been addressed through qualitative and quantitative methodologies able to support the operational optimisation in the industrial domain.

The methods and approaches here described have been tested on the energy sector, but they are of wider application and their extension to other domains will be the way forward.

ACKNOWLEDGMENT

This publication has emanated from research supported in part by a research grant from Science Foundation Ireland (SFI) under Grant Number 14/IFB/2718 and by the SAFERA ERA-NET partnership through the Project PROAGE.

REFERENCES

Bagheri, B., Yang, S., Kao, H.-A., & Lee, J. (2015). Cyber-physical Systems Architecture for Self-Aware Machines in Industry 4.0 Environment. *IFAC-PapersOnLine*, *48*(3), 1622–1627. doi:10.1016/j.ifacol.2015.06.318

Baldissone, G., Cavaglià, G., & Demichela, M. (2014). Are intensified processes safer and more reliable than traditional processes? an emblematic case study. *Chemical Engineering Transactions*, *36*, 415–420.

Baldissone, G., Demichela, M., Camuncoli, G., & Comberti, L. (2017). Formaldehyde production plant modification: Risk based decision making. *Chemical Engineering Transactions*, *57*, 703–708.

Baldissone, G., Demichela, M., & Comberti, L. (2019). Multivariable Based Decision-making for the Maintenance Strategy of Process Equipment. *Chemical Engineering Transactions*, *74*, 643–648.

Baldissone, G., Fissore, D., & Demichela, M. (2016). Catalytic after-treatment of lean VOC–air streams: Process intensification vs. plant reliability. *Process Safety and Environmental Protection*, *100*, 208–219. doi:10.1016/j.psep.2016.01.012

Chiu, Y.-C., Cheng, F.-T., & Huang, H.-C. (2017). Developing a factory-wide intelligent predictive maintenance system based on Industry 4.0. *Zhongguo Gongcheng Xuekan*, *40*(7), 562–571. doi:10.1080/02533839.2017.1362357

Clementel, S., & Galvagni, R. (1984). The use of the event tree in the design of nuclear power plants. *Environment International*, *10*(5), 377–382. doi:10.1016/0160-4120(84)90045-X

Darabnia, B., & Demichela, M. (2013a). Data field for decision making in maintenance optimization: An opportunity for energy saving. *Chemical Engineering Transactions*, *33*, 367–372.

Darabnia, B., & Demichela, M. (2013b). Maintenance an opportunity for energy saving. *Chemical Engineering Transactions*, *32*, 259–264.

Das, A., & Bhattacharya, S. C. (1990). Circulating fluidised-bed combustion. *Applied Energy*, *37*(3), 227–246.

Demichela, M. (2014). Integrated dynamic decision analysis: A method for PSA in dynamic process system. *Safety, Reliability and Risk Analysis: Beyond the Horizon - Proceedings of the European Safety and Reliability Conference*, 1969-1975.

Demichela, M., Pirani, R., & Leva, M. C. (2014). Human factor analysis embedded in risk assessment of industrial machines: Effects on the safety integrity level. *International Journal of Performability Engineering*, *10*(5), 487–496.

Dong, W., Moan, T., & Gao, Z. (2012). Fatigue reliability analysis of the jacket support structure for offshore wind turbine considering the effect of corrosion and inspection. *Reliability Engineering & System Safety*, *106*, 11–27.

Ferreiro, S., Konde, E., Fernández, S., & Prado, A. (2016,). Industry 4.0: predictive intelligent maintenance for production equipment. *European Conference of the prognostics and health management society*, 1-8.

Galvagni, R., & Clementel, S. (1989). Risk analysis as an instrument of design. In C. Maurizio & N. Antonio (Eds.), *Safety design criteria for industrial plants* (Vol. 1). Boca Raton: CRC.

Gerbec, M., Baldissone, G., & Demichela, M. (2016). Design of procedures for rare, new or complex processes: Part 2 – Comparative risk assessment and CEA of the case study. *Safety Science, 100*(B), 203-215.

ISO 55001:2014, Asset management — Management systems — Requirements

ISO 55002:2014, Asset management — Management systems — Guidelines on the application of ISO 55001

Kezunovic, M., Xie, L., & Grijalva, S. (2013). The Role of Big Data in Improving Power System Operation and Protection. *2013 IREP Symposium-Bulk Power System Dynamics and Control –IX (IREP)*.

Lee, J., Ardakani, H. D., Yang, S., & Bagheri, B. (20015). Industrial big data analytics and cyber-physical systems for future maintenance & service innovation. Procedia CIRP, 38, 3–7.

Lefton, S. A., & Besuner, P. (2006). The cost of cycling coal fired power plants. *Coal Power Magazine*, *2006*, 16–20.

Leva, M. C., Baldissone, G., Caso, R., Demichela, M., Lawlor, L., & Mcaleer, B. (2018). Cost benefit evaluation of maintenance options for aging equipment using monetised risk values: A practical application. *Procedia Manufacturing*, *19*, 119–126.

Mc Nett, W. (2016). *Deferred maintenance and risk assessment: Technical analysis is critical to the process, 2016*. Available on line at: https://www.plantengineering.com/single-article/deferred-maintenance-and-risk-assessment-technical-analysis-is-critical-to-the-process/ae7fd166735e91e4d9b405aeae82355f.html

Millwater, H. R., & Wu, Y.-T. (1993). Computational Structural Reliability Analysis of a turbine blade. *Proceedings of the international gas turbine and aeroengine congress and exposition*. 10.1115/93-GT-237

Moan, T. (2005). Reliability-based management of inspection, maintenance and repair of offshore structures. *Structure and Infrastructure Engineering*, *1*(1), 33–62.

Troy, N., Denny, E., & O'Malley, M. (2010). Base-load cycling on a system with significant wind penetration. *IEEE Transactions on Power Systems*, *25*(2), 1088–1097.

Wan, J., Tang, S., Li, D., Wang, S., Liu, C., Abbas, H., & Vasilakos, A. V. (2017). A Manufacturing Big Data Solution for Active Preventive Maintenance. *IEEE Transactions on Industrial Informatics*, *13*(4), 2039–2047. doi:10.1109/TII.2017.2670505

Chapter 15
Supporting Maintenance and Mandatory Inspections Through Digital Technologies on Lifting Equipment

Maria Grazia Gnoni
University of Salento, Italy

Valerio Elia
University of Salento, Italy

Sara Anastasi
Department of Technological Innovation and Safety Equipment, Products and Anthropic Settlements, Italian Workers' Compensation Authority (INAIL), Rome, Italy

Luigi Monica
Department of Technological Innovation and Safety Equipment, Products and Anthropic Settlements, Italian Workers' Compensation Authority (INAIL), Rome, Italy

ABSTRACT

In this chapter, the authors present a critical analysis about the current maintenance and inspection process carried out on hazardous lifting equipment. In Italy, a mandatory audit schema is working requesting a periodic interaction between owners of the lifting equipment and inspectors. The current condition has been analyzed aiming to evaluate potential points of criticalities. A smart platform integrating physical devices—based on internet of things technologies, mobile, and cloud applications—has been developed in order to provide companies and inspectors with a reliable and modular tool to organize, certify, and trace maintenance activities developed on the specific equipment. The final purpose is to guarantee a high level of safety for this type of hazardous equipment.

DOI: 10.4018/978-1-7998-3904-0.ch015

BACKGROUND

Maintenance of hazardous equipment represents a complex task as this process is strictly connected to the overall safety level of this component. By adopting a preventive maintenance policy, equipment is maintained on a schedule over time using inspections, calibrations, repairs and other regular service checks to reduce the chances of failure in the future. Otherwise, predictive maintenance relies on installed or embedded devices to monitor a machine's actual condition. Regular (preventive and/or predictive) maintenance of equipment – especially hazardous ones - is an essential activity to both provide company productivity as well as machine and worker safety. Insufficient/inadequate maintenance can cause serious (and potentially deadly) accidents or health problems. Among several typologies of equipment normally used in a company, one category is currently characterized by a high level of risk (Jannadi and Bu-Khamsin, 2002): lifting equipment. Lifting equipment includes a wide range of equipment – which are characterized by several different components- from loader cranes, mobile cranes, forklifts, mobile elevating platforms, to winches and, hoists. First of all, this equipment is wide spread in several sectors as they are flexible and adaptable tools for handling products as well as moving workers: thus, they are used in different workplaces from construction sites to industrial facilities. As they are easily transferable based on their specific feature, they could move from one workplace to another one as they are rented by different companies or they are used in different locations of the same companies. This condition determines an uncertainty in the actual level of use of the equipment as they could work under different organizational and environmental conditions. Furthermore, based on injuries data, these types of equipment represent nowadays a relevant source of hazard at workplace (Aneziris et al, 2008, Raviv et al, 2017). Historical data (Shepperd et al., 2000; Anastasi et al., 2018) often describe how the improper use of heavy equipment has been the cause of fatal or serious accidents.

Finally, the number of components and accessories in a lifting equipment is usually high and a standardization is not often applied. Thus, evaluating the actual condition of a lifting equipment is usually a complex task involving both maintenance and safety issues. In order to reduce hazards due to the use of this equipment, several national legislations have defined a mandatory audit schemes aiming that all work equipment be maintained in an efficient state, in efficient order and in good repair. The Italian scheme is based on the integration of preventive and predictive maintenance policies for evaluating the actual condition of the equipment: fixed frequencies of controls, visual inspections, tests and an examination of the documentation certifying the developed activities are the basic pillars of this schema. The type and frequencies of the checks to be performed are indicated in the instructions provided by the equipment manufacturer. The mandatory schema requires that all these interventions are reported in a special register (the so called control registry) aiming to provide a reliable evidence of the check developed during the use phase as well as providing the actual "picture" of the equipment status. Therefore, employer must organize internal maintenance activities in order to guarantee the safety of their equipment and at the same time provide information on this process to public/private inspectors during periodic checks. Currently, the audit process is developed through a paper-based system. Several criticalities could be outlined by analyzing historical experience in inspection and maintenance of hazardous lifting equipment which falls under a mandatory audit schema.

The first issue to be discussed – if a paper based system is used to trace the inspection and maintenance process developed on a lifting equipment - is the potential loss of information about activities developed. This condition could occur more often due to the increasing diffusion of renting contracts in this sector. Under a renting contract, maintenance activities over the lifting equipment are usually developed by different users (e.g. customers). Each user could also apply different maintenance strategies- and, often, the owner (i.e. the rental company) could not completely be informed about real activities developed for inspection and maintenance. Based on these conditions, information required during the inspection process could be not completely reliable and available thus contributing to estimate correctly equipment safety levels. Moreover, due to easiness in moving several types of lifting equipment, another potential criticality could be the reliable identification of each single equipment as documents (instruction, EC declaration of conformity and reports of periodic checks) could be lost. This is a very hazardous condition, as the reliability of information provided is not guaranteed, and consequently, the safety level of the equipment. In addition, this lack of information could be also due to the possibility of developing the inspection by different subjects, i.e. Public authorities and Authorized Private Bodies: as information from previous inspections are in paper-based system, the probability of losing such data increases if inspectors do not share them. By focusing specifically on maintenance activity, the main problem is that the maintenance activities managed directly by the employer must be coordinated with those of mandatory inspection (Walker, 2004). As defined previously, different maintenance policies could be adopted for the lifting equipment: a fixed periodic period is defined by the legislation under instructions provided by the manufacturer; different policies (proactive, predictive, etc.) aiming to increase also the productivity of the equipment could be also applied by the company to optimize its internal service level. Thus, a lack of coordination could determine an increased cost for company, an uncertainty in safety level of the machinery as non-compliance during inspection activities could increase. The lack of a standard procedures and an effective tracing system determines an unavailability of relevant information that shall be critical also for the safety of the machinery (Koehn, 1995; Dekker, 2014). All these occurred criticalities outline how the hazardous lifting equipment could contribute to reduce the safety level of workers.

CURRENT LEVEL ADOPTION OF IOT TECHNOLOGIES TO SUPPORT INSPECTION AND MAINTENANCE PROCESSES

Recently, This last option is wide spreading thanks to the diffusion of a new technological paradigm defined as Internet of Things (IOT). IOT represents the transition from computer networks to a network of objects where each object – a drill, a machine tool, a washing machine - becomes equipped with its own digital identity and communicates with the environment: the goal is then to make the objects more functional and/or providing new economic potential in terms of both direct and indirect. Several recent papers proposed applications of IOT technologies for supporting a more efficiency and effectiveness in maintenance process (March and Scudder, 2017; Ayad, 2018). Recent studies have faced with the integration of these technologies for supporting maintenance. Kanawaday and Sane (2017) adopted machine learning technique to elaborate data acquired by installing IOT devices. Jung et al. (2017) described an application of IOT sensors in vibration for improving the measurement accuracy and lowering the total cost of the data acquisition process. Civerchia et al. (2017) discussed the performance of an integrated IOT system - based on battery-powered sensing devices – to monitor machineries. Differently, form these studies, Gore et al. (2018) discussed an IOT based system to support identification and traceability of machinery also for improving maintenance processes. Hegedűs et al (2018) proposed an IOT based platform where IOT devices are fully integrated with software tools for supporting proactive maintenance strategies. At the same time, the use of these technologies is enabling the availability of a huge quantity of data that could be useful for optimizing maintenance processes (Truong, 2018).

THE PROPOSED PROTOYPE SYSTEM

The prototype is based on three components. Specific features of each component are detailed as follows. The first component is a *cloud based software* aiming to sharing information about the equipment lifecycle in a dynamic and reliable way. Installation data as well as data about all mandatory maintenance activities developed by owners for a specific hazardous device in an equipment could be managed by this tool. Information in the cloud software is provided both by duty holders and by inspectors, who periodically verify the actual condition of such an equipment. The second one is the *smart label* based on IOT technologies: it must be attached to each lifting equipment at its first use. Smart labels communicate information with users – i.e. equipment owner or inspector - trough commercial mobile devices. The IOT technology adopted is the Near Field Communication (NFC) based on proximity communication The main goal of this technology is to add "features" to objects aiming to provide automatic information to people interacting with them. Smart labels are able to exchange itself information and also with the surrounding environment thanks to specific enabling technologies. Finally, the third component is the *mobile APP* used to "read" information from smart labels at workplace and, consequently, providing information sharing between companies and inspectors regarding all equipment history from its first installation.

CONCLUSION

The study proposes a smart tool to support companies in tracing all maintenance activities developed on hazardous lifting equipment, which requires mandatory periodic inspection under an audit schema. The prototype is based on three components: smart labels attached to the equipment, a mobile Application for acquiring data directly in the place where the equipment is installed and a cloud software to manage all information about maintenance activities developed on the equipment. The proposed tool aims to overcome two current criticalities outlined for the mandatory inspection process in hazardous equipment: uncertainty in equipment identification and lack of information during mandatory inspection activities. The first problem is solved by creating a reliable information system based on IOT smart labels, where identification could be carried out easily – trough commercial mobile devices - and in a reliable way. Next, information is stored and updated in a dynamic way through a cloud web based software that manages different accesses based on the user type (company or inspector). The software allows different levels of complexity from a basic level – where functionalities as well as information to be added are reduced at minimum – to a higher one where information to be added in the cloud is higher as well as the service provided as on output. The proposed tool provides support both to company employer and inspectors for each own specific activities. In addition, a coordination of different maintenance policies (corrective, preventive, proactive, etc.) could be also realized by adopting the proposed tool as they are all managed in the same information system.

ACKNOWLEDGMENT

The study has been developed under the research project called "Smartbench" ID15 funded by INAIL under the BRIC 2016-2018 tender.

REFERENCES

Ayad, S., Terrissa, L. S., & Zerhouni, N. (2018, March). An IoT approach for a smart maintenance. In *2018 International Conference on Advanced Systems and Electric Technologies (IC_ASET)* (pp. 210-214). IEEE. 10.1109/ASET.2018.8379861

Civerchia, F., Bocchino, S., Salvadori, C., Rossi, E., Maggiani, L., & Petracca, M. (2017). Industrial Internet of Things monitoring solution for advanced predictive maintenance applications. *Journal of Industrial Information Integration*, *7*, 4–12. doi:10.1016/j.jii.2017.02.003

Dekker, S. W. (2014). The bureaucratization of safety. *Safety Science*, *70*, 348–357. doi:10.1016/j.ssci.2014.07.015

Gore, R. N., Kour, H., & Gandhi, M. (2018). IoT based equipment identification and location for maintenance in large deployment industrial plants. In *2018 10th International Conference on Communication Systems & Networks (COMSNETS)* (pp. 461-463). IEEE. 10.1109/COMSNETS.2018.8328244

Hegedűs, C., Varga, P., & Moldován, I. (2018). The MANTIS architecture for proactive maintenance. In *2018 5th International Conference on Control, Decision and Information Technologies (CoDIT)* (pp. 719-724). IEEE. 10.1109/CoDIT.2018.8394904

Jung, D., Zhang, Z., & Winslett, M. (2017). Vibration analysis for iot enabled predictive maintenance. In *2017 IEEE 33rd International Conference on Data Engineering (ICDE)* (pp. 1271-1282). IEEE. 10.1109/ICDE.2017.170

Kanawaday, A., & Sane, A. (2017). Machine learning for predictive maintenance of industrial machines using IoT sensor data. In *2017 8th IEEE International Conference on Software Engineering and Service Science (ICSESS)* (pp. 87-90). IEEE. 10.1109/ICSESS.2017.8342870

Koehn, E. E., Kothari, R. K., & Pan, C. S. (1995). Safety in developing countries: Professional and bureaucratic problems. *Journal of Construction Engineering and Management*, *121*(3), 261–265. doi:10.1061/(ASCE)0733-9364(1995)121:3(261)

March, S. T., & Scudder, G. D. (2017). Predictive maintenance: Strategic use of IT in manufacturing organizations. *Information Systems Frontiers*, 1–15.

Truong, H. L. (2018). Integrated analytics for IIoT predictive maintenance using IoT big data cloud systems. In *2018 IEEE International Conference on Industrial Internet (ICII)* (pp. 109-118). IEEE. 10.1109/ICII.2018.00020

Walker, D., & Tait, R. (2004). Health and safety management in small enterprises: An effective low cost approach. *Safety Science*, *42*(1), 69–83. doi:10.1016/S0925-7535(02)00068-1

Compilation of References

Abbaszadeh, M. (2019). *U.S. Patent Application No. 15/988,515*. Washington, DC: US Patent Office.

Abrahão, F. T. M., Filho, J. N. M., Neto, L. P. D., & de Mesquita, A. C. P. (2019). Development of the AeroLogLabTOOL®. *SPOLM2019*, 14. https://www.marinha.mil.br/spolm/sites/www.marinha.mil.br.spolm/files/DEVELOPMENT%20OF%20 THE%20AEROLOGLABTOOL.pdf

Ackert, S. P. (2010). *Basics of aircraft maintenance programs for financies: evaluations and insights of commercial aircraft programs*. Accessed from: www.aircraftmonitor.com

Act, A. (1996). Health insurance portability and accountability act of 1996. *Public Law*, *104*, 191.

Aggarwal, C. C., & Zhai, C. (2012). *Mining text data*. Springer Science & Business Media. doi:10.1007/978-1-4614-3223-4

Alaloul, W. S., Liew, M. S., Zawawi, N. A. W. A., & Kennedy, I. B. (2019). Industrial Revolution 4.0 in the construction industry: Challenges and opportunities for stakeholders. *Ain Shams Engineering Journal*.

Alam, N., Vats, P., & Kashyap, N. (2018). Internet of Things: A literature review. *2017 Recent Developments in Control. Automation and Power Engineering, RDCAPE*, *2017*(May), 192–197. doi:10.1109/RDCAPE.2017.8358265

Al-Badi, A. H., Ahshan, R., Hosseinzadeh, N., Ghorbani, R., & Hossain, E. (2020). Survey of Smart Grid Concepts and Technological Demonstrations Worldwide Emphasizing on the Oman Perspective. *Applied System Innovation*, *3*(1), 5. doi:10.3390/asi3010005

Alcácer, V., & Cruz-Machado, V. (2019). Scanning the Industry 4.0: A Literature Review on Technologies for Manufacturing Systems. *Engineering Science and Technology, an International Journal*. doi:10.1016/j.jestch.2019.01.006

Alert, I. C. S. (2013). *Industrial Control Systems Cyber Emergency Response Team*. Alert 13-164-01. Cited 19.01. 2015, https://ics-cert. uscert. gov/alerts/ICS-ALERT-13-164-01

Almada-Lobo, F. (2016). The Industry 4.0 revolution and the future of Manufacturing Execution Systems (MES). *Journal of Innovation Management*, *3*(4), 16–21. doi:10.24840/2183-0606_003.004_0003

Al-Najjar, B., Algabroun, H., & Jonsson, M. (2018). Maintenance 4.0 to fulfill the demands of Industry 4.0 and Factory of the Future. *Journal of Engineering Research and Application*, *8*(11), 20–31. Www.Ijera.Com. doi:10.9790/9622-0811022031

Alqahtani, A. Y., Gupta, S. M., & Nakashima, K. (2018, December). Warranty and maintenance analysis of sensor embedded products using internet of things in industry 4.0. *International Journal of Production Economics*, *208*, 483–499. doi:10.1016/j.ijpe.2018.12.022

Alsina, E. F., Chica, M., Trawiński, K., & Regattieri, A. (2018). On the use of machine learning methods to predict component reliability from data-driven industrial case studies. *International Journal of Advanced Manufacturing Technology*, *94*(5-8), 2419–2433. doi:10.100700170-017-1039-x

Alter, S. (2008). Defining information systems as work systems: Implications for the IS field. *European Journal of Information Systems*, *17*(5), 448–469. doi:10.1057/ejis.2008.37

Ansari, F., Khobreh, M., Seidenberg, U., & Sihn, W. (2018). A problem-solving ontology for human-centered cyber physical production systems. *CIRP Journal of Manufacturing Science and Technology*, *22*, 91–106. doi:10.1016/j.cirpj.2018.06.002

Arifiani, L., & Arifiani, L. (2019). The effect of disruption technology, opportunities and challenges of telecommunication industry 4.0 in Indonesia. *International Journal of Recent Technology and Engineering*, *7*(6), 808–819. doi:10.35940/ijeat.F1040.0986S319

Arslan, M., Cruz, C., & Ginhac, D. (2019). Semantic trajectory insights for worker safety in dynamic environments. *Automation in Construction*, *106*, 102854. doi:10.1016/j.autcon.2019.102854

Arthur, W. B. (2017). Where is technology taking the economy? *McKinsey Quarterly*. https://usblearn.belpark.sun.ac.za/pluginfile.php/44861/mod_resource/content/0/7b - Arthur %282017%29 Where is technology taking the economy.pdf

AS 7770:2018. (2018). *Rail cyber security*. Rail Industry Safety and Standards Board.

ASD/AIA. (2018, July). *ASD/AIA SX000i – International guide for the use of the S-Series Integrated Logistics Support (ILS) specifications*. https://www.sx000i.org/

Ashraf, W. (2018). *A model to assess and customize computerized maintenance management systems complies with industry 4.0 vision and requirements: A case study in food processing plant*. Academic Press.

Avizienis, A., Laprie, J.-C., & Randell, B. (2001). *Fundamental Concepts of Dependability*. Research Report No 1145, LAAS-CNRS.

Ayad, S., Terrissa, L. S., & Zerhouni, N. (2018, March). An IoT approach for a smart maintenance. In *2018 International Conference on Advanced Systems and Electric Technologies (IC_ASET)* (pp. 210-214). IEEE. 10.1109/ASET.2018.8379861

Backes, M., Miwa, F., Okajima, C., Souza, A., Jr., & Tkacz, D. (2017). *Digital records: Improving efficiency in commercial aviation*. https://repositorio.itl.org.br/jspui/handle/123456789/11

Bagheri, B., Yang, S., Kao, H., & Lee, J. (2015). Cyber-physical systems architecture for self-aware machines in industry 4.0 environment. *IFAC-PapersOnLine*, *48*(3), 1622–1627. doi:10.1016/j.ifacol.2015.06.318

Bakuei, M., Flores, R., Kropotov, V., & Yarochkin, F. (2017). *Securing Smart Factories*. Academic Press.

Baldissone, G., Cavaglià, G., & Demichela, M. (2014). Are intensified processes safer and more reliable than traditional processes? an emblematic case study. *Chemical Engineering Transactions*, *36*, 415–420.

Baldissone, G., Comberti, L., Bosca, S., & Murè, S. (2019). The analysis and management of unsafe acts and unsafe conditions. data collection and analysis. *Safety Science*, *119*, 240–251. doi:10.1016/j.ssci.2018.10.006

Baldissone, G., Demichela, M., Camuncoli, G., & Comberti, L. (2017). Formaldehyde production plant modification: Risk based decision making. *Chemical Engineering Transactions*, *57*, 703–708.

Baldissone, G., Demichela, M., & Comberti, L. (2019). Multivariable Based Decision-making for the Maintenance Strategy of Process Equipment. *Chemical Engineering Transactions*, *74*, 643–648.

Baldissone, G., Fissore, D., & Demichela, M. (2016). Catalytic after-treatment of lean VOC–air streams: Process intensification vs. plant reliability. *Process Safety and Environmental Protection, 100*, 208–219. doi:10.1016/j.psep.2016.01.012

Bantin, C. C., & Siu, J. (2011). Designing a secure data communications system for automatic train control. *Proceedings of the Institution of Mechanical Engineers. Part F, Journal of Rail and Rapid Transit, 225*(4), 395–402. doi:10.1243/09544097JRRT390

Bastos, P., Lopes, I., & Pires, L. (2014). Application of data mining in a maintenance system for failure prediction. *Safety, Reliability and Risk Analysis: Beyond the Horizon: 22nd European Safety and Reliability, 1*, 933-940.

Batista, N., Melício, R., & Mendes, V. (2017). Services enabler architecture for smart grid and smart living services providers under industry 4.0. *Energy and Building, 141*, 16–27. doi:10.1016/j.enbuild.2017.02.039

Behzadan, A., Dong, S., & Kamat, V. (2015). Augmented reality visualization: A review of civil infrastructure system applications. *Advanced Engineering Informatics, 29*(2), 252–267. doi:10.1016/j.aei.2015.03.005

Behzadan, A., & Kamat, V. (2013). Enabling discovery-based learning in construction using telepresent augmented reality. *Automation in Construction, 33*, 3–10. doi:10.1016/j.autcon.2012.09.003

Ben-Daya, M., Kumar, U., & Murthy, D. P. (2016). *Introduction to maintenance engineering: Modelling, optimization and management*. John Wiley & Sons. doi:10.1002/9781118926581

Berawi, M. A. (2018). Utilizing Big Data in Industry 4.0: Managing Competitive Advantages and Business Ethics. *International Journal of Technology, 9*(3), 430. doi:10.14716/ijtech.v9i3.1948

Berry, M. J. A., & Linoff, G. (1997). *Data Mining Techniques. For Marketing, Sales and Customer Support.* Wiley Computer Publishing.

Bierer, A., Götze, U., Köhler, S., & Lindner, R. (2016). Control and Evaluation Concept for Smart MRO Approaches. *Procedia CIRP, 40*, 699–704. doi:10.1016/j.procir.2016.01.157

Billinghurst, M., Clark, A., & Lee, G. (2014). A survey of augmented reality. *Foundations and Trends in Human-Computer Interaction, 8*(2–3), 73–272. doi:10.1561/1100000049

Bimber, O., & Raskar, R. (2004). *Spatial Augmented Reality. Merging Real and Virtual Worlds*. Wellesley, MA: K Peters Ltd.

Bi, Z., & Cochran, D. (2014). Big data analytics with applications. *Journal of Management Analytics, 1*(4), 249–265. doi:10.1080/23270012.2014.992985

Blanchard, B. S., Fabrycky, W. J., & Fabrycky, W. J. (2011). *Systems engineering and analysis* (Vol. 5). Prentice Hall.

Blanco-Novoa, O., Fernandez-Carames, T. M., Fraga-Lamas, P., & Vilar-Montesinos, M. A. (2018). A Practical Evaluation of Commercial Industrial Augmented Reality Systems in an Industry 4.0 Shipyard. *IEEE Access: Practical Innovations, Open Solutions, 6*, 8201–8218. doi:10.1109/ACCESS.2018.2802699

Bloomfield, R., Bendele, M., Bishop, P., Stroud, R., & Tonks, S. (2016, June). The risk assessment of ERTMS-based railway systems from a cyber security perspective: Methodology and lessons learned. In *International Conference on Reliability, Safety, and Security of Railway Systems* (pp. 3-19). Springer. 10.1007/978-3-319-33951-1_1

Bokrantz, J., Skoogh, A., Berlin, C., & Stahre, J. (2017). Maintenance in digitalised manufacturing: Delphi-based scenarios for 2030. *International Journal of Production Economics, 191*, 154–169. doi:10.1016/j.ijpe.2017.06.010

Borchellini, R., Cardu, M., Colella, F., Labagnara, D., Martinetti, A., Patrucco, M., . . . Verda, V. (2013). A Prevention through Design Approach for the Environmental S&H Conditions and the Ventilation System at an Italian Underground Quarry. *Chemical Engineering Transactions, AIDIC - The Italian Association of Chemical Engineering, 32*, 181-186. doi:10.3303/CET1332031

Borchellini, R., Cardu, M., Colella, F., Labagnara, D., Martinetti, A., Patrucco, M., ... Verda, V. (2013). A Prevention through Design Approach for the Environmental S&H Conditions and the Ventilation System at an Italian Underground Quarry. *Chemical Engineering Transactions*, *32*, 181–186.

Borchiellinia, R., Cardub, M., Colella, F., Labagnara, D., Martinetti, A., Patrucco, M., ... Vittorio, V. (2013). A prevention through design approach for the environmental S&H conditions and the ventilation system at an Italian underground quarry. *AIDIC Conference Series, 11*, 51-60.

Bose, S., Vahabzadeh, S., & Bandyopadhyay, A. (2013). Bone tissue engineering using 3D printing. *Biochemical Pharmacology, 16*(12), 496–504. doi:10.1016/j.mattod.2013.11.017

Bošnjak, S. M., Savićević, S. D., Gnjatović, N. B., Milenović, I. L. J., & Pantelić, M. P. (2015). Disaster of the bucket wheel excavator caused by extreme environmental impact: Consequences, rescue and reconstruction. *Engineering Failure Analysis*, *56*, 360–374. doi:10.1016/j.engfailanal.2015.01.002

Bottani, E., & Vignali, G. (2019). Augmented reality technology in the manufacturing industry: A review of the last decade. *IISE Transactions*, *51*(3), 284–310. doi:10.1080/24725854.2018.1493244

Bowles, J. B., & Pelaez, C. E. (1995). Fuzzy logic prioritization of failures in a system failure mode effects and criticality analysis. *Reliability Engineering & System Safety*, *50*(2), 203–213. doi:10.1016/0951-8320(95)00068-D

Braband, J. (2017, November). Cyber security in railways: Quo vadis? In *International conference on reliability, safety and security of railway systems* (pp. 3-14). Springer.

Bragatto, P. A., Agnello, P., Ansaldi, S. M., & Pittiglio, P. (2010). The digital representation of safety systems at "Seveso" plants and its potential for improving risk management. *Journal of Loss Prevention in the Process Industries*, *23*(5), 601–612. doi:10.1016/j.jlp.2010.05.013

Bream, R. (2006). Lawsuit poses further risk to BP's image. *Financial Times*.

Brettel, M., Friederichsen, N., Keller, M., & Rosenberg, M. (2014). How Virtualization, Decentralization and Network Building Change the Manufacturing Landscape: An Industry 4.0 Perspective. *International Journal of Science, Engineering and Technology*, *8*(1), 37–44.

Campbell, R. J. (2011, June). *The smart grid and cybersecurity: Regulatory policy and issues*. Congressional Research Service, Library of Congress.

Cao, G., Xu, C. A., & Shi, H. a (2013). Design of a management of change platform based on Web and RFID. In *1st CCPS Asia-Pacific Conference on Process Safety 2013* (pp. 784-792). Qingdao, China: AIChE.

Cao, J., & Zhang, S. (2016). *The Application and Design of Big Data in Operation and Maintenance of Industry 4.0.pdf*. Academic Press.

Cappelli, D. M., Moore, A. P., & Trzeciak, R. F. (2012). *The CERT guide to insider threats: how to prevent, detect, and respond to information technology crimes (Theft, Sabotage, Fraud)*. Addison-Wesley.

Case, D. U. (2016). Analysis of the cyber attack on the Ukrainian power grid. *Electricity Information Sharing and Analysis Center, 388.*

CCPS. (2013). *Guidelines for Managing Process Safety Risks During Organizational Change*. Hoboken, NJ: Wiley.

Cegelec. (2020). *Integrated Logistic Support – Maintaining systems in Operating and Security Conditions*. Cegelec Defense. https://www.cegelec-defense.com/en/infrastructures-networks/moc-msc/

Cellary, W. (2019). Non-technical Challenges of Industry 4.0. In Advances in Information and Communication Technology (Vol. 568, pp. 245–258). Springer International Publishing. doi:10.1007/978-3-030-28464-0_1

CEN. (2017). *NEN-EN 50126-1 Railway Applications - The specification and Demonstration of Reliability, Availability, Maintainability and Safety (RAMS)- Part 1: Generic RAMS process* (Vol. 1). Brussels: CEN-CENELEC Management Center.

Cen, E. (2001). *13306: 2001-maintenance terminology. European Standard*. European Committee for Standardization.

Ceruti, A., Marzocca, P., Liverani, A., & Bil, C. (2019). Maintenance in Aeronautics in an Industry 4.0 Context: The Role of Augmented Reality and Additive Manufacturing. *Journal of Computational Design and Engineering*.

Ceruti, A., Marzocca, P., Liverani, A., & Bil, C. (2019). Maintenance in Aeronautics in an Industry 4.0 Context: The Role of Augmented Reality and Additive Manufacturing. *Journal of Computational Design and Engineering, S2288430018302781*(4), 516–526. doi:10.1016/j.jcde.2019.02.001

Cezarino, L. O., Liboni, L. B., Oliveira Stefanelli, N., Oliveira, B. G., & Stocco, L. C. (2019). Diving into emerging economies bottleneck: Industry 4.0 and implications for circular economy. *Management Decision*, MD-10-2018-1084. doi:10.1108/MD-10-2018-1084

Cha, M., Han, S., Lee, J., & Choi, B. (2012). A virtual reality based fire training simulator integrated with fire dynamics data. *Fire Safety Journal*, *50*, 12–24. doi:10.1016/j.firesaf.2012.01.004

Chemweno, P., Pintelon, L., Jongers, L., & Muchiri, P. (2016). *i-RCAM: Intelligent expert system for root cause analysis in maintenance decision making*. Paper presented at the 2016 IEEE International Conference on Prognostics and Health Management (ICPHM). 10.1109/ICPHM.2016.7542830

Chemweno, P., Morag, I., Sheikhalishahi, M., Pintelon, L., Muchiri, P., & Wakiru, J. (2016). Development of a novel methodology for root cause analysis and selection of maintenance strategy for a thermal power plant: A data exploration approach. *Engineering Failure Analysis*, *66*, 19–34. doi:10.1016/j.engfailanal.2016.04.001

Chemweno, P., Pintelon, L., Muchiri, P. N., & Van Horenbeek, A. (2018). Risk assessment methodologies in maintenance decision making: A review of dependability modelling approaches. *Reliability Engineering & System Safety*, *173*, 64–77. doi:10.1016/j.ress.2018.01.011

Chemweno, P., Pintelon, L., Van Horenbeek, A., & Muchiri, P. (2015). Development of a risk assessment selection methodology for asset maintenance decision making: An analytic network process (ANP) approach. *International Journal of Production Economics*, *170*, 663–676. doi:10.1016/j.ijpe.2015.03.017

Chen, J., & Zhu, Q. (2017, July). Interdependent strategic cyber defense and robust switching control design for wind energy systems. In *2017 IEEE Power & Energy Society General Meeting* (pp. 1-5). IEEE.

Chena, G. L., Wu, W. C., Ling, W. Y., Yang, S. C., & Tang, S. M. (2011). Explicit knowledge and tacit knowledge sharing. *International Conference on Management and Service Science, MASS 2011*, 1–4. 10.1109/ICMSS.2011.5998951

Chen, B., Wan, J., Shu, L., Li, P., Mukherjee, M., & Yin, B. (2017). Smart factory of industry 4.0: Key technologies, application case, and challenges. *IEEE Access: Practical Innovations, Open Solutions*, *6*, 6505–6519. doi:10.1109/ACCESS.2017.2783682

Chen, G., He, W., Leung, L. C., Lan, T., & Han, Y. (2017). Assigning licenced technicians to maintenance tasks at aircraft maintenance base: A bi-objective approach and a Chinese airline application. *International Journal of Production Research, 55*(19), 5550–5563. doi:10.1080/00207543.2017.1296204

Chen, M., Mao, S., & Liu, Y. (2014). Big data: A survey. *Mobile Networks and Applications, 19*(2), 171–209. doi:10.100711036-013-0489-0

Chen, X., Xu, L., Wang, Y., Wang, H., Wang, F., Zeng, X., ... Egge, J. (2015). Development of a surgical navigation system based on augmented reality using an optical see-through head-mounted display. *Journal of Biomedical Informatics, 55*, 124–131. doi:10.1016/j.jbi.2015.04.003 PMID:25882923

Chiu, Y.-C., Cheng, F.-T., & Huang, H.-C. (2017). Developing a factory-wide intelligent predictive maintenance system based on Industry 4.0. *Zhongguo Gongcheng Xuekan, 40*(7), 562–571. doi:10.1080/02533839.2017.1362357

Chong, L., Ramakrishna, S., & Singh, S. (2018). A review of digital manufacturing-based hybrid additive manufacturing processes. *International Journal of Advanced Manufacturing Technology, 95*(5-8), 2281–2300. doi:10.100700170-017-1345-3

Chung, C. A. (2004). *Simulation Modelling Handbook: A practical approach*. Academic Press.

Cimini, C., Pirola, F., Pinto, R., & Cavalieri, S. (2020). A human-in-the-loop manufacturing control architecture for the next generation of production systems. *Journal of Manufacturing Systems, 54*, 258–271. doi:10.1016/j.jmsy.2020.01.002

Cipresso, P., Giglioli, I. A. C., Raya, M. A., & Riva, G. (2018). The Past, Present, and Future of Virtual and Augmented Reality Research: A Network and Cluster Analysis of the Literature. *Frontiers in Psychology, 9*, 2086. doi:10.3389/fpsyg.2018.02086 PMID:30459681

Civerchia, F., Bocchino, S., Salvadori, C., Rossi, E., Maggiani, L., & Petracca, M. (2017). Industrial Internet of Things monitoring solution for advanced predictive maintenance applications. *Journal of Industrial Information Integration, 7*, 4–12. doi:10.1016/j.jii.2017.02.003

Clark, J., & Goulder, R. (2002, July). Human systems integration (HSI) ensuring design & development meet human performance capability early in acquisition process. *Program Management*, 88–91.

Clarke, I. (2019). Common Failures in Management of Change. *Chemical Engineering Transactions, 77*, 439–444.

Clementel, S., & Galvagni, R. (1984). The use of the event tree in the design of nuclear power plants. *Environment International, 10*(5), 377–382. doi:10.1016/0160-4120(84)90045-X

Colakovic, A., & Hadzialic, M. (2018). Internet of Things (IoT): A review of enabling technologies, challenges, and open research issues. *Computer Networks, 144*, 17–39. doi:10.1016/j.comnet.2018.07.017

Columbus, L. (2016). Roundup of internet of things forecasts and market estimates, 2016. Forbes Magazine.

Comberti, L., Baldissone, G., Bosca, S., Demichela, M., Murè, S., Petruni, A., . . . Cencetti, S. (2015). Comparison of two methodologies for occupational accidents pre-cursors data collection. In *Safety and Reliability of Complex Engineered Systems - Proceedings of the 25th European Safety and Reliability Conference, ESREL 2015* (pp. 3237-3244). London: CRC Press/Balkema - Taylor & Francis Group. 10.1201/b19094-426

Conner, K. R., & Prahalad, C. K. (1996). A Resource-Based Theory of the Firm: Knowledge Versus Opportunism. *Organization Science, 7*(5), 477–501. doi:10.1287/orsc.7.5.477

Corallo, A., Lazoi, M., & Lezzi, M. (2020). Cybersecurity in the context of industry 4.0: A structured classification of critical assets and business impacts. *Computers in Industry, 114*, 103165. doi:10.1016/j.compind.2019.103165

Cui, Y., Kara, S., & Chan, K. C. (2020). Manufacturing big data ecosystem: A systematic literature review. *Robotics and Computer-integrated Manufacturing*, *62*, 101861. doi:10.1016/j.rcim.2019.101861

Cylus. (2020). Retrieved from https://cylus.com/

Da Xu, L., Xu, E. L., & Li, L. (2018). Industry 4.0: State of the art and future trends. *International Journal of Production Research*, *56*(8), 2941–2962. doi:10.1080/00207543.2018.1444806

Dalmarco, G., Ramalho, F. R., Barros, A. C., & Soares, A. L. (2019). Providing industry 4.0 technologies: The case of a production technology cluster. *The Journal of High Technology Management Research*, *30*(2), 100355. doi:10.1016/j.hitech.2019.100355

Darabnia, B., & Demichela, M. (2013). Data field for decision making in maintenance optimization: An opportunity for energy saving. *Chemical Engineering Transactions*, *32*, 259–264.

Darabnia, B., & Demichela, M. (2013b). Maintenance an opportunity for energy saving. *Chemical Engineering Transactions*, *32*, 259–264.

Das, A., & Bhattacharya, S. C. (1990). Circulating fluidised-bed combustion. *Applied Energy*, *37*(3), 227–246.

Datta, S. (2017). Emergence of Digital Twins Is this the march of reason? *Journal of Innovation Management*, *5*(3), 14–33. doi:10.24840/2183-0606_005.003_0003

Davis, F. D., Bagozzi, R. P., & Warshaw, P. R. (1989). User Acceptance of Computer Technology: A Comparison of Two Theoretical Models. *Management Science*, *35*(8), 982–1003. doi:10.1287/mnsc.35.8.982

Davos. (2016). Emerging economies are less likely to benefit from fourth industrial revolution. *The Economic Times*.

De Pace, F., Manuri, F., Sanna, A., & Zappia, D. (2019). A Comparison Between Two Different Approaches for a Collaborative Mixed-Virtual Environment in Industrial Maintenance. *Frontiers in Robotics and AI*, *6*(March), 1–14. doi:10.3389/frobt.2019.00018

Defense Acquisition University. (2011, December 1). *Integrated Product Support (IPS) Element Guidebook*. https://www.dau.edu/tools/t/Integrated-Product-Support-(IPS)-Element-Guidebook-

Dekker, S. W. (2014). The bureaucratization of safety. *Safety Science*, *70*, 348–357. doi:10.1016/j.ssci.2014.07.015

Demichela, M. (2014). Integrated dynamic decision analysis: A method for PSA in dynamic process system. *Safety, Reliability and Risk Analysis: Beyond the Horizon - Proceedings of the European Safety and Reliability Conference*, 1969-1975.

Demichela, M., & Baldissone, G. (2019). Process risk assessment: From the basics to new frontiers. In M. C. Leva, T. Kontogiannis, M. Gerbec, & O. Aneziris (Eds.), *Total Safety and the Productivity Challenge* (pp. 43–68). Abingdon, UK: Routledge. doi:10.4324/9781315108100-3

Demichela, M., Baldissone, G., & Camuncoli, G. (2017). Risk-Based Decision Making for the Management of Change in Process Plants: Benefits of Integrating Probabilistic and Phenomenological Analysis. *Industrial & Engineering Chemistry Research*, *56*(50), 14873–14887. doi:10.1021/acs.iecr.7b03059

Demichela, M., Baldissone, G., & Darabnia, B. (2018). Using Field Data for Energy Efficiency Based on Maintenance and Operational Optimisation. A Step towards PHM in Process Plants. *Processes (Basel, Switzerland)*, *6*(3), 1–15. doi:10.3390/pr6030025

Demichela, M., Pirani, R., & Leva, M. C. (2014). Human factor analysis embedded in risk assessment of industrial machines: Effects on the safety integrity level. *International Journal of Performability Engineering*, *10*(5), 487–496.

Denis, M. Wm. (2008, September 8). *MRO Technologies & Services: Point of View*. https://www.slideshare.net/guesta9496c4/aviation-mro-it-emergence-of-saas-and-convergence-of-bpo

Department for transport. (2016). *Rail Cyber Security Guidance to Industry*. Retrieved from https://www.rssb.co.uk/Library/improving-industry-performance/2016-02-cyber-security-rail-cyber-security-guidance-to-industry.pdf

Dhillon, B. S. (2008). *Mining Equipment Reliability, Maintainability and Safety*. Springer. doi:10.1007/978-1-84800-288-3

Diaz, V. G. (2019). Handbook of Research on Industrial Advancement in Scientific Knowledge. doi:10.4018/978-1-5225-7152-0

Díaz, M., Martín, C., & Rubio, B. (2016). State-of-the-art, challenges, and open issues in the integration of Internet of things and cloud computing. *Journal of Network and Computer Applications*, *67*, 99–117. doi:10.1016/j.jnca.2016.01.010

Diez-Olivan, A., Del Ser, J., Galar, D., & Sierra, B. (2019). Data fusion and machine learning for industrial prognosis: Trends and perspectives towards industry 4.0. *Information Fusion*, *50*, 92–111. doi:10.1016/j.inffus.2018.10.005

Dilberoglu, U. M., Gharehpapagh, B., Yaman, U., & Dolen, M. (2017). The Role of Additive Manufacturing in the Era of Industry 4.0. *Procedia Manufacturing*, *11*(June), 545–554. doi:10.1016/j.promfg.2017.07.148

Dinis, D., Barbosa-Póvoa, A., & Teixeira, Â. P. (2019). Valuing data in aircraft maintenance through big data analytics: A probabilistic approach for capacity planning using Bayesian networks. *Computers & Industrial Engineering*, *128*, 920–936. doi:10.1016/j.cie.2018.10.015

Djatna, T., & Alitu, I. (2015). An application of association rule mining in total productive maintenance strategy: An analysis and modelling in wooden door manufacturing industry. *Procedia Manufacturing*, *4*, 336–343. doi:10.1016/j.promfg.2015.11.049

Dong, W., Moan, T., & Gao, Z. (2012). Fatigue reliability analysis of the jacket support structure for offshore wind turbine considering the effect of corrosion and inspection. *Reliability Engineering & System Safety*, *106*, 11–27.

Donovan, P. O., Leahy, K., Bruton, K., & Sullivan, D. T. J. O. (2015). An industrial big data pipeline for data - driven analytics maintenance applications in large - scale smart manufacturing facilities. *Journal of Big Data*. doi:10.118640537-015-0034-z

Ebramhimipour, V., & Suzuki, K. (2006). A synergetic approach for assessing and improving equipment performance in offshore industry based on dependability. *Reliability Engineering & System Safety*, *91*(1), 10–19. doi:10.1016/j.ress.2004.11.008

Eckhart, M., & Ekelhart, A. (2018). Towards Security-Aware Virtual Environments for Digital Twins. In *Proceedings of the 4th ACM Workshop on Cyber-Physical System Security* (pp. 61–72). Incheon, South Korea: Association for Computing Machinery. 10.1145/3198458.3198464

Edwards, T., Bayoumi, A., & Lester Eisner, M. (2017). Internet of Things – A Complete Solution for Aviation's Predictive Maintenance. In Y. Bahei-El-Din & M. Hassan (Eds.), *Advanced Technologies for Sustainable Systems: Selected Contributions from the International Conference on Sustainable Vital Technologies in Engineering and Informatics, BUE ACE1 2016, 7-9 November 2016, Cairo, Egypt* (pp. 167–177). Springer International Publishing. 10.1007/978-3-319-48725-0_16

EN 50159:2010 (or IEC 62280). (2010). *Railway applications – signalling, telecommunication and processing systems – safety communication in transmission systems*. Brussels: European Committee for Electrotechnical Standardization.

EN 60300-3-14. (2004). 60300 (3-14): Dependability Management–Part 3-14: Application Guide–Maintenance and maintenance support.

Enisa. (2018). *Industry 4.0 Cybersecurity: Challenges & Recommendations*. Retrieved from https://www.enisa.europa.eu/publications/industry-4-0-cybersecurity-challenges-and-recommendations

Erkoyuncu, J. A., Khan, S., Eiroa, A. L., Butler, N., Rushton, K., & Brocklebank, S. (2017). Perspectives on trading cost and availability for corrective maintenance at the equipment type level. *Reliability Engineering & System Safety*, *168*, 53–69. doi:10.1016/j.ress.2017.05.041

Esposito, M., Lazoi, M., Margarito, A., & Quarta, L. (2019). Innovating the Maintenance Repair and Overhaul Phase through Digitalization. *Aerospace*, *6*(5), 53. doi:10.3390/aerospace6050053

Ezenwa, O. (2018). Effect of competitive intelligence on competitive advantage in Innoson technical and industry limited, Enugu state, Nigeria. *International Journal of Business, Economics & Management*. doi:10.31295/ijbem.v1n1.25

Fahey, R. (2013). *Human factors in information security management systems*. Academic Press.

Famurewa, S. M., Zhang, L., & Asplund, M. (2017). Maintenance analytics for railway infrastructure decision support. *Journal of Quality in Maintenance Engineering*, *23*(3), 310–325. doi:10.1108/JQME-11-2016-0059

Fantini, P., Pinzone, M., & Taisch, M. (2018). Placing the operator at the centre of industry 4.0 design: Modelling and assessing human activities within cyber-physical systems. *Computers & Industrial Engineering*.

Farid, S., Shirazi, S., Gharehkhani, S., Mehrali, M., Yarmand, H., Simon, H., ... Osman, A. (2015). A review on powder-based additive manufacturing for tissue engineering : Selective laser sintering and inkjet 3D printing. *Science and Technology of Advanced Materials*, *16*, 1–20. doi:10.1088/1468-6996/16/3/033502 PMID:27877783

Federal Aviation Administration. (2019). *Flight Standards – Draft Advisory Circulars (ACs) Open for Comment*. https://www.faa.gov/aircraft/draft_docs/media/afx/AC_43-218_Coord_Copy.pdf

Feiner, S. (1994). Redefining the user interface: Augmented reality. *ACM SIGGRAPH. Course Notes*, *2*, 1–18.

Felsberger, L., Todd, B., & Kranzlmüller, D. (2019). Cost and Availability Improvements for Fault-Tolerant Systems Through Optimal Load-Sharing Policies. *Procedia Computer Science*, *151*(2018), 592–599. doi:10.1016/j.procs.2019.04.080

Fernández-Caramés, T.M., Fraga-Lamas, P., Suárez-Albela, M., & Vilar-Montesinos, M. (2018). *A Fog Computing and Cloudlet Based Augmented Reality System for the Industry 4.0 Shipyard.pdf*. Academic Press.

Fernández-Caramés, T. M., Blanco-Novoa, O., Froiz-Míguez, I., & Fraga-Lamas, P. (2019). Towards an Autonomous Industry 4.0 Warehouse: A UAV and Blockchain-Based System for Inventory and Traceability Applications in Big Data-Driven Supply Chain Management. *Sensors (Basel)*, *19*(10), 2394. doi:10.339019102394 PMID:31130644

Ferreiro, S., Konde, E., Fernández, S., & Prado, A. (2016,). Industry 4.0: predictive intelligent maintenance for production equipment. *European Conference of the prognostics and health management society*, 1-8.

Ferrin, F. (1999). Update on optical systems for military head-mounted displays. In *Proc. SPIE 3689, Helmet- and Head-Mounted Displays IV* (pp. 178–185). Orlando, FL: SPIE. doi:10.1117/12.352829

Fischer, E.A. (2016). *Cybersecurity Issues and challenges: in brief*. Academic Press.

Fontaine, O., Martinetti, A., & Mateou, S. (2016). Remote Pilot Aircraft System (RPAS): Just culture, human factors and learnt lessons. *Chemical Engineering Transactions*, *53*, 205–210.

Fontaine, O., Martinetti, A., & Michaelides-Mateou, S. (2016). Remote pilot aircraft system (RPAS): Just culture, human factors and learnt lessons. *Chemical Engineering Transactions*, *52*, 205–2010.

Fontaine, O., Martinetti, A., & Michaelides-Mateou, S. (2016). Remote pilot aircraft system (RPAS): Just culture, human factors and learnt lessons. In *Chemical engineering transactions, AIDIC* (pp. 205–210). The Italian Association of Chemical Engineering. doi:10.3303/CET1653035

Froes, F. H., & Boyer, R. (2019). *Additive manufacturing for the aerospace industry.*

Froese, M. (2016). *Talking with turbines through the Internet of Things.* Retrieved from https://www.windpowerengineering.com/talking-turbines-internet-things/

Fusko, M., Rakyta, M., Krajcovic, M., Dulina, L., Gaso, M., & Grznar, P. (2018). *Basics of designing maintenance processes in Industry 4.0.* Academic Press.

Galar, D., & Kans, M. (2017). *The impact of maintenance 4.0 and big data analytics within strategic asset management.* Paper presented at the Maintenance Performance and Measurement and Management 2016 (MPMM 2016), Luleå, Sweden.

Galar, D., Gustafson, A., Tormos, B., & Berges, L. (2012). Maintenance Decision Making based on different types of data fusion. *Ekspolatacja i Niezawodnosc - Maintenance and Reliability, 14*(2), 135–144.

Galar, D., Gustafson, A., Tormos Martínez, B. V., & Berges, L. (2012). Maintenance decision making based on different types of data fusion. *Eksploatacja i Niezawodnosc-Maintenance and Reliability, 14*(2), 135–144.

Galar, D., Kans, M., & Schmidt, B. (2016). Big data in asset management: Knowledge discovery in asset data by the means of data mining. *Proceedings of the 10th World Congress on Engineering Asset Management (WCEAM 2015),* 161-171. 10.1007/978-3-319-27064-7_16

Galati, F., & Bigliardi, B. (2019). Industry 4.0: Emerging themes and future research avenues using a text mining approach. *Computers in Industry, 109,* 100–113. doi:10.1016/j.compind.2019.04.018

Galvagni, R., & Clementel, S. (1989). Risk analysis as an instrument of design. In C. Maurizio & N. Antonio (Eds.), *Safety design criteria for industrial plants* (Vol. 1). Boca Raton: CRC.

Gambetti, F., Casalli, A., & Chisari, V. (2013). Why sometimes may be neglected management of change? *Chemical Engineering Transactions, 31,* 553–558.

Ganguly, P., Nasipuri, M., & Dutta, S. (2019, August). Challenges of the Existing Security Measures Deployed in the Smart Grid Framework. In *2019 IEEE 7th International Conference on Smart Energy Grid Engineering (SEGE)* (pp. 1-5). IEEE. 10.1109/SEGE.2019.8859917

Garrido-Hidalgo, C., Olivares, T., Ramirez, F. J., & Roda-Sanchez, L. (2019). An end-to-end Internet of Things solution for Reverse Supply Chain Management in Industry 4.0. *Computers in Industry, 112,* 103127. doi:10.1016/j.compind.2019.103127

Gattullo, M., Scurati, G. W., Fiorentino, M., Uva, A. E., Ferrise, F., & Bordegoni, M. (2019). Towards augmented reality manuals for industry 4.0: A methodology. *Robotics and Computer-integrated Manufacturing, 56,* 276–286. doi:10.1016/j.rcim.2018.10.001

Gaub, H. (2016). *Customization of mass-produced parts by combining injection molding and additive manufacturing with Industry 4.0 technologies.* Reinforced Plastics. doi:10.1016/j.repl.2015.09.004

Gdalevitch, M. (2000, November 1). *MSG-3, The Intelligent Maintenance.* Aviation Pros. https://www.aviationpros.com/engines-components/article/10388498/msg3-the-intelligent-maintenance

Gera, M., & Goel, S. (2015). Data mining-techniques, methods and algorithms: A review on tools and their validity. *International Journal of Computers and Applications, 113*(18).

Gerbec, M. (2017). Safety change management – A new method for integrated management of organizational and technical changes. *Safety Science, 100*(Part B), 225-234.

Gerbec, M., Baldissone, G., & Demichela, M. (2016). Design of procedures for rare, new or complex processes: Part 2 – Comparative risk assessment and CEA of the case study. *Safety Science, 100*(B), 203-215.

Gerbec, M., & Levovnik, D. (2019). Management of technical and organizational changes in major hazard industries. In M. C. Leva, T. Kontogiannis, M. Gerbec, & O. Aneziris (Eds.), *Total Safety and the Productivity Challenge* (pp. 226–250). Abingdon, UK: Routledge. doi:10.4324/9781315108100-12

Gerdes, M. (2019). *Health Monitoring for Aircraft Systems using Decision Trees and Genetic Evolution* (Doctoral dissertation). Luleå, Sweden: Luleå University of Technology, Graphic Production.

Gharbi, A., Briceno, S. I., & Mavris, D. N. (2018). *Enabling the Digital Thread in Commercial Aircraft Companies. In 2018 Aviation Technology, Integration, and Operations Conference*. American Institute of Aeronautics and Astronautics. doi:10.2514/6.2018-4008

Gibbons, J. (2009). Debunking the myths surrounding sustainability. *The Irish Times.*

Gilchrist, A. (2016). *Introducing Industry 4.0. In Industry 4.0* (pp. 195–215). Berkeley, CA: Apress. doi:10.1007/978-1-4842-2047-4

Gill, S. S., & Kaplas, M. (2009). Comparative study of 3D printing technologies for rapid casting of aluminium alloy. *Materials and Manufacturing Processes, 24*(12), 1405–1411. doi:10.1080/10426910902997571

Gits, C. (1992). Design of maintenance concepts. *International Journal of Production Economics, 24*(3), 217–226. doi:10.1016/0925-5273(92)90133-R

Goel, P., Datta, A., & Mannan, M. S. (2017). Application of big data analytics in process safety and risk management. In *2017 IEEE International Conference on Big Data (Big Data)* (pp. 1143-1152). Boston, MA: IEEE. 10.1109/BigData.2017.8258040

Gölzer, P., & Fritzsche, A. (2017). Data-driven operations management: Organisational implications of the digital transformation in industrial practice. *Production Planning and Control, 28*(16), 1332–1343. doi:10.1080/09537287.2017.1375148

Gore, R. N., Kour, H., & Gandhi, M. (2018). IoT based equipment identification and location for maintenance in large deployment industrial plants. In *2018 10th International Conference on Communication Systems & Networks (COMSNETS)* (pp. 461-463). IEEE. 10.1109/COMSNETS.2018.8328244

Grabowski, M., Rowen, A., & Rancy, J. (2018). Evaluation of wearable immersive augmented reality technology in safety critical systems. *Safety Science, 103*, 23–32. doi:10.1016/j.ssci.2017.11.013

Grody, A. D. (2018). Rebuilding Financial Industry Infrastructure. *Journal of Risk Management in Financial Institutions, 11*(1), 34–46.

Gunduz, M. Z., & Das, R. (2020). Cyber-security on smart grid: Threats and potential solutions. *Computer Networks, 169*, 107094. doi:10.1016/j.comnet.2019.107094

Guo, Y., Ten, C. W., Hu, S., & Weaver, W. W. (2015). Preventive maintenance for advanced metering infrastructure against malware propagation. *IEEE Transactions on Smart Grid, 7*(3), 1314–1328. doi:10.1109/TSG.2015.2453342

Hadjina, M., & Matulja, T. (2018). *Toward shipbuilding 4.0-an Industry 4.0 changing the face of the shipbuilding industry.* Academic Press.

Hallaway, D., Feiner, S., & Hollerer, T. (2010). Bridging the gaps: Hybrid tracking for adaptive mobile augmented reality. *Applied Artificial Intelligence*, *18*(6), 477–500. doi:10.1080/08839510490462768

Han Siong, P., Chin, K. Y., Bakar, H. T. A., Ling, C. H., Kidam, K., Ali, M. W., ... Kamarden, H. (2017). The contribution of management of change to process safety accident in the chemical process industry. *Chemical Engineering Transactions*, *56*, 1363–1368.

Haroun, A. E., & Duffuaa, S. O. (2009). Maintenance Organization. In M. Ben-Daya, S. O. Duffuaa, A. Raouf, J. Knezevic, & D. Ait-Kadi (Eds.), *Handbook of maintenance management and engineering* (pp. 4–5). Springer. doi:10.1007/978-1-84882-472-0_1

He, Han, & Gu. (2018). *Cost-oriented predictive maintenance based on mission reliability state for cyber manufacturing systems*. Academic Press.

Hegedűs, C., Varga, P., & Moldován, I. (2018). The MANTIS architecture for proactive maintenance. In *2018 5th International Conference on Control, Decision and Information Technologies (CoDIT)* (pp. 719-724). IEEE. 10.1109/CoDIT.2018.8394904

Hélie, S., & Sun, R. (2010). Incubation, insight, and creative problem solving: a unified theory and a connectionist model. *Psychological review* (Vol. 117). doi:10.1037/a0019532

Henderson, S., & Feiner, S. (2011). Exploring the Benefits of Augmented Reality Documentation for Maintenance and Repair. *IEEE Transactions on Visualization and Computer Graphics*, *17*(10), 1355–1368. doi:10.1109/TVCG.2010.245 PMID:21041888

Henshel, D., Cains, M. G., Hoffman, B., & Kelley, T. (2015). Trust as a human factor in holistic cyber security risk assessment. *Procedia Manufacturing*, *3*, 1117–1124. doi:10.1016/j.promfg.2015.07.186

Henshel, D., Sample, C., Cains, M., & Hoffman, B. (2016). *Integrating cultural factors into human factors framework and ontology for cyber attackers*. In D. Nicholson (Ed.), *Advances in Human Factors in Cybersecurity. Advances in Intelligent Systems and Computing* (pp. 123–137). Cham: Springer International Publishing.

He, R., Chen, G., Dong, C., Sun, S., & Shen, X. (2019). Data-driven digital twin technology for optimized control in process systems. *ISA Transactions*, *95*, 221–234. doi:10.1016/j.isatra.2019.05.011 PMID:31109723

Heritage, I. (2019). Protecting Industry 4.0: Challenges and solutions as IT, OT and IP converge. *Network Security*, *2019*(10), 6–9. doi:10.1016/S1353-4858(19)30120-5

Hermann, M., Pentek, T., & Otto, B. (2016). Design principles for industrie 4.0 scenarios. *Proceedings of the Annual Hawaii International Conference on System Sciences*, 3928–3937. 10.1109/HICSS.2016.488

Hoff, R. (2013). MoC scoping—Ensuring that MoC action items are correctly and completely described. *Journal of Loss Prevention in the Process Industries*, *26*(3), 499–510. doi:10.1016/j.jlp.2012.07.007

Holler, J., Tsiatsis, V., Mulligan, C., Avesand, S., Karnouskos, S., & Boyle, D. (2015). *From machine-to-machine to the internet of things*. Academic Press.

Holmberg, K., Adgar, A., Arnaiz, A., Jantunen, E., Mascolo, J., & Mekid, S. (Eds.). (2010). *E-maintenance*. Springer London. doi:10.1007/978-1-84996-205-6

Hopali, E., & Vayvay, Ö. (2018). Industry 4.0 as the last industrial revolution and its opportunities for developing countries. *Analyzing the Impacts of Industry 4.0 in Modern Business Environments*, 65–80. doi:10.4018/978-1-5225-3468-6.ch004

Huang, K., Blue, J., Weng, H., & Liu, S. (2018). *Single machine scheduling with consideration of preventive maintenance and machine health.* Academic Press.

Huang, Z., Hui, P., Peylo, C., & Chatzopoulos, D. (2013). Mobile augmented reality survey: a bottom-up approach. *The Computing Research Repository (CoRR)*, 1309.4413.

Huang, J., Kong, L., Chen, G., Wu, M. Y., Liu, X., & Zeng, P. (2019). Towards secure industrial iot: Blockchain system with credit-based consensus mechanism. *IEEE Transactions on Industrial Informatics*, *15*(6), 3680–3689. doi:10.1109/TII.2019.2903342

Hunsaker, B. T., & Youngdahl, W. E. (2018). *Intel: AI and Industry 4.0 Strategy.* Thunderbird School of Global Management, HBR.

Hutchins, E. M., Cloppert, M. J., & Amin, R. M. (2011). Intelligence-driven computer network defense informed by analysis of adversary campaigns and intrusion kill chains. *Leading Issues in Information Warfare & Security Research*, *1*(1), 80.

IAEA-TECDOC-658, Safety related maintenance in the framework of the reliability cantered maintenance concept, International Atomic Energy Agency.

Iliffe, R. E., Chung, P. W. H., & Kletz, T. A. (1999). More Effective Permit-to-Work Systems. *Process Safety and Environmental Protection*, *77*(2), 69–76. doi:10.1205/095758299529839

Indra. (2020). *Logistics & Maintenance.* Indra. https://www.indracompany.com/en/logistics-maintenance

International Council on Systems Engineering. (2010). *Systems Engineering Handbook Volume 3.1 Appendix M.* Author.

International Electrotechnical Commission. International standard - Dependability management ISO-IEC 300. https://www.iec.ch/

International, S. A. E. (2019). *Standard Practice for Human Systems Integration (SAE6906).* Warrendale, PA: SAE. https://saemobilus.sae.org/content/sae6906

ISO 55001:2014, Asset management — Management systems — Requirements

ISO 55002:2014, Asset management — Management systems — Guidelines on the application of ISO 55001

ISO/IEC 27032. (2012). Information technology–Security techniques–Guidelines for cybersecurity.

Ivanov, D., Sethi, S., Dolgui, A., & Sokolov, B. (2018). A survey on control theory applications to operational systems, supply chain management, and Industry 4.0. *Annual Reviews in Control*, *46*, 134–147. doi:10.1016/j.arcontrol.2018.10.014

Ivezić, D., Tanasijević, M., & Ignjatović, D. (2008). Fuzzy Approach to Dependability Performance Evaluation. *Quality and Reliability Engineering International*, *24*(7), 779–792. doi:10.1002/qre.926

Ivezić, D., Tanasijević, M., Jovančić, P., & Đurić, R. (2019). A Fuzzy Expert Model for Availability Evaluation. *Proceedings of the 2019 20th International Carpathian Control Conference (ICCC).* 10.1109/CarpathianCC.2019.8766031

Jacobs, S., Grunert, R., Mohr, F. W., & Falk, V. (2008). 3D-Imaging of cardiac structures using 3D heart models for planning in heart surgery : A preliminary study. *Interactive Cardiovascular and Thoracic Surgery*, *7*(1), 6–9. doi:10.1510/icvts.2007.156588 PMID:17925319

Jarmakiewicz, J., Parobczak, K., & Maślanka, K. (2017). Cybersecurity protection for power grid control infrastructures. *International Journal of Critical Infrastructure Protection*, *18*, 20–33. doi:10.1016/j.ijcip.2017.07.002

Jasiulewicz-Kaczmarek, M., & Gola, A. (2019). Maintenance 4.0 Technologies for Sustainable Manufacturing-an Overview. *IFAC-PapersOnLine*, *52*(10), 91–96. doi:10.1016/j.ifacol.2019.10.005

Jindal, A., Marnerides, A. K., Scott, A., & Hutchison, D. (2019, June). Identifying Security Challenges in Renewable Energy Systems: A Wind Turbine Case Study. In *Proceedings of the Tenth ACM International Conference on Future Energy Systems* (pp. 370-372). 10.1145/3307772.3330154

Ji, Z., Yanhong, Z., Baicun, W., & Jiyuan, Z. (2019). Human–Cyber–Physical Systems (HCPSs) in the Context of New-Generation Intelligent Manufacturing. *Engineering*, *5*(4), 624–636. doi:10.1016/j.eng.2019.07.015

Johansson, J. (2019). *Countermeasures Against Coordinated Cyber-Attacks Towards Power Grid Systems: A systematic literature study*. Academic Press.

Jovančić, P. (2016). Asset management and condition monitoring on maintenance of mining equipment lignite mines. *13th International symposium Continuous Surface Mining ISCSM 2016, 12-14 September 2016, Belgrade – Serbia, Metropol Palace, Conference Proceedings*, 197-207.

Jovančić, P., Ignjatović, D., & Tanasijević, M. (2012a). Modern concepts of mining equipment maintenance. In *10th International Symposia Mechanization and Automation in Mining and Energetics MAREN2012*. University of Belgrade.

Jovančić, P., Ignjatović, D., & Tanasijević, M. (2012b). Proactive monitoring system for basic mining equipment at open pit mines of Electric Power Industry of Serbia. *XXI International Congress on Maintenance and Asset Management – EUROMAINTENANCE 2012, May 14-16, 2012 – Sava Centar, Belgrade, Serbia, Conference Proceedings*, 622-630.

Jung, D., Zhang, Z., & Winslett, M. (2017). Vibration analysis for iot enabled predictive maintenance. In *2017 IEEE 33rd International Conference on Data Engineering (ICDE)* (pp. 1271-1282). IEEE. 10.1109/ICDE.2017.170

Junior, J. A. G., Busso, C. M., Gobbo, S. C. O., & Carreão, H. (2018). Making the links among environmental protection, process safety, and industry 4.0. *Process Safety and Environmental Protection*, *117*, 372–382. doi:10.1016/j.psep.2018.05.017

Kaasinen, E., Schmalfuß, F., Özturk, C., Aromaa, S., Boubekeur, M., Heilala, J., ... Mehta, R. (2020). Empowering and engaging industrial workers with Operator 4.0 solutions. *Computers & Industrial Engineering*, *139*, 105678. doi:10.1016/j.cie.2019.01.052

Kadir, B. A., Broberg, O., & da Conceição, C. S. (2019). Current research and future perspectives on human factors and ergonomics in Industry 4.0. *Computers & Industrial Engineering*, *137*, 106004. doi:10.1016/j.cie.2019.106004

Kagermann, H., Wahlster, W., & Helbig, J. (2013). Recommendations for Implementing the Strategic Initiative INDUSTRIE 4.0 - Final Report of the Industrie 4.0 Working Group. *Industrie*, 82. doi:10.13140/RG.2.1.1205.8966

Kakabadse, N. K., Kouzmin, A., & Kakabadse, A. (2001). Fromtacit knowledge to knowledgemanagement: Leveraging invisible assets. *Knowledge and Process Management*, *8*(3), 137–154. doi:10.1002/kpm.120

Kamal, M., & Rizza, G. (2019). *Design for metal additive manufacturing for aerospace applications. Additive Manufacturing for the Aerospace Industry*. Elsevier Inc. doi:10.1016/b978-0-12-814062-8.00005-4

Kamarul Bahrin, M. A., Othman, M. F., Nor Azli, N. H., & Talib, M. F. (2016). Industry 4.0: A review on industrial automation and robotic. *Jurnal Teknologi*, *78*(6-13), 6–13. doi:10.11113/jt.v78.9285

Kamat, V., & El-Tawil, S. (2007). Evaluation of Augmented Reality for Rapid Assessment of Earthquake-Induced Building Damage. *Journal of Computing in Civil Engineering*, *21*(5), 303–310. doi:10.1061/(ASCE)0887-3801(2007)21:5(303)

Kanawaday, A., & Sane, A. (2017). Machine learning for predictive maintenance of industrial machines using IoT sensor data. In *2017 8th IEEE International Conference on Software Engineering and Service Science (ICSESS)* (pp. 87-90). IEEE. 10.1109/ICSESS.2017.8342870

Kans, M., Galar, D., & Thaduri, A. (2016). Maintenance 4.0 in railway transportation industry. In Lecture Notes in Mechanical Engineering, F4, 317–331. doi:10.1007/978-3-319-27064-7_30

Karakuş, G., Karşıgil, E., & Polat, L. (2019). The Role of IoT on Production of Services: A Research on Aviation Industry. In N. M. Durakbasa, & M. G. Gencyilmaz (Eds.), *Proceedings of the International Symposium for Production Research 2018* (pp. 503–511). Springer International Publishing. 10.1007/978-3-319-92267-6_43

Karimnia, H., & Bagloo, H. (2015). Optimum mining method selection using fuzzy analytical hierarchy process-Qapiliq salt mine. *Iran. Int. J. Min. Sci. Technol., 25*(2), 225–230. doi:10.1016/j.ijmst.2015.02.010

Karim, R., Westerberg, J., Galar, D., & Kumar, U. (2016). Maintenance analytics–the new know in maintenance. *IFAC-PapersOnLine, 49*(28), 214–219. doi:10.1016/j.ifacol.2016.11.037

Kato, H., & Billinghurst, M. (1999). Marker tracking and HMD calibration for a video-based augmented reality conferencing system. In *Proceedings 2nd IEEE and ACM International Workshop on Augmented Reality (IWAR'99)* (pp. 85-94). San Francisco, CA: IEEE. 10.1109/IWAR.1999.803809

Kearney, K. (2019). Four Ways Blockchain will Change Aircraft Maintenance [Web Blog]. Honeywell Aerospace. https://aerospace.honeywell.com/content/aero/en/us/home/learn/about-us/blogs/2019/08/four-ways-blockchain-will-change-aircraft-maintenance.html

Keil, S. (2017). Design of a cyber-physical production system for semiconductor manufacturing design. *Proceedings of the Hamburg International Conference of Logistics (HICL)*, 318–340.

Keren, N., West, H. H., & Mannan, M. S. (2002). Benchmarking MoC practices in the process industries. *Process Safety Progress, 21*(2), 103–112. doi:10.1002/prs.680210205

Kergroach, S. (2017). Industry 4.0: New Challenges and Opportunities for the Labour Market. *Foresight and STI Governance, 11*(4), 6–8. doi:10.17323/2500-2597.2017.4.6.8

Kezunovic, M., Xie, L., & Grijalva, S. (2013). The Role of Big Data in Improving Power System Operation and Protection. *2013 IREP Symposium-Bulk Power System Dynamics and Control –IX (IREP).*

Khan, A. Y., Latif, R., Latif, S., Tahir, S., Batool, G., & Saba, T. (2019). Malicious Insider Attack Detection in IoTs Using Data Analytics. *IEEE Access : Practical Innovations, Open Solutions, 8*, 11743–11753. doi:10.1109/ACCESS.2019.2959047

Khas, K. S., Pandey, P. M., & Ray, A. R. (2015). Design and development of a device to measure the deformities of clubfoot. *Proceedings of the Institution of Mechanical Engineers. Part H, Journal of Engineering in Medicine, 229*(3), 194–204. doi:10.1177/0954411915574758 PMID:25833995

Kim, D., Lee, H., & Kwak, J. (2017). Standards as a driving force that influences emerging technological trajectories in the converging world of the Internet and things: An investigation of the M2M/IoT patent network. *Research Policy, 46*(7), 1234–1254. doi:10.1016/j.respol.2017.05.008

King, R., & Curran, K. (2019). Predictive Maintenance for Vibration-Related failures in the Semi-Conductor Industry. *Journal of Computer Engineering & Information Technology, 8*(1), 1.

Kinnison, H. A. (2004). *Aviation maintenance management*. McGraw-Hill.

Kissel, R. (2012). Glossary of key information security terms (draft). National Institute of Standards and Technology. http://dx.doi.org/.IR,7298 doi:10.6028/NIST

Klathae, V. (2019). *The Predictable Maintenance 4.0 by Applying Digital Technology A Case Study of Heavy Construction Machinery.pdf*. Academic Press.

Klir, G. J., & Yuan, B. (1995). *Fuzzy Sets and Fuzzy Logic, Theory and Applications*. New York: Prentice Hall.

Knofius, N., van der Heijden, M. C., & Zijm, W. H. M. (2019). Consolidating spare parts for asset maintenance with additive manufacturing. *International Journal of Production Economics, 208*, 269–280. doi:10.1016/j.ijpe.2018.11.007

Koehn, E. E., Kothari, R. K., & Pan, C. S. (1995). Safety in developing countries: Professional and bureaucratic problems. *Journal of Construction Engineering and Management, 121*(3), 261–265. doi:10.1061/(ASCE)0733-9364(1995)121:3(261)

Komonen, K. (2002). *A cost model of industrial maintenance for profitability analysis and benchmarking*. Academic Press.

Koulali, M., Koulali, S., Tembine, H., & Kobbane, A. (2018). Industrial Internet of Things-Based Prognostic Health Management: A Mean-Field Stochastic Game Approach. *IEEE Access: Practical Innovations, Open Solutions, 6*, 54388–54395. doi:10.1109/ACCESS.2018.2871859

Kour, R., Aljumaili, M., Karim, R., & Tretten, P. (2019). eMaintenance in railways: Issues and challenges in cybersecurity. *Proceedings of the Institution of Mechanical Engineers. Part F, Journal of Rail and Rapid Transit, 233*(10), 1012–1022. doi:10.1177/0954409718822915

Kour, R., Thaduri, A., & Karim, R. (2020). Railway Defender Kill Chain to Predict and Detect Cyber-Attacks. *Journal of Cyber Security and Mobility, 9*(1), 47–90. doi:10.13052/jcsm2245-1439.912

Kour, R., Thaduri, A., Singh, S., & Martinetti, A. (2019). *Big data analytics for maintaining transportation systems. In Transportation systems* (pp. 73–91). Springer.

Kovacs, O. (2018). The dark corners of industry 4.0 – Grounding economic governance 2.0. *Technology in Society, 55*, 140–145. doi:10.1016/j.techsoc.2018.07.009

Kraft, J., & Kuntzagk, S. (2017). *Engine Fleet-Management: The Use of Digital Twins From a MRO Perspective*. V001T01A007-V001T01A007. doi:10.1115/GT2017-63336

Kroes, M. J., Watkins, W. A., Delp, F., & Sterkenburg, R. (2013). *Aircraft Maintenance and Repair* (7th ed.). McGraw-Hill Education.

Kulev, N., Reuter, A., Eichhorn, O., Engler, E., & Wrede, C. (2019). Non-resilient behavior of offshore wind farms due to cyber-physical attacks. *REA Symposium on Resilience Engineering Embracing Resilience.*

Kumar, G. E. P., Baskaran, K., Blessing, R. E., & Lydia, M. (2016, December). Securing the smart grid network: A review. In *2016 IEEE International Conference on Computational Intelligence and Computing Research (ICCIC)* (pp. 1-6). IEEE. 10.1109/ICCIC.2016.7919570

Kumar, U. (2008). *System maintenance: Trends in management and technology. In Handbook of performability engineering* (pp. 773–787). Springer.

Kumar, U., & Galar, D. (2018). *Maintenance in the era of industry 4.0: Issues and challenges. In Quality, IT and business operations* (pp. 231–250). Springer. doi:10.1007/978-981-10-5577-5_19

Kumar, V., Verma, P., Sharma, R. R. K., & Khan, A. F. (2017). Conquering in emerging markets: Critical success factors to enhance supply chain performance. *Benchmarking: An International Journal, 24*(3), 570–593. doi:10.1108/BIJ-05-2016-0078

Kunst, R., Avila, L., Binotto, A., Pignaton, E., Bampi, S., & Rochol, J. (2019). Improving devices communication in Industry 4.0 wireless networks. *Engineering Applications of Artificial Intelligence, 83*, 1–12. doi:10.1016/j.engappai.2019.04.014

Kuo, C. C., Shyu, J. Z., & Ding, K. (2019). Industrial revitalization via industry 4.0–A comparative policy analysis among China, Germany and the USA. *Global Transitions, 1*, 3–14. doi:10.1016/j.glt.2018.12.001

Kwok, P. K., Yan, M., Chan, B. K. P., & Lau, H. Y. K. (2019). Crisis management training using discrete-event simulation and virtual reality techniques. *Computers & Industrial Engineering*, *135*, 711–722. doi:10.1016/j.cie.2019.06.035

Labagnara, D., Martinetti, A., & Patrucco, M. (2013). Tunneling operations, occupational S&H and environmental protection: A Prevention through Design approach. *American Journal of Applied Sciences*, *11*(11), 1371–1377. doi:10.3844/ajassp.2013.1371.1377

Lasi, H., Fettke, P., Kemper, H.-G., Feld, T., & Hoffmann, M. (2014). Industry 4.0. *Business & Information Systems Engineering*, *6*(4), 239–242. doi:10.100712599-014-0334-4

Laszka, A., Dubey, A., Walker, M., & Schmidt, D. (2017). Providing privacy, safety, and security in IoT-based transactive energy systems using distributed ledgers. *7th International Conference on the Internet of Things, IoT 2017*. 10.1145/3131542.3131562

Leahy, K., Hu, R. L., Konstantakopoulos, I. C., Spanos, C. J., & Agogino, A. M. (2016, June). Diagnosing wind turbine faults using machine learning techniques applied to operational data. In *2016 IEEE International Conference on Prognostics and Health Management (ICPHM)* (pp. 1-8). IEEE. 10.1109/ICPHM.2016.7542860

Lee, J., Ardakani, H. D., Yang, S., & Bagheri, B. (20015). Industrial big data analytics and cyber-physical systems for future maintenance & service innovation. Procedia CIRP, 38, 3–7.

Lee, J., Kao, H., & Yang, S. (2014). *Service innovation and smart analytics for Industry 4.0 and big data environment.* doi:10.1016/j.procir.2014.02.001

Lee, C.-H., Shin, H.-S., Tsourdos, A., & Skaf, Z. (2018). New Application of Data Analysis Using Aircraft Fault Record Data. *Journal of Aerospace Information Systems*, *15*(5), 297–306. doi:10.2514/1.I010577

Lee, E. (2015). The Past, Present and Future of Cyber-Physical Systems: A Focus on Models. *Sensors (Basel)*, *15*(3), 4837–4869. doi:10.3390150304837 PMID:25730486

Lee, H. (2005). A cost/benefit model for investments in inventory and preventive maintenance in an imperfect production system. *Computers & Industrial Engineering*, *48*(1), 55–68. doi:10.1016/j.cie.2004.07.008

Lee, J., & Bagheri, B. (2015). *Cyber-physical systems in future maintenance. In 9th WCEAM Research Papers* (pp. 299–305). Cham: Springer. doi:10.1007/978-3-319-15536-4_25

Lee, J., Bagheri, B., & Kao, H. A. (2015). A cyber-physical systems architecture for industry 4.0-based manufacturing systems. *Manufacturing Letters*, *3*, 18–23. doi:10.1016/j.mfglet.2014.12.001

Lee, J., Cameron, I., & Hassall, M. (2019). Improving process safety: What roles for Digitalization and Industry 4.0. *Process Safety and Environmental Protection*, *132*, 325–339. doi:10.1016/j.psep.2019.10.021

Lee, J., & Wang, H. (2008). *New technologies for maintenance. In Complex system maintenance handbook* (pp. 49–78). Springer. doi:10.1007/978-1-84800-011-7_3

Lefton, S. A., & Besuner, P. (2006). The cost of cycling coal fired power plants. *Coal Power Magazine*, *2006*, 16–20.

Leonard-barton, D. (1992). Core Capabilities and Core Rigidities : A Paradox in Managing New Product Development. *Strategic Management Journal*, *13*(S1), 111–125. doi:10.1002mj.4250131009

Leva, M. C., Baldissone, G., Caso, R., Demichela, M., Lawlor, L., & Mcaleer, B. (2018). Cost benefit evaluation of maintenance options for aging equipment using monetised risk values: A practical application. *Procedia Manufacturing*, *19*, 119–126.

Levovnik, D., & Gerbec, M. (2018). Operational readiness for the integrated management of changes in the industrial organizations – Assessment approach and results. *Safety Science*, *107*, 119–129. doi:10.1016/j.ssci.2018.04.006

Lewis, T., Aggarwal, R., Rajaretnam, N., Grantcharov, T., & Darzi, A. (2011). Training in surgical oncology – The role of VR simulation. *Surgical Oncology*, *20*(3), 134–139. doi:10.1016/j.suronc.2011.04.005 PMID:21605972

Leyh, C., Martin, S., & Schaffer, T. (2017). Industry 4.0 and Lean Production-A matching relationship? An analysis of selected Industry 4.0 models. *Proceedings of the 2017 Federated Conference on Computer Science and Information Systems, FedCSIS 2017*, *11*, 989–993. 10.15439/2017F365

Lezzi, M., Lazoi, M., & Corallo, A. (2018). Cybersecurity for Industry 4.0 in the current literature: A reference framework. *Computers in Industry*, *103*, 97–110. doi:10.1016/j.compind.2018.09.004

Li, Z. (2018). *Deep Learning Driven Approaches for Predictive Maintenance*. Academic Press.

Liao, Deschamps, Loures, & Ramos. (2017). Past, present and future of Industry 4.0 - a systematic literature review and research agenda proposal. International Journal of Production Research, 55(12), 3609–3629. doi:10.1080/00207543.2017.1308576

Lin, C. C., Deng, D. J., Chen, Z. Y., & Chen, K. C. (2016). Key design of driving industry 4.0: Joint energy-efficient deployment and scheduling in group-based industrial wireless sensor networks. *IEEE Communications Magazine*, *54*(10), 46–52. doi:10.1109/MCOM.2016.7588228

Linnimaa, S. (2016). *Information management in aircraft maintenance*. https://lutpub.lut.fi/handle/10024/121873

Liu, S., Hua, H., & Cheng, D. (2010). A Novel Prototype for an Optical See-Through Head-Mounted Display with Addressable Focus Cues. *IEEE Transactions on Visualization and Computer Graphics*, *16*(3), 381–393. doi:10.1109/TVCG.2009.95 PMID:20224134

Liu, T., Abd-Elrahman, A., Morton, J., & Wilhelm, V. L. (2018). Comparing fully convolutional networks, random forest, support vector machine, and patch-based deep convolutional neural networks for object-based wetland mapping using images from small unmanned aircraft system. *GIScience & Remote Sensing*, *55*(2), 243–264. doi:10.1080/15481603.2018.1426091

Liu, Z., Meyendorf, N., & Mrad, N. (2018). The role of data fusion in predictive maintenance using digital twin. *AIP Conference Proceedings*, *1949*(1), 020023. doi:10.1063/1.5031520

Li, X., Liang, X., Lu, R., Shen, X., Lin, X., & Zhu, H. (2012). Securing smart grid: Cyber attacks, countermeasures, and challenges. *IEEE Communications Magazine*, *50*(8), 38–45. doi:10.1109/MCOM.2012.6257525

Li, Z., Wang, Y., & Wang, K.-S. (2017). Intelligent predictive maintenance for fault diagnosis and prognosis in machine centers: Industry 4.0 scenario. *Advances in Manufacturing*, *5*(4), 377–387. doi:10.100740436-017-0203-8

Longo, F., Nicoletti, L., & Padovano, A. (2019). Emergency preparedness in industrial plants: A forward-looking solution based on industry 4.0 enabling technologies. *Computers in Industry*, *105*, 99–122. doi:10.1016/j.compind.2018.12.003

Lucas, J., & Thabet, W. (2008). Implementation and evaluation of a VR task-based training tool for conveyor belt safety training. *Journal of Information Technology in Construction*, *13*(40), 637–659.

Luthra, S., & Mangla, S. K. (2018). Evaluating challenges to Industry 4.0 initiatives for supply chain sustainability in emerging economies. *Process Safety and Environmental Protection*, *117*, 168–179. doi:10.1016/j.psep.2018.04.018

Lutz, R. (2018). Safe-AR: Reducing Risk While Augmenting Reality. In *IEEE 29th International Symposium on Software Reliability Engineering (ISSRE)* (pp. 70-75). Memphis, TN: IEEE.

Lu, Y. (2017). Industry 4.0: A survey on technologies, applications and open research issues. *Journal of Industrial Information Integration*, *6*, 1–10. doi:10.1016/j.jii.2017.04.005

Mack, D. L., Biswas, G., Koutsoukos, X. D., & Mylaraswamy, D. (2016). Learning bayesian network structures to augment aircraft diagnostic reference models. *IEEE Transactions on Automation Science and Engineering*, *14*(1), 358–369. doi:10.1109/TASE.2016.2542186

Madni, A. M., & Sievers, M. (2013). System of Systems Integration: Key Considerations and Challenges. *Systems Engineering*, *17*(3), 330–347. doi:10.1002ys.21272

Malone, S. (2016). *Using an expanded cyber kill chain model to increase attack resiliency*. Black Hat, US.

Manda, M. I., & Ben Dhaou, S. (2019). Responding to the challenges and opportunities in the 4th Industrial revolution in developing countries. *Proceedings of the 12th International Conference on Theory and Practice of Electronic Governance - ICEGOV2019*, 244–253. 10.1145/3326365.3326398

Mangla, S. K., Luthra, S., Jakhar, S., Gandhi, S., Muduli, K., & Kumar, A. (2020). A step to clean energy - Sustainability in energy system management in an emerging economy context. *Journal of Cleaner Production*, 242.

Mannan, S. (2012). *Lees' Loss Prevention in the Process Industries* (4th ed.). Oxford, UK: Elsevier.

Maquee, A., Shojaie, A. A., & Mosaddar, D. (2012). Clustering and association rules in analyzing the efficiency of maintenance system of an urban bus network. *International Journal of System Assurance Engineering and Management*, *3*(3), 175–183. doi:10.100713198-012-0121-x

March, S. T., & Scudder, G. D. (2017). Predictive maintenance: Strategic use of IT in manufacturing organizations. *Information Systems Frontiers*, 1–15.

Marques, H. C., & Giacotto, A. (2019). *Prescriptive Maintenance: Building Alternative Plans for Smart Operations*. doi:10.3384/ecp19162027

Martinetti, A., Braaksma, A. J. J., & van Dongen, L. A. M. (2017). Beyond RAMS Design: Towards an Integral Asset and Process Approach. In L. Redding, R. Roy, & A. Shaw (Eds.), Advances in Through-life Engineering Services (pp. 417-428). Cham, Switzerland: Springer.

Martinetti, A., Braaksma, A. J. J., & van Dongen, L. A. M. (2017). Beyond RAMS Design: Towards an Integral Asset and Process Approach. In L. Redding, R. Roy, & A. Shaw (Eds.), Advances in Through-life Engineering Services (pp. 417-428). Springer. doi:10.1007/978-3-319-49938-3_25

Martinetti, A., 't Hart, K., Damgrave, R., van Dongen, L. A. M., Turkenburg, R., & Nouwens, A. (2018). There is no spoon: Applying virtual reality for maintenance training of rolling stock technicians. *International Journal of System of Systems Engineering*, *8*(4), 398–415. doi:10.1504/IJSSE.2018.094563

Martinetti, A., Chatzimichailidou, M., Maida, L., & van Dongen, L. A. M. (2018). Safety I-II, Resilience and Antifragility Engineering: A debate explained through an accident occurred on a Mobile Elevating Work Platform. *International Journal of Occupational Safety and Ergonomics*, *25*(1), 66–75. doi:10.1080/10803548.2018.1444724 PMID:29473459

Martinetti, A., Costa Marques, H., Singh, S., & van Dongen, L. (2019). Reflections on the Limited Pervasiveness of Augmented Reality in Industrial Sectors. *Applied Sciences (Basel, Switzerland)*, *9*(16), 3382. doi:10.3390/app9163382

Masoni, R., Ferrise, F., Bordegoni, M., Gattullo, M., Uva, A. E., Fiorentino, M., ... Di Donato, M. (2017). Supporting Remote Maintenance in Industry 4.0 through Augmented Reality. *Procedia Manufacturing*, *11*, 1296–1302. doi:10.1016/j.promfg.2017.07.257

Massaro, A., Maritati, V., Galiano, A., Birardi, V., & Pellicani, L. (2018). ESB platform integrating KNIME data mining tool oriented on industry 4.0 based on artificial neural network predictive maintenance. *Int.J.Artif.Intell Appl, 9*, 1–17.

Masson, É., & Gransart, C. (2017, May). Cyber security for railways–a huge challenge–Shift2Rail perspective. In *International workshop on communication technologies for vehicles* (pp. 97-104). Springer. 10.1007/978-3-319-56880-5_10

Mc Nett, W. (2016). *Deferred maintenance and risk assessment: Technical analysis is critical to the process, 2016*. Available on line at: https://www.plantengineering.com/single-article/deferred-maintenance-and-risk-assessment-technical-analysis-is-critical-to-the-process/ae7fd166735e91e4d9b405aeae82355f.html

McNamara, J. E. (2014). Interface Standards. *Technical Aspects of Data Communication*, 17–36. doi:10.1016/b978-1-4831-8400-5.50007-3

Megahed, F. M., & Jones-Farmer, L. A. (2013). A Statistical Process Monitoring Perspective on "Big Data". *Frontiers in Statistical Quality Control*, 1–21.

Mehrpouya, M., Dehghanghadikolaei, A., Fotovvati, B., Vosooghnia, A., Emamian, S. S., & Gisario, A. (2019). The Potential of Additive Manufacturing in the Smart Factory Industrial 4.0: A Review. *Applied Sciences (Basel, Switzerland), 9*(18), 3865. doi:10.3390/app9183865

Mell, P., & Grance, T. (2011). *The NIST definition of cloud computing*. Academic Press.

Melnyk, S. A., Flynn, B. B., & Awaysheh, A. (2018). The best of times and the worst of times: Empirical operations and supply chain management research. *International Journal of Production Research, 56*(1–2), 164–192. doi:10.1080/00207543.2017.1391423

Merlino, G., Bruneo, D., Distefano, S., Longo, F., Puliafito, A., & Al-Anbuky, A. (2015). A Smart City Lighting Case Study on an OpenStack-Powered Infrastructure. *Sensors (Basel), 15*(7), 16314–16335. doi:10.339015071631 4PMID:26153775

Michael, K. (1991). Information as Thing. *Journal of the American Society for Information Science, 42*(5), 351–360. doi:10.1002/(SICI)1097-4571(199106)42:5<351::AID-ASI5>3.0.CO;2-3

Milgram, P., Takemura, H., Utsumi, A., & Kishino, F. (1994). Augmented reality: a class of displays on the reality-virtuality continuum. In *Photonics for industrial applications* (pp. 282–292). Boston, MA: SPIE.

Millwater, H. R., & Wu, Y.-T. (1993). Computational Structural Reliability Analysis of a turbine blade. *Proceedings of the international gas turbine and aeroengine congress and exposition*. 10.1115/93-GT-237

Miodragović R., Tanasijević M., Mileusnić Z., Jovančić P., (2012) Effectiveness assessment of agricultural machinery based on fuzzy sets theory. *Expert Systems with Applications, 39*(10), 8940–8946.

Mitsuhara, H., Tanimura, C., Nemoto, J., & Shishibori, M. (2019). Failure-enhanced evacuation training using a VR-based disaster simulator: A comparative experiment with simulated evacuees. *Procedia Computer Science, 159*, 1670–1679. doi:10.1016/j.procs.2019.09.337

Miyashita, T., Meier, P., Tachikawa, T., Orlic, S., Eble, T., Scholz, V., ... Lieberknecht, S. (2008). An Augmented Reality Museum Guide. In *IEEE International Symposium on Mixed and Augmented Reality 2008* (pp. 103-106). Cambridge, UK: IEEE. 10.1109/ISMAR.2008.4637334

Moan, T. (2005). Reliability-based management of inspection, maintenance and repair of offshore structures. *Structure and Infrastructure Engineering, 1*(1), 33–62.

Modernizing MRO. (2018, January 2). *Aerospace America*. https://aerospaceamerica.aiaa.org/departments/modernizing-mro/

Modular. (2018). In *Merriam-Webster.* Merriam-Webster Inc.

Moharana, U., Sarmah, S., & Rathore, P. (2019). Application of data mining for spare parts information in maintenance schedule: A case study. *Journal of Manufacturing Technology Management, 11*(7), 21. doi:10.1108/JMTM-09-2018-0303

Moktadir, M. A., Ali, S. M., Kusi-Sarpong, S., & Shaikh, M. A. A. (2018). Assessing challenges for implementing Industry 4.0: Implications for process safety and environmental protection. *Process Safety and Environmental Protection, 117*, 730–741. doi:10.1016/j.psep.2018.04.020

Monferini, A., Konstandinidou, M., Nivolianitou, Z., Weber, S., Kontogiannis, T., Kafka, P., ... Demichela, M. (2013). A compound methodology to assess the impact of human and organizational factors impact on the risk level of hazardous industrial plants. *Reliability Engineering & System Safety, 119*, 280–289. doi:10.1016/j.ress.2013.04.012

Monostori, L. (2014). Cyber-physical production systems: Roots, expectations and R&D challenges. *Procedia CIRP, 17*, 9–13. doi:10.1016/j.procir.2014.03.115

Monshizadeh, M., Yan, Z., Hippeläinen, L., & Khatri, V. (2015, September). Cloudification and security implications of TaaS. In *2015 World Symposium on Computer Networks and Information Security (WSCNIS)* (pp. 1-8). IEEE.

Mourtzis, D., Doukas, M., & Bernidaki, D. (2014). Simulation in manufacturing: Review and challenges. *Procedia CIRP, 25*, 213–229. doi:10.1016/j.procir.2014.10.032

Mueller, E., Chen, X.-L., & Riedel, R. (2017). Challenges and Requirements for the Application of Industry 4.0: A Special Insight with the Usage of Cyber-Physical System [English Edition]. *Chinese Journal of Mechanical Engineering, 30*(5), 1050–1057. doi:10.100710033-017-0164-7

Muhuri, P. K., Shukla, A. K., & Abraham, A. (2019). Industry 4.0: A bibliometric analysis and detailed overview. *Engineering Applications of Artificial Intelligence, 78*, 218–235. doi:10.1016/j.engappai.2018.11.007

Müller, J. M. (2020). Industry 4.0 in the Context of the Triple Bottom Line of Sustainability. In Customer Satisfaction and Sustainability Initiatives in the Fourth Industrial Revolution (pp. 1–20). IGI Global.

Muniz, J., Dias Batista, E., & Nakano, D. (2013). Engaging environments: Tacit knowledge sharing on the shop floor. *Journal of Knowledge Management, 17*(2), 290–306. doi:10.1108/13673271311315222

Mun, J., Yun, B. G., Ju, J., & Chang, B. M. (2015). Indirect additive manufacturing based casting of a periodic 3D cellular metal - Flow simulation of molten aluminum alloy. *Journal of Manufacturing Processes, 17*, 28–40. doi:10.1016/j.jmapro.2014.11.001

Nee, A., Ong, S., Chryssolouris, G., & Mourtzis, D. (2012). Augmented reality applications in design and manufacturing. *CIRP Annals - Manufacturing Technology, 61*(2), 657-679.

Negahban, A., & Smith, J. S. (2014). Simulation for manufacturing system design and operation: Literature review and analysis. *Journal of Manufacturing Systems, 33*(2), 241–261. doi:10.1016/j.jmsy.2013.12.007

Neumann, U., & Majoros, A. (1998). Cognitive, performance, and systems issues for augmented reality applications in manufacturing and maintenance. In *Proceedings. IEEE 1998 Virtual Reality Annual International Symposium* (pp. 4-11). Atlanta, GA: IEEE. 10.1109/VRAIS.1998.658416

Neumann, U., & You, S. (1999). Natural feature tracking for augmented reality. *IEEE Transactions on Multimedia, 1*(1), 53–64. doi:10.1109/6046.748171

Ngo, T. D., Kashani, A., Imbalzano, G., Nguyen, K. T. Q., & Hui, D. (2018). Additive manufacturing (3D printing): A review of materials, methods, applications and challenges. *Composites Part B, 143*, 172–196. doi:10.1016/j.compositesb.2018.02.012

Nimmagadda, S. L., Reiners, T., & Wood, L. (2018). On big data-guided upstream business research and its knowledge management. *Journal of Business Research, 89*, 143–158. doi:10.1016/j.jbusres.2018.04.029

Nkamnebe, A. D. (2011). Sustainability marketing in the emerging markets: Imperatives, challenges, and agenda setting. *International Journal of Emerging Markets, 6*(3), 217–232. doi:10.1108/17468801111144058

Nokia. (2019). *IoT & Analytics for railways*. Retrieved from https://onestore.nokia.com/asset/206522

Nonaka, I., & von Krogh, G. (2009). Perspective—Tacit Knowledge and Knowledge Conversion: Controversy and Advancement in Organizational Knowledge Creation Theory. *Organization Science, 20*(3), 635–652. doi:10.1287/orsc.1080.0412

NS. (2016). *Back on Track The passenger as our first, second and third priority NS Strategic Document*. Academic Press.

Odelius, J., Famurewa, S. M., Forslöf, L., Casselgren, J., & Konttaniemi, H. (2017). Industrial internet applications for efficient road winter maintenance. *Journal of Quality in Maintenance Engineering, 23*(3), 355–367. doi:10.1108/JQME-11-2016-0071

Olshannikova, E., Ometov, A., Koucheryavy, Y., & Olsson, T. (2015). Visualizing Big Data with augmented and virtual reality: Challenges and research agenda. *Journal of Big Data, 2*(1), 22. doi:10.118640537-015-0031-2

Onik, M. M. H., Kim, C.-S., & Yang, J. (2019). Personal Data Privacy Challenges of the Fourth Industrial Revolution. *21st International Conference on Advanced Communication Technology, ICACT 2019*, 635–638. 10.23919/ICACT.2019.8701932

Oztemel, E., & Gursev, S. (2020). Literature review of Industry 4.0 and related technologies. *Journal of Intelligent Manufacturing, 31*(1), 127–182. doi:10.100710845-018-1433-8

Paelke, V. (2014). Augmented reality in the smart factory: Supporting workers in an industry 4.0. environment. In *Proceedings of the 2014 IEEE emerging technology and factory automation (ETFA)* (pp. 1-4). IEEE. 10.1109/ETFA.2014.7005252

Palattella, M. R., Dohler, M., Grieco, A., Rizzo, G., Torsner, J., Engel, T., & Ladid, L. (2016). Internet of Things in the 5G Era: Enablers, Architecture, and Business Models. *IEEE Journal on Selected Areas in Communications, 34*(3), 510–527. doi:10.1109/JSAC.2016.2525418

Palmarini, R., Erkoyuncu, J. A., Roy, R., & Torabmostaedi, H. (2018). A systematic review of augmented reality applications in maintenance. *Robotics and Computer-Integrated Manufacturing, 49*(March 2017), 215–228. doi:10.1016/j.rcim.2017.06.002

Pandey, P. M., & Singh, G. (2019). Effect of unit cell shape and strut size on flexural properties of ordered Copper foam. *International Journal of Engineering Science, 12*(3), 73–79.

Papic, L., & Milovanovic, Z.N. (2007). *Systems Maintainability and Reliability*. The Research Center of Dependability and Quality Management DQM.

Papic, L., Aronov, J., & Pantelic, M. (2009). Safety based maintenance concept. *International Journal of Reliability Quality and Safety Engineering, 16*(6), 533–549. doi:10.1142/S0218539309003563

Parida, A., & Kumar, U. (2009). Maintenance Productivity and Performance Measurement. In M. Ben-Daya, S. O. Duffuaa, A. Raouf, J. Knezevic, & D. Ait-Kadi (Eds.), *Handbook of maintenance management and engineering* (pp. 17–37). Springer. doi:10.1007/978-1-84882-472-0_2

Park, J., Yang, X., Cho, Y. K., & Seo, J. (2017). Improving dynamic proximity sensing and processing for smart work-zone safety. *Automation in Construction*, *84*, 111–120. doi:10.1016/j.autcon.2017.08.025

Patra, S., & Young, V. (2016). A Review of 3D Printing Techniques and the Future in Biofabrication of Bioprinted Tissue. *Cell Biochemistry and Biophysics*, *74*(2), 93–98. doi:10.100712013-016-0730-0 PMID:27193609

Peinl, R. (2017). Knowledge management 4.0 - Lessons learned from IT trends. *CEUR Workshop Proceedings*, *1821*(July), 112–117.

Pereira, A. C., & Romero, F. (2017). A review of the meanings and the implications of the Industry 4.0 concept. *Procedia Manufacturing*, *13*, 1206–1214. doi:10.1016/j.promfg.2017.09.032

Peres, R. S., Dionisio Rocha, A., Leitao, P., & Barata, J. (2018). IDARTS – Towards intelligent data analysis and real-time supervision for industry 4.0. *Computers in Industry*, *101*, 138–146. doi:10.1016/j.compind.2018.07.004

Permatasari, C. I., Yuniaristanto, Sutopo, W., & Hisjam, M. (2019). Aircraft maintenance manpower shift planning with multiple aircraft maintenance licenced. *IOP Conference Series. Materials Science and Engineering*, *495*, 012023. doi:10.1088/1757-899X/495/1/012023

Petrović, D. V., Tanasijević, M., Milić, V., Lilić, N., Stojadinović, S., & Svrkota, I. (2014). Risk assessment model of mining equipment failure based on fuzzy logic. *Expert Systems with Applications*, *41*(18), 8157–8164. doi:10.1016/j.eswa.2014.06.042

Pfeffer, J. (2010). Building Sustainable Organizations: The Human Factor Executive Overview. *The Academy of Management Perspectives*, *24*(1), 34–45.

Polanyi, M. (1966). *The tacit dimension*. Chicago: University of Chicago Press.

Poomathi, N., Singh, S., Prakash, C., Patil, R. V., Perumal, P. T., & Barathi, V. A. … Maheshwari, N. U. (2018). Bioprinting in ophthalmology: Current advances and future pathways. *Rapid Prototyping Journal*. doi:10.1108/RPJ-06-2018-0144

Prakash, C., Singh, S., Pabla, B. S., Sidhu, S. S., & Uddin, M. S. (2018). Bio-inspired low elastic biodegradable Mg-Zn-Mn-Si-HA alloy fabricated by spark plasma sintering. *Materials and Manufacturing Processes*, *00*(00), 1–12. doi: 10.1080/10426914.2018.1512117

Preuveneers, D., & Ilie-Zudor, E. (2017). The intelligent industry of the future: A survey on emerging trends, research challenges and opportunities in Industry 4.0. *Journal of Ambient Intelligence and Smart Environments*, *9*(3), 287–298. doi:10.3233/AIS-170432

PrivilegedA. T. R. (2019). Retrieved from https://www.beyondtrust.com/resources/whitepapers/privileged-access-threat-report

Qi, Q., & Tao, F. (2018). *Digital Twin and Big Data Towards Smart Manufacturing and Industry 4.0 360 Degree Comparison*. Academic Press.

Qian, Y., Dersch, M. S., Gao, Z., & Edwards, J. R. (2019). Railroad infrastructure 4.0: Development and application of an automatic ballast support condition assessment system. *Transportation Geotechnics*, *19*, 19–34. doi:10.1016/j.trgeo.2019.01.002

Rabah, S., Assila, A., Khouri, E., Maier, F., Ababsa, F., Bourny, V., ... Mérienne, F. (2018). Towards improving the future of manufacturing through digital twin and augmented reality technologies. *Procedia Manufacturing*, *17*, 460–467. doi:10.1016/j.promfg.2018.10.070

Rachburee, N., Arunrerk, J., & Punlumjeak, W. (2018). Failure Part Mining Using an Association Rules Mining by FP-Growth and Apriori Algorithms: Case of ATM Maintenance in Thailand. In *IT Convergence and Security 2017* (pp. 19–26). Springer. doi:10.1007/978-981-10-6451-7_3

Rail Delivery Group. (2017). *Rail Cyber Security Strategy*. Retrieved from https://www.raildeliverygroup.com/component/arkhive/?task=file.download&id=469772253

Rajabalinejad, M. (2018). System Integration : Challenges and Opportunities. In *13th System of Systems Engineering Conference* (pp. 6–11). IEEE.

Rakyta, M., Fusko, M., Herčko, J., Závodská, Ľ., & Zrnić, N. (2016). Proactive approach to smart maintenance and logistics as a auxiliary and service processes in a company. *Journal of Applied Engineering Science*, 433–442.

Rakyta, M., Fusko, M., Herčko, J., Závodská, Ľ., & Zrnić, N. (2016). Proactive approach to smart maintenance and logistics as a auxiliary and service processes in a company. *Journal of Applied Engineering Science*, *14*(4), 433–442. doi:10.5937/jaes14-11664

Rauschnabel, P. A., Brem, A., & Ro, Y. K. (2015). *Augmented Reality Smart Glasses: Definition, Conceptual Insights, and Managerial Importance*. Dearborn, MI: The University of Michigan-Dearborn, College of Business.

Rauschnabel, P. A., & Ro, Y. K. (2016). Augmented reality smart glasses: An investigation of technology acceptance drivers. *International Journal of Technology Marketing*, *11*(2), 123–148. doi:10.1504/IJTMKT.2016.075690

Ravna, R. (2016). *Industry 4.0 and Maintenance*. Academic Press.

Reis, M., & Gins, G. (2017). Industrial process monitoring in the big data/industry 4.0 era: From detection, to diagnosis, to prognosis. *Processes*, *5*(3), 35. doi:10.3390/pr5030035

Reitmayr, G., & Schmalstieg, D. (2003). Data Management Strategies for Mobile Augmented Reality. In *Proceedings of the International Workshop on Software Technology for Augmented Reality Systems* (pp. 47-52). Tokyo, Japan: IEEE.

Reuss, P., Stram, R., Althoff, K.-D., Henkel, W., & Henning, F. (2018). Knowledge engineering for decision support on diagnosis and maintenance in the aircraft domain. In *Synergies Between Knowledge Engineering and Software Engineering* (pp. 173–196). Springer. doi:10.1007/978-3-319-64161-4_9

Reyes Garcia, J. R., Martinetti, A., Jauregui Becker, J. M., Singh, S., & van Dongen, L. A. M. (2019). Towards an Industry 4.0-Based Maintenance Approach in the Manufacturing Processes. In V. González-Prida Diaz & J. P. Zamora Bonilla (Eds.), *Handbook of Research on Industrial Advancement in Scientific Knowledge* (pp. 135–159). IGI Global. doi:10.4018/978-1-5225-7152-0.ch008

Ríos, J., Morate, F. M., Oliva, M., & Hernández, J. C. (2016). Framework to support the aircraft digital counterpart concept with an industrial design view. *International Journal of Agile Systems and Management*, *9*(3), 212–231. doi:10.1504/IJASM.2016.079934

Roblek, V., Meško, M., & Krapež, A. (2016). A Complex View of Industry 4.0. *SAGE Open*, *6*(2), 215824401665398. doi:10.1177/2158244016653987

Rodrigues, D., & Lavorato, P. (2016). Maintenance, Repair and Overhaul (MRO) Fundamentals and Strategies: An Aeronautical Industry Overview. *International Journal of Computers and Applications*, *135*(12), 21–29. doi:10.5120/ijca2016908563

Rodrigues, R. S., Balestrassi, P. P., Paiva, A. P., Garcia-Diaz, A., & Pontes, F. (2012). Aircraft interior failure pattern recognition utilizing text mining and neural networks. *Journal of Intelligent Information Systems*, *38*(3), 741–766. doi:10.100710844-011-0176-1

Rødseth, H., Schjølberg, P., & Marhaug, A. (2017). Deep digital maintenance. *Advances in Manufacturing*, *5*(4), 299–310. doi:10.100740436-017-0202-9

Rosen, D. W. (2014). Research supporting principles for design for additive manufacturing: This paper provides a comprehensive review on current design principles and strategies for AM. *Virtual and Physical Prototyping*, *9*(4), 225–232. doi:10.1080/17452759.2014.951530

Roy, R., Stark, R., Tracht, K., Takata, S., & Mori, M. (2016). Continuous maintenance and the future – Foundations and technological challenges. *CIRP Annals - Manufacturing Technology*, *65*(2), 667–688. doi:10.1016/j.cirp.2016.06.006

Roy, R., Stark, R., Tracht, K., Takata, S., & Mori, M. (2018). *Continuous maintenance and the future – Foundations and technological challenges.pdf*. Academic Press.

Ruiz-Sarmiento, J. R., Monroy, J., Moreno, F. A., Galindo, C., Bonelo, J. M., & Gonzalez-Jimenez, J. (2020). A predictive model for the maintenance of industrial machinery in the context of industry 4.0. *Engineering Applications of Artificial Intelligence*, *87*, 103289. doi:10.1016/j.engappai.2019.103289

Rzevski, G., Knezevic, J., Skobelev, P., Borgest, N., & Lakhin, O. (2016). Managing aircraft lifecycle complexity. *International Journal of Design & Nature and Ecodynamics*, *11*(2), 77–87. doi:10.2495/DNE-V11-N2-77-87

Saaty, T. L. (1980). *Decision Making: The Analytical Hierarchy Process*. New York: McGraw-Hill.

Sadeghi, A. R., Wachsmann, C., & Waidner, M. (2015). Security and privacy challenges in industrial Internet of Things. *Proceedings - Design Automation Conference*, 1–6. 10.1145/2744769.2747942

Sahay, A. (2012). *Leveraging Information Technology for Optimal Aircraft Maintenance, Repair and Overhaul (MRO)*. Elsevier. doi:10.1533/9780857091437

Sakas, G. (2002). Trends in medical imaging: from 2D to 3D. *Computers & Graphics*, *26*(4), 577–587. doi:10.1016/S0097-8493(02)00103-6

Santos, M. Y., Oliveira e Sá, J., Andrade, C., Vale Lima, F., Costa, E., Costa, C., ... Galvão, J. (2017). A Big Data system supporting Bosch Braga Industry 4.0 strategy. *International Journal of Information Management*, *37*(6), 750–760. doi:10.1016/j.ijinfomgt.2017.07.012

Saucedo-Martínez, J. A., Pérez-Lara, M., Marmolejo-Saucedo, J. A., Salais-Fierro, T. E., & Vasant, P. (2018). Industry 4.0 framework for management and operations: A review. *Journal of Ambient Intelligence and Humanized Computing*, *9*(3), 789–801. doi:10.100712652-017-0533-1

Saweros, E., & Song, Y.-T. (2019). Connecting heterogeneous electronic health record systems using tangle. In *13th International Conference on Ubiquitous Information Management and Communication, IMCOM 2019* (Vol. 935, pp. 858–869). 10.1007/978-3-030-19063-7_68

Scheifele, S., Friedrich, J., & Lechler, A. (2014), Self-configuring control system for a modular production system. *2nd International Conference on System-Integrated Intelligence: Challenges for Product and Production Engineering Flexible*, 1-8.

Schulz, K. F., & Freund, D. (2019). A multichain architecture for distributed supply chain design in industry 4.0. In *21st International Conference on Business Information Systems, BIS 2018* (*Vol. 339*, pp. 277–288). 10.1007/978-3-030-04849-5_25

Scurati, G. W., Gattullo, M., Fiorentino, M., Ferrise, F., Bordegoni, M., & Uva, A. E. (2018). Converting maintenance actions into standard symbols for Augmented Reality applications in Industry 4.0. *Computers in Industry*, *98*, 68–79. doi:10.1016/j.compind.2018.02.001

Seneviratne, D., Ciani, L., Catelani, M., & Galar, D. (2018). Smart maintenance and inspection of linear assets: An industry 4.0 approach. Acta Imeko.

Seneviratne, D., Ciani, L., Catelani, M., & Galar, D. (2018). Smart maintenance and inspection of linear assets: An Industry 4.0 approach. *Acta IMEKO*, *7*(1), 50–56. doi:10.21014/acta_imeko.v7i1.519

Shafiq, S. I., Sanin, C., Toro, C., & Szczerbicki, E. (2015). Virtual engineering object (VEO): Toward experience-based design and manufacturing for industry 4.0. *Cybernetics and Systems*, *46*(1-2), 35–50. doi:10.1080/01969722.2015.1007734

Sheth, J. N. (2011). Impact of Emerging Markets on Marketing: Rethinking Existing Perspectives and Practices. *Journal of Marketing*, *75*(4), 166–182. doi:10.1509/jmkg.75.4.166

Shift2rail report. (2017). *CYbersecurity in the RAILway sector D2.1 – Safety and Security requirements of Rail transport system in multi-stakeholder environments.*

Shift2Rail. (2016). *Cybersecurity in the railway sector.* Retrieved from https://shift2rail.org/project/cyrail/

Shin, D., & Dunston, P. (2008). Identification of application areas for Augmented Reality in industrial construction based on technology suitability. *Automation in Construction*, *17*(7), 882–894. doi:10.1016/j.autcon.2008.02.012

Shrouf, F., Ordieres, J., & Miragliotta, G. (2014). *Smart factories in Industry 4.0: A review of the concept and of energy management approached in production based on the Internet of Things paradigm. In 2014 IEEE international conference on industrial engineering and engineering management* (pp. 697–701). IEEE.

Simon, J., Trojanova, M., Zbihlej, J., & Sarosi, J. (2018). Mass customization model in food industry using industry 4.0 standard with fuzzy-based multi-criteria decision making methodology. *Advances in Mechanical Engineering*, *10*(3). doi:10.1177/1687814018766776

Singh, A., Payal, A., & Bharti, S. (2019). A walkthrough of the emerging IoT paradigm: Visualizing inside functionalities, key features, and open issues. *Journal of Network and Computer Applications*, *143*(February), 111–151. doi:10.1016/j.jnca.2019.06.013

Singh, G., & Pandey, P. M. (2018). Design and Analysis of Long-Stepped Horn for Ultrasonic Assisted Sintering. *21st International Conference on Advances in Materials and Processing Technology (AMPT).*

Singh, G., & Pandey, P. M. (2019a). Experimental investigations into mechanical and thermal properties of rapid manufactured copper parts. *Proceedings of the Institution of Mechanical Engineers. Part C, Journal of Mechanical Engineering Science*, *0*(0), 1–14. doi:10.1177/0954406219875483

Singh, G., & Pandey, P. M. (2019b). Rapid manufacturing of copper components using 3D printing and ultrasonic assisted pressureless sintering: Experimental investigations and process optimization. *Journal of Manufacturing Processes*, *43*, 253–269. doi:10.1016/j.jmapro.2019.05.010

Singh, G., & Pandey, P. M. (2019c). Topological ordered copper graphene composite foam: Fabrication and compression properties study. *Materials Letters*, *257*, 126712. doi:10.1016/j.matlet.2019.126712

Singh, G., & Pandey, P. M. (2019d). Ultrasonic Assisted Pressureless Sintering for rapid manufacturing of complex copper components. *Materials Letters*, *236*, 276–280. doi:10.1016/j.matlet.2018.10.123

Singh, G., & Pandey, P. M. (2019e). Uniform and graded copper open cell ordered foams fabricated by rapid manufacturing: Surface morphology, mechanical properties and energy absorption capacity. *Materials Science and Engineering A*, *761*(June), 138035. doi:10.1016/j.msea.2019.138035

Singh, G., & Pandey, P. M. (2020). Neck growth kinetics during ultrasonic-assisted sintering of copper powder. *Proceedings of the Institution of Mechanical Engineers. Part C, Journal of Mechanical Engineering Science*, *0*(0), 1–11. doi:10.1177/0954406220904108

Singh, J. P., & Pandey, P. M. (2018). Fabrication and assessment of mechanical properties of open cell porous regular interconnected metallic structure through rapid manufacturing route. *Rapid Prototyping Journal*, *24*(1), 138–149. doi:10.1108/RPJ-04-2015-0043

Singh, S., Galar, D., Baglee, D., & Björling, S. E. (2014). Self-maintenance techniques: A smart approach towards self-maintenance system. *International Journal of System Assurance Engineering and Management*, *5*(1), 75–83. doi:10.100713198-013-0200-7

Singh, S., Prakash, C., & Ramakrishna, S. (2019). 3D printing of polyether-ether-ketone for biomedical applications. *European Polymer Journal*, *114*(February), 234–248. doi:10.1016/j.eurpolymj.2019.02.035

Singh, S., Ramakrishna, S., & Singh, R. (2017). Material issues in additive manufacturing: A review. *Journal of Manufacturing Processes*, *25*, 185–200. doi:10.1016/j.jmapro.2016.11.006

Singh, S., Singh, N., Gupta, M., Prakash, C., & Singh, R. (2019). Mechanical feasibility of ABS/HIPS-based multi-material structures primed by low-cost polymer printer. *Rapid Prototyping Journal*, *25*(1), 152–161. doi:10.1108/RPJ-01-2018-0028

Singh, S., & Singh, R. (2015). Wear modelling of Al-Al2O3functionally graded material prepared by FDM assisted investment castings using dimensionless analysis. *Journal of Manufacturing Processes*, *20*, 507–514. doi:10.1016/j.jmapro.2015.01.007

Singh, S., & Singh, R. (2016). Fused deposition modelling based rapid patterns for investment casting applications : A review. *Rapid Prototyping Journal*, *22*(2), 123–143. doi:10.1108/RPJ-02-2014-0017

Sipsas, K., Alexopoulos, K., Xanthakis, V., & Chryssolouris, G. (2016). Collaborative maintenance in flow-line manufacturing environments: An Industry 4.0 approach. *Procedia CIRP*, *55*, 236–241. doi:10.1016/j.procir.2016.09.013

Sisinni, E., Saifullah, A., Han, S., Jennehag, U., & Gidlund, M. (2018). Industrial internet of things: Challenges, opportunities, and directions. *IEEE Transactions on Industrial Informatics*, *14*(11), 4724–4734. doi:10.1109/TII.2018.2852491

Snowden, D. (2002). Complex acts of knowing: Paradox and descriptive self-awareness. *Journal of Knowledge Management*, *6*(2), 100–111. doi:10.1108/13673270210424639

SonicWall Cyber Threat Report. (2019). Retrieved from https://cdw-prod.adobecqms.net/content/dam/cdw/on-domain-cdw/brands/sonicwall/2019-sonicwall-cyber-threat-report.pdf

SS-CC. A. P. T. A. (2015). Securing control and communications systems in rail transit environments. Washington, DC: American Public Transportation Association.

Staggs, J., Ferlemann, D., & Shenoi, S. (2017). Wind farm security: Attack surface, targets, scenarios and mitigation. *International Journal of Critical Infrastructure Protection*, *17*, 3–14. doi:10.1016/j.ijcip.2017.03.001

Steinsland, K. (2018). *A model to develop and implement a predictive maintenance programme complying with Industry 4.0 vision: A case study in Skretting AS*. Academic Press.

Stock, T., & Seliger, G. (2016). Opportunities of Sustainable Manufacturing in Industry 4.0. *Procedia CIRP*, *40*, 536–541. doi:10.1016/j.procir.2016.01.129

Stork, S., & Schubö, A. (2010). Human cognition in manual assembly: Theories and applications. *Advanced Engineering Informatics*, *24*(3), 320–328. doi:10.1016/j.aei.2010.05.010

Strandberg, K. (1991). IEC 300: The Dependability Counterpart of ISO 9000. *Reliability and Maintainability Symposium*, 463-467. 10.1109/ARMS.1991.154481

Sun, T.-L., & Salgado, G. A. M. (2017). *Sustainable Data Collection Framework: Real-Time, Online Data Visualization.* Paper presented at the International Conference on Sustainable Design and Manufacturing. 10.1007/978-3-319-57078-5_6

Tack, P., Victor, J., Gemmel, P., & Annemans, L. (2016). 3D - printing techniques in a medical setting : A systematic literature review. *Biomedical Engineering Online*, *15*(1), 1–21. doi:10.118612938-016-0236-4 PMID:27769304

Tanasijević, M. (2016). A fuzzy-based decision support model for evaluation of mining machinery. *Full Papers Proceeding of International Conference ''48th International October Conference on Mining and Metallurgy*, 15-18.

Tanasijević, M., Ivezić, D., Jovančić, P., Ćatić, D., & Zlatanović, D. (2013). Study of Dependability Evaluation for Multi-hierarchical Systems Based on Max–Min Composition. *Quality and Reliability Engineering International*, *29*(3), 317–326. doi:10.1002/qre.1383

Tarnowski, I. (2017). How to use cyber kill chain model to build cybersecurity? *European Journal of Higher Education IT*. Available: http://www. eunis. org/download/TNC2017/TNC17-IreneuszTarnowski-cybersecurity. pdf

Tatic, D., & Tešic, B. (2017). The application of augmented reality technologies for the improvement of occupational safety in an industrial environment. *Computers in Industry*, *85*, 1–10. doi:10.1016/j.compind.2016.11.004

Tech, I. T. (2020). *Integrated Logistics – IT Tech Direct*. https://www.ittechdirect.net/integrated-logistics/

Tedeschi, S., Rodrigues, D., Emmanouilidis, C., Erkoyuncu, J., Roy, R., & Starr, A. (2018). A cost estimation approach for IoT modular architectures implementation in legacy systems. *Procedia Manufacturing*, *19*, 103–110. doi:10.1016/j.promfg.2018.01.015

Teodorović, D., & Vukadinović, K. (1998). *Traffic Control and Transport Planning: A Fuzzy Sets and Neural Networks Approach*. Boston: Kluwer Academic Publishers. doi:10.1007/978-94-011-4403-2

Thaduri, A., Famurewa, S. M., Verma, A. K., & Kumar, U. (2019). Process mining for maintenance decision support. In *System performance and management analytics* (pp. 279–293). Springer.

Thaduri, A., Galar, D., & Kumar, U. (2015). Railway assets: A potential domain for big data analytics. *Procedia Computer Science*, *53*, 457–467. doi:10.1016/j.procs.2015.07.323

Thaduri, A., Galar, D., Kumar, U., & Verma, A. K. (2016). *Context-based maintenance and repair shop suggestion for a moving vehicle. In Current trends in reliability, availability, maintainability and safety* (pp. 67–81). Springer. doi:10.1007/978-3-319-23597-4_6

Thaduri, A., Kumar, U., & Verma, A. K. (2017). Computational intelligence framework for context-aware decision making. *International Journal of System Assurance Engineering and Management*, *8*(4), 2146–2157. doi:10.100713198-014-0320-8

Thai, V. P., Zhong, W., Pham, T., Alam, S., & Duong, V. 2019, April. Detection, Tracking and Classification of Aircraft and Drones in Digital Towers Using Machine Learning on Motion Patterns. In *2019 Integrated Communications, Navigation and Surveillance Conference (ICNS)* (pp. 1-8). IEEE.

Thales. (2020). *Railway Digitalization: Cybersecurity*. Retrieved from https://www.thalesgroup.com/en/spain/magazine/railway-digitalization-cybersecurity

Thames, L., & Schaefer, D. (2017). *Cybersecurity for industry 4.0*. New York: Springer. doi:10.1007/978-3-319-50660-9

Thoben, K., Wiesner, S., & Wuest, T. (2017). "Industrie 4.0" and smart manufacturing-a review of research issues and application examples. *International Journal of Automotive Technology*, *11*(1), 4–16.

Tortorella, G. L., & Fettermann, D. (2018). Implementation of industry 4.0 and lean production in brazilian manufacturing companies. *International Journal of Production Research*, *56*(8), 2975–2987. doi:10.1080/00207543.2017.1391420

Trappey, A. J., Trappey, C. V., Govindarajan, U. H., Chuang, A. C., & Sun, J. J. (2017). A review of essential standards and patent landscapes for the internet of things: A key enabler for industry 4.0. *Advanced Engineering Informatics*, *33*, 208–229. doi:10.1016/j.aei.2016.11.007

Tretten, P., & Karim, R. (2016). *Project: iMain 'Methodologies and tools for the sustainable, predictive maintenance of production equipment'*. Academic Press.

Tronchoni, D., & Brennan, G. (2017). Industry 4.0: A Key Enabler of the Circular Economy. *Climate Innovation Insights*, *2*(7), 2–5. http://edinn.com/dwn/Insights_2-7_Industry_4 0_EU.pdf

Troy, N., Denny, E., & O'Malley, M. (2010). Base-load cycling on a system with significant wind penetration. *IEEE Transactions on Power Systems*, *25*(2), 1088–1097.

Truong, H. L. (2018). Integrated analytics for IIoT predictive maintenance using IoT big data cloud systems. In *2018 IEEE International Conference on Industrial Internet (ICII)* (pp. 109-118). IEEE. 10.1109/ICII.2018.00020

Turnbull, A., Carroll, J., Koukoura, S., & McDonald, A. (2019). Prediction of wind turbine generator bearing failure through analysis of high-frequency vibration data and the application of support vector machine algorithms. *Journal of Engineering (Stevenage, England)*, *2019*(18), 4965–4969. doi:10.1049/joe.2018.9281

U.S. Department of Transportation. (2003). *Guidance on Aviation Rules and Statutes*. Author.

Ucler, C., & Gok, O. (2015). Innovating General Aviation MRO's through IT: The Sky Aircraft Management System - SAMS. *Procedia: Social and Behavioral Sciences*, *195*, 1503–1513. doi:10.1016/j.sbspro.2015.06.452

Ulz, T., Pieber, T., Steger, C., Haas, S., Bock, H., & Matischek, R. (2017). Bring your own key for the industrial Internet of Things. *2017 IEEE International Conference on Industrial Technology, ICIT 2017*, 1430–1435. 10.1109/ICIT.2017.7915575

United States Department of Energy. (2003). *"Grid 2030" A National Vision for Electricity's Second 100 Years*. Office of Electric Transmission and Distribution. Retrieved from https://www.energy.gov/sites/prod/files/oeprod/DocumentsandMedia/Electric_Vision_Document.pdf

Utzig, S., Kaps, R., Azeem, S. M., & Gerndt, A. (2019). Augmented Reality for Remote Collaboration in Aircraft Maintenance Tasks. *2019 IEEE Aerospace Conference*, 1–10. 10.1109/AERO.2019.8742228

Vacík, E., & Špaček, M. (2018). Process innovation in service sector matching industry 4.0 environment. *Annals of DAAAM and Proceedings of the International DAAAM Symposium*, *29*(1), 352–360. doi:10.2507/29th.daaam.proceedings.05

van Beek, M. V. (2015). Proposal for a Directive of the European Parliament and of the Council on Improving the Gender Balance among Non-Executive Directors of Companies Listed on Stock Exchanges and Related Measures: Is It Courageous and Visionary Or Is It. *RRDE*, 109.

Vanderroost, M., Ragaert, P., Verwaeren, J., De Meulenaer, B., De Baets, B., & Devlieghere, F. (2017). The digitization of a food package's life cycle: Existing and emerging computer systems in the logistics and post-logistics phase. *Computers in Industry*, *87*, 15–30. doi:10.1016/j.compind.2017.01.004

Vatn, J. (2018). Industry 4.0 and real-time synchronization of operation and maintenance. *Safety and Reliability - Safe Societies in a Changing World - Proceedings of the 28th International European Safety and Reliability Conference, ESREL 2018*, 681–686. https://www.scopus.com/inward/record.uri?eid=2-s2.0-85058076521&partnerID=40&md5=027b2998282b9231868068dfaf3e7000

Verma, P., & Sharma, R. R. K. (2016). Structure, Evaluation System, and Employee Resistance: Impediments to Horizontal Strategy Implementation in Diversified Corporations. *Sixth International Conference on Industrial Engineering and Operations Management (IEOM).*

Verma, P., Sharma, R. R. K., & Kumar, V. (2017). The sustainability issues of diversified firms in emerging economies context: A theoretical model and propositions. *International Journal of Process Management and Benchmarking*, *7*(2), 224. doi:10.1504/IJPMB.2017.083107

Vyatkin, V., Salcic, Z., Roop, P., & Fitzgerald, J. (2007). Now That's Smart! *IEEE Industrial Electronics Magazine*, *1*(4), 17–29. doi:10.1109/MIE.2007.909540

Waeyenbergh, G., & Pintelon, L. (2002). A framework for maintenance concept development. *International Journal of Production Economics*, *77*(3), 299–313. doi:10.1016/S0925-5273(01)00156-6

Wagner, D., & Schmalstieg, D. (2006). Handheld Augmented Reality Displays. In *IEEE Virtual Reality Conference (VR 2006)* (pp. 321-321). Alexandria, VA: IEEE. 10.1109/VR.2006.67

Walker, D., & Tait, R. (2004). Health and safety management in small enterprises: An effective low cost approach. *Safety Science*, *42*(1), 69–83. doi:10.1016/S0925-7535(02)00068-1

Wan, S., Li, D., Gao, J., & Li, J. (2019). A knowledge based machine tool maintenance planning system using case-based reasoning techniques. *Robotics and Computer-Integrated Manufacturing*, *58*, 80–96. doi:10.1016/j.rcim.2019.01.012

Wang, Z., & Li, Y. (2017). *Equipment Maintenance Support Decision Method Research Based on Big Data.* Paper presented at the International Conference on Geo-Spatial Knowledge and Intelligence.

Wang, J. (2000). A subjective modelling tool applied to formal ship safety assessment. *Ocean Engineering*, *27*(10), 1019–1035. doi:10.1016/S0029-8018(99)00037-2

Wang, J., Erkoyuncu, J., & Roy, R. (2018). A Conceptual Design for Smell Based Augmented Reality: Case Study in Maintenance Diagnosis. *Procedia CIRP*, *78*, 109–114. doi:10.1016/j.procir.2018.09.067

Wang, J., Yang, J. B., & Sen, P. (1995). Safety analyses and synthesis using fuzzy sets and evidential reasoning. *Reliability Engineering & System Safety*, *47*(2), 103–118. doi:10.1016/0951-8320(94)00053-Q

Wang, S., Wan, J., Zhang, D., Li, D., & Zhang, C. (2016). Towards smart factory for industry 4.0: A self-organized multi-agent system with big data based feedback and coordination. *Computer Networks*, *101*(June), 158–168. doi:10.1016/j.comnet.2015.12.017

Wang, X., Jiang, M., Zhou, Z., Gou, J., & Hui, D. (2017). 3D printing of polymer matrix composites : A review and prospective. *Composites. Part B, Engineering*, *110*, 442–458. doi:10.1016/j.compositesb.2016.11.034

Wang, X., Ong, S., & Nee, A. (2016). A comprehensive survey of augmented reality assembly research. *Advances in Manufacturing*, *4*(1), 1–22. doi:10.100740436-015-0131-4

Wan, J., Tang, S., Li, D., Wang, S., Liu, C., Abbas, H., & Vasilakos, A. V. (2017). A manufacturing big data solution for active preventive maintenance. *IEEE Transactions on Industrial Informatics*, *13*(4), 2039–2047. doi:10.1109/TII.2017.2670505

Webster, G. D., Harris, R. L., Hanif, Z. D., Hembree, B. A., Grossklags, J., & Eckert, C. (2018). Sharing is Caring: Collaborative Analysis and Real-Time Enquiry for Security Analytics. *11th IEEE International Congress on Conferences on Internet of Things, 14th IEEE International Conference on Green Computing and Communications, 11th IEEE International Conference on Cyber, Physical and Social Computing, 4th IEEE International Conference*, 1402–1409. 10.1109/Cybermatics_2018.2018.00240

Weibull, W. (1951). A statistical distribution function of wide applicability. *Journal of Applied Mechanics - Transaction ASME, 18*(3), 293–297.

Welte, T. M., Foros, J., Nielsen, M. H., & Adsten, M. (2018). *MonitorX – Experience from a Norwegian-Swedish research project on industry 4.0 and digitalization applied to fault detection and maintenance of hydropower plants*. Academic Press.

Whitman, M. E., & Mattord, H. J. (2012). *Hands-on information security lab manual*. Cengage Learning.

Willett, K. D. (2008). *Information assurance architecture*. Auerbach Publications. doi:10.1201/9780849380686

Wilson, J. R. (2014). Fundamentals of systems ergonomics/human factors. *Applied Ergonomics, 45*(1), 5–13. doi:10.1016/j.apergo.2013.03.021 PMID:23684119

Winter, S. (1987). Knowledge and Competence as Strategic Assets. The Strategic Management of Intellectual Capital. doi:10.1016/b978-0-7506-9850-4.50013-0

Winters, R., Adae, I., & Silipo, R. (2014). *Anomaly detection in predictive maintenance anomaly detection with time series analysis*. Paper presented at the Knime.

Wits, W. W., García, J. R. R., & Becker, J. M. J. (2016). How Additive Manufacturing Enables more Sustainable End-user Maintenance, Repair and Overhaul (MRO) Strategies. *Procedia CIRP, 40*, 693–698. doi:10.1016/j.procir.2016.01.156

Wong, K. V., & Hernandez, A. (2012). A Review of Additive Manufacturing. *International Scholarly Research Network, 2012*, 1–10. doi:10.5402/2012/208760

World Economic Forum. (2017). Harnessing the fourth industrial revolution for sustainable emerging cities. *Fourth Industrial Revolution for the Earth Series*. http://www3.weforum.org/docs/WEF_Harnessing_the_4IR_for_Sustainable_Emerging_Cities.pdf

Yan, J., Meng, Y., Lu, L., & Li, L. (2017). *Industrial Big Data in an Industry 4.0 Environment : Challenges, Schemes, and Applications for Predictive Maintenance. 5*. Academic Press.

Yang, J.B., Wang, Y.M., Xu, D.L., & Chin, K.S. (2006) The evidential reasoning approach for MADA under both probabilistic and fuzzy uncertainties. *European Journal of Operational Research, 71*(1), 309-343.

Yi, L., Gläßner, C., & Aurich, J. C. (2019). How to integrate additive manufacturing technologies into manufacturing systems successfully: A perspective from the commercial vehicle industry. *Journal of Manufacturing Systems, 53*, 195–211. doi:10.1016/j.jmsy.2019.09.007

Ying-Ming, W., & Kwai-Sang, C. (2009). Risk evaluation in failure mode and effects analysis using fuzzy weighted geometric mean. *Expert Systems with Applications, 36*(2), 1195–1207. doi:10.1016/j.eswa.2007.11.028

Yoo, J. H., Lee, H. S., & Ko, J. W. (2011). Risk assessment and drawing information system based change management. *Korean Journal of Chemical Engineering, 28*(2), 323–331. doi:10.100711814-010-0389-y

Young, T., Fehskens, M., Pujara, P., Burger, M., & Edwards, G. (2010). *Utilizing data mining to influence maintenance actions*. Paper presented at the 2010 IEEE AUTOTESTCON. 10.1109/AUTEST.2010.5613610

Yu, S., Wang, C., Ren, K., & Lou, W. (2010, March). Achieving secure, scalable, and fine-grained data access control in cloud computing. In 2010 Proceedings IEEE INFOCOM (pp. 1-9). IEEE. doi:10.1109/INFCOM.2010.5462174

Zadeh, L. A. (1996). Fuzzy Logic = Computing with words. *IEEE Transactions on Fuzzy Systems, 4*(2), 103–111. doi:10.1109/91.493904

Zhang, G., Wang, Y., Mu, B., & Wang, T. (2019). Big Data and Safety Management Methods: The Reduction Model of Hot Work Number. In *2019 IEEE 4th International Conference on Big Data Analytics (ICBDA)* (pp. 140-143). Suzhou, China: IEEE.

Zhang, B., & Wang, J. (1994). Knowledge information and human function in manufacturing systems. *Chinese Journal of Mechanical Engineering*, *30*(5), 61–65.

Zhang, F., Du, B., Zhang, L., & Xu, M. (2016). Weakly supervised learning based on coupled convolutional neural networks for aircraft detection. *IEEE Transactions on Geoscience and Remote Sensing*, *54*(9), 5553–5563. doi:10.1109/TGRS.2016.2569141

Zhang, P., White, J., & Schmidt, D. (2018). Architectures and patterns for leveraging high-frequency, low-fidelity data in the healthcare domain. *6th IEEE International Conference on Healthcare Informatics, ICHI 2018*, 463–464. 10.1109/ICHI.2018.00104

Zhang, W., Yang, D., & Wang, H. (2019). Data-Driven Methods for Predictive Maintenance of Industrial Equipment: A Survey. *IEEE Systems Journal*, *13*(3), 2213–2227. doi:10.1109/JSYST.2019.2905565

Zhang, X., & Liang, E. (2019). Metal additive manufacturing in aircraft: Current application, opportunities and challenges. *IOP Conference Series. Materials Science and Engineering*, *493*, 012032. doi:10.1088/1757-899X/493/1/012032

Zhao, L., Zhou, Y., Matsuo, I., Korkua, S. K., & Lee, W. J. (2019, May). The Design of a Holistic IoT-Based Monitoring System for a Wind Turbine. In *2019 IEEE/IAS 55th Industrial and Commercial Power Systems Technical Conference (I&CPS)* (pp. 1-7). IEEE. 10.1109/ICPS.2019.8733375

Zhou, K., Liu, T., & Zhou, L. (2015). Industry 4.0: Towards future industrial opportunities and challenges. *2015 12th International Conference on Fuzzy Systems and Knowledge Discovery (FSKD)*, 2147-2152.

Zhou, K., Liu, T., & Zhou, L. (2015). Industry 4.0: Towards future industrial opportunities and challenges. *2015 12th International Conference on Fuzzy Systems and Knowledge Discovery (FSKD)*, 2147–2152. 10.1109/FSKD.2015.7382284

Zhou, J. P., Li, Y., Zhou, B., Wang, J., Zang, L., & Meng, L. (2018). MengToward new-generation intelligent manufacturing. *Engineering*, *4*(1), 11–20. doi:10.1016/j.eng.2018.01.002

Zissis, D., & Lekkas, D. (2012). Addressing cloud computing security issues. *Future Generation Computer Systems*, *28*(3), 583–592. doi:10.1016/j.future.2010.12.006

Zou, C., Deng, H., & Qiu, Q. (2013). Design and implementation of hybrid cloud computing architecture based on cloud bus. *2013 IEEE 9th International Conference on Mobile Ad-Hoc and Sensor Networks*, 289-293. 10.1109/MSN.2013.72

About the Contributors

Alberto Martinetti (1985) is an assistant professor in the chair of Maintenance Engineering within the department of Design, Production and Management. He is track coordinator of the Maintenance Engineering and Operations specialisation in Mechanical Engineering at the University of Twente and has worked for the Polytechnic of Turin and for the University of Turin. He holds a Master's degree in Geo-resources and Geo-technologies Engineering (2009) and a PhD degree in Environmental and Land / Safety and Health at the Polytechnic of Turin (2013) on the Prevention through Design approach in mining activities.

Sarbjeet Singh is working at Division of Operation and Maintenance, Luleå University of Technology, Sweden and Mechanical Engineering Department, Government College of Engineering &Technology, (University of Jammu) Jammu, India.

Micaela Demichela is an Associate Professor at the Department of Applied Science and Technology of Politecnico di Torino. PhD in Chemical Engineering, both for teaching and research she operates in the field of process and occupational safety and environment and territories protection. She is in the Board of Directors of the ESReDa Association (European Safety, Reliability and Data) and in the Scientific Board of the R3C (Responsible Risk Resilience Centre) Inter Department Centre at Politecnico di Torino.

* * *

Yawar Abbas is a Ph.D. Candidate in faculty of Engineering Technology at the University of Twente. Yawar completed his BSc in Mechanical Engineering at the Polytechnic of Turin and his MSc in Mechanical Engineering at the University of Twente.

Fernando Abrahão graduated from the Brazilian Air Force Academy Aviation Officer Training Course (1988), Master's Degree in Maintenance Management, Logistics Management, and Procurement Process Logistics Management - US Air Force Institute of Technology (1998) and Ph.D. in Transportation Engineering from the Polytechnic School of the University of São Paulo (2005). He was Manager of Acquisition and Life Cycle Management Programs for the Aeronautical Systems at the SDDP / COPAC Development and Programs Sub-Directorate and Professor of the Specialization Course in Logistics at the Aeronautics Logistics Institute, both of the Brazilian Air Force. He was the Dean of Administration at ITA, DCTA, São Jose dos Campos (2014/2015). He is currently Chief Coordinator of the ITA Logistics Engineering Laboratory - AeroLogLab-ITA.

Gabriele Baldissone is a Chemical Engineer wit a PhD in Metrology. He is Research Assistant at the Department of Applied Science and Technology of Politecnico di Torino. Within the research group SAfeR (Safety, Reliability and Risks) he manage and operates research activities related to the modeling of complex socio-technical systems in the process and manufacturing environment.

Priyanka C. Bhatt is currently a Ph.D. Research Fellow in the Department of Information Management, College of Informatics, at Chaoyang University of Technology, Taiwan. She has extensive experience of more than 5 years in the area of Computer Engineering and Information Science. She has worked with organizations and institutes of national importance like Times Group, Greater Noida; SIDBI, IIT Kanpur; Homi Bhabha Centre for Science Education – Tata Institute of Fundamental Research, Mumbai; and National Medical Library, New Delhi. 'e-services' initiated by her at National Medical Library, New Delhi received appreciation from the Honorable Minister of Health and Family Welfare, Govt. Of India, Dr. J.P. Nadda in 2016. She has more than five research contributions in international conferences and has published two books. Her research interests include technology management, social network analysis, scientometrics, patent statistics, artificial neural networks, etc.

Gianfranco Camuncoli is a Chemical Engineer, with a Master Degree in Industrial Safety and Risk Analysis. He is the CEO of an Italian service company, ARIA srl, whose activity is devoted to the risk assessment for the benefit of industrial and occupational safety. Recent activities deal with the implementation of Industry 4.0 enabling technologies to enhance safety at the workplace and safety and reliability of process plants and machineries in manufacturing.

Peter Chemweno is an Assistant Professor in Manufacturing Systems at the University of Twente, in the Netherlands. His current research is aligned towards developing decision support models of among other aspects, asset management, optimizing manufacturing systems, and for safety assessment of collaborative robot systems. His interests extends to healthcare, where he focuses on developing decision support models for optimally managing diagnostic and treatment devices. Specifically, he is researching on robust protocols for assessing clinical hazards embedded in the patient's care pathway. Peter received his Ph.D. in Mechanical Engineering in 2016 from University of Leuven, Belgium. His postdoctoral research work focused on developing safety assessment models of healthcare and collaborative robot systems, which emphasize systematic approaches for hazard analysis and risk assessment.

Lorenzo Comberti is a Environmental and geology Engineer with a PhD in Geotechnical Engineering. He is Research Fellow at the Department of Applied Science and Technology of Politecnico di Torino. Within the research group SAfeR (Safety, Reliability and Risks) he manage and operates research activities related to the modeling within the manufacturing environment of human and organisational factors inter-relation with safety and quality.

Aleksandar Cvjetic received his diploma in Mining Engineering from University of Belgrade, Faculty of Mining and Geology in 1992. In 2010 he defended his PhD thesis also at the University of Belgrade, in the field of Occupational Health and Safety. He started his career as Teaching Assistant at the Faculty of Mining and Geology, where he currently works as Associate Professor teaching couple courses, one of which is related to Occupational Health and Safety in Mining. He is Experts of Ministry of Energy, Development and Environmental Protection of Republic of Serbia, the Chairman of the Commission for

Personal Protective Equipment - Institute for Standardization of Republic of Serbia, one of the members of the Commission for protection against noise and one of the members of the Commission for Optics, Photonics and Eye protection in the same Institute. From the 2017 he is the Vice President of Commission for gaining Chartered Engineer Degree in Mining Engineering – Republic of Serbia - Autonomous Province of Vojvodina - Provincial Secretariat for Energy, Construction and Transport. Throughout his career he fostered collaboration with mining industry both locally and abroad, as well as developing collaboration with regional and European Universities.

Vinayak A. Drave is working as Assistant Professor, Department of Operations, Jindal Global Business School, O.P. Jindal Global University, India & Visiting Researcher, Department and Graduate Institute of Business Administration, College of Management, Chaoyang University of Technology, Taichung, Taiwan, ROC.

Valerio Elia is Assistant Professor at University of Salento (Italy). He received his degree in Physics at University of Lecce in 1993, with a dissertation on "Measurement of total and differential cross sections of Hadron production of J/ψ at FNAL/E771". He received his PhD in Physics at University of Bari with a dissertation on "Perspectives for the violation of CPT and T at KLOE". He carries out his research and teaching activities at the Department of Engineering for Innovation of the University of Salento. He is the author of papers at national and international level on books, conference proceedings, and journals.

Stephen Famurewa is working as Railway Track Specialist at Trafikverket, Lulea, Sweden and Associate Senior Lecturer, Division: Operation, Maintenance and Acoustics, Luleå University of Technology, Luleå, Sweden

Lex Frunt is Head of Quality Management at Nederlandse Spoorwegen.

Sung-Chi Hsu is working in the Department of Civil Engineering, Chaoyang University of Technology, Taiwan.

Dejan Ivezić is the full professor at the University of Belgrade - Faculty of Mining and Geology. He graduated at the Faculty of Mechanical Engineering of the University of Belgrade in 1994. He obtained magister degree and PhD in Technical Sciences at the same Faculty in 1999 and 2004, respectively. He has worked at the Faculty of Mining and Geology since 1997. He teaches graduate and postgraduate courses in gas engineering, energy modelling and control of energy processes. His research interests include sustainable development, environmental protection concerning negative impact of energy activities, energy efficiency and renewable energy sources utilization and conservation of natural resources. He is an author or co-author of over 60 scientific papers, two university books and two scientific monographs. He took part in numerous energy and climate change projects funded by the EU, Serbian ministries of energy and environmental protection, Serbian municipalities and others. He drafted Energy Sector Development Strategy of Republic of Serbia up to 2025 with Projections to 2030. He is the member of Scientific Committee for Energy, Mining and Energy Efficiency at the Ministry of Education, Science and Technological Development.

Predrag D. Jovančić acquired his PhD from the University of Belgrade, Faculty of Mining and Geology, in 2007. He is a full professor at this institution. Prof. Jovančić is the head of the Mine mechanization chair. Has published over a hundred papers in journals and conferences. Also, there are over a hundred projects and studies prepared for the mining and mechanical industry. His current areas of interest include the operation and maintenance of mining machinery, and the role of vibrations as an overriding parameter in defining the condition and performance of drive assemblies and support structures.

Ravdeep Kour is a Ph.D. student in the Division of Operation and Maintenance Engineering at Luleå University of Technology, Sweden. She received Bachelor's degree in Information Technology and Master's degree in Computer Science Engineering from India, in 2004 and 2012 respectively. She worked as Assistant Professor in India from 2004 to 2012 and worked in Luleå Technical University, Lulea, Sweden as Research Engineer from 2012 to 2014. She worked on European Union and Swedish Railway Projects. Her total academic and research work experience is 15 years. Her research interests are cybersecurity in the context of IT and OT technologies, security risk assessment, machine learning, cloud computing, and big data analytics.

Vikas Kukshal is working as an Assistant Professor in the Department of Mechanical Engineering at National Institute of Technology, Uttarakhand, India.

Vimal Kumar is an Assistant Professor at Chaoyang University of Technology, Taichung, Taiwan (R.O.C.) in the Department of Information Management. He completed his Postdoctoral Research at Chaoyang University of Technology, Taichung, Taiwan (R.O.C.) in the Department of Business Administration in the domain of Technological Innovation and Patent Analysis. He has served as an Assistant Professor under TEQIP III, an initiative of MHRD, Govt. of India at AEC Guwahati in the Department of Industrial and Production Engineering. Prior to joining AEC, he served as Assistant Professor at MANIT, Bhopal in the Department of Management Studies and also served as Visiting Faculty at IMT Nagpur. He obtained his PhD in the domain of TQM and Manufacturing Strategy in the year 2017 and Masters in Supply Chain Management from the Department of Industrial & Management Engineering, IIT Kanpur in the year 2012. He graduated (B.Tech) in Manufacturing Technology from JSS Academy of Technical Education Noida, in the year 2010. He has published nineteen articles in reputable international journals and presented eighteen papers at international conferences. His research paper entitled "Time Table Scheduling for Educational Sector on an E-Governance Platform: A Solution from an Analytics Company" has been selected for best paper award in the International Conference on Industrial Engineering and Operations Management (IEOM) held in Bandung, Indonesia, March 6-8, 2018. He was also invited to serve as session chair of session on "Energy Related Awareness" held on 19th September, 2018 at iCAST 2018, IEEE International Conference on Awareness Science and Technology and "Lean Six Sigma" at the International Conference on Industrial Engineering & Operations Management (IEOM-2018) at Bandung, Indonesia and "Quality Control & Management" at the International Conference on Industrial Engineering & Operations Management (IEOM-2016) at Kuala Lumpur, Malaysia. He is a contributing author in journals including CSREM, IJPPM, IJQRM, IJPMB, IJPQM, IJBIS, AJOR, The TQM Journal, and Benchmarking: An International Journal, etc. and also a guest reviewer of a reputable journal like IJQRM, TQM & Business Excellence, The TQM Journal, Benchmarking: An International Journal, Journal of Asia Business Studies, and JSIT.

Arun Kumar is pursuing PhD From Department of Mechanical Engineering, Institute of Technology Delhi, India. He has good industrial experience and worked with Metallurgical and Material Handling group of Larsin and Toubro Constructions as Assistant Manager.

Kuei Kuei Lai is working in the Department of Business Administration, Chaoyang University of Technology, Taichung City, Taiwan.

Maria Chiara Leva (PI in ESHI) is the co-chair of the technical committee on Human Factors for the European Safety and Reliability Association (ESRA), former chair of the Irish Ergonomics Society and co-chair of the Symposium on Human Mental Workload. In 2016 she was awarded a Female Founder Competitive Start Fund by the National Digital Research Centre and Enterprise Ireland for her Campus Company 'Tosca Human Factors Solutions'. The company is a spin out of one of the EU project Dr Leva led as a PI, She currently holds a scientific advisory role in the business. Dr Leva has more than 60 publications on Human Factors (HF), Operational Risk Assessment and Safety Management in Science and Engineering Journals. She is a Lecturer in TU Dublin and adjunct professor for Risk Assessment and Safety Management in the School of Engineering, associated PI in the Science and Technology in Advanced Manufacturing research centre and in the Centre for Innovative Human systems in Trinity College Dublin. (Maria Chiara is a PI in the Environmental Sustainability and Health Institute).

Henrique Marques graduated in CFOAV from the Brazilian Air Force Academy (1991), Masters and Doctorate in Electronics and Computer Engineering from the Aeronautics Institute of Technology - ITA. He was responsible for the initial structuring of the ITA's Command and Control Laboratory and is currently assisting in the structuring of the ITA's Logistics Engineering Laboratory (AeroLogLab-ITA). Has experience in Command and Control, acting on the following subjects: simulation, command and control, joint operations, high level data fusion, probabilistic ontologies. Currently conducting research in the areas of e-Maintenance, Integrated Vehicle Health Management (IVHM), Prognostics and Health Management (PHM) and VirtualReality/Augmented Reality/Mixed Reality for applications in the area of operation and maintenance of aerospace systems.

Vladimir Milisavljević received his diploma in Mining Engineering from University of Belgrade, Faculty of Mining and Geology in 1994. In 2011 he defended his PhD thesis also at the University of Belgrade, in the field of Mining Mechanization. He started his career as Teaching Assistant at the Faculty of Mining and Geology, where he currently works as Associate Professor teaching courses related to Underground Mining Mechanization and Safe Operation of Mining Equipment. Since 2018 he is member of the Supervising Board of Faculty of Mining and Geology. Throughout his career he fostered collaboration with mining industry both locally and abroad, as well as developing collaboration with regional and European Universities.

Luigi Monica is technologist and responsible of the technical-scientific section "Technical Assessments" of the Italian Workers' Compensation Authority (INAIL) Department of Technological Innovation and Safety of Plants, Products and Anthropic settlements, with tasks of coordination of conformity assessment activities of machines, plants, appliances and products to the safety requirements prescribed by the provisions applicable laws (Machinery Directive, PED Directive, etc.), in support of the Authorities market surveillance. Mechanical engineer and PhD in "Production Systems and Industrial Plants".

It carries out research and study mainly on some thematic areas such as: risk assessment methodologies, machine and equipment safety of work and production plants, risk management and technological innovation.

Vijay Pal is Working as Assistant Professor, Department of Mechanical Engineering, IIT Jammu, India

Pulak M. Pandey completed his B.Tech. degree from H.B.T.I. Kanpur in 1993 securing first position and got Master's degree from IIT Kanpur in 1995 in Manufacturing Science specialization. He served H.B.T.I. Kanpur as faculty member for approximately 8 years and also completed Ph.D. in the area of Additive Manufacturing/3D Printing from IIT Kanpur in 2003. He joined IIT Delhi as a faculty member in 2004 and is presently serving as Professor. In IIT Delhi, Dr Pandey diversified his research areas in the field of micro and nano finishing, micro-deposition and also continued working in the area of 3D Printing. He supervised 30 PhDs and more than 34 MTech theses in last 10 years and also filed 17 Indian patent applications. He has approximately 154 international journal papers and 45 international/national refereed conference papers to his credit. These papers have been cited for more than 4403 times with h-index as 33. He received Highly Commended Paper Award by Rapid Prototyping Journal for the paper "Fabrication of three dimensional open porous regular structure of PA 2200 for enhanced strnegth of scaffold using selective laser sintering" published in 2017. Many of his B.Tech. and M.Tech. supervised projects have been awarded by IIT Delhi. He is recipient of Outstanding Young Faculty Fellowship (IIT Delhi) sponsored by Kusuma Trust, Gibraltar and J.M. Mahajan outstanding teacher award of IIT Delhi. His students have won GYTI (Gandhian Young Technological Innovation Award) in 2013, 2015, 2017 and 2018.

Amar Patnaik is working as an Associate Professor in the Department of Mechanical Engineering at Malaviya National Institute of Technology Jaipur, India.

Liliane Pintelon is a Professor at the KU Leuven (CIB), where she teaches logistics courses, including maintenance management. Her research interests are in asset management, both in industry and healthcare.

Mohammad Rajabalinejad obtained a master degree on safety of civil infrastructures in 2001. After the graduation, he worked in several national projects in design, system control and later on inspection for about 6 years. In 2006, he started my research on the use of prior information for reliability of infrastructures and accomplished a PhD from TUDelft in 3 years. He continued with this research during my postdoc at Ecole Polytechnique de Montreal, Canada and conducted studies on European and American pilot sites. Then, he started as an assistant professor in TUDelft focusing on the use of information to manage uncertainties in the course of product design. Moving to UTwente, his research attained focus on infrastructures and their engineering. He is serving two journals as associate editor, and he has organized or chaired special sessions in different conferences. As educator he is interested in design of smart products, system design, engineering and safety. He is also a member of the standardization committee for Safety of Machinery (NEC 44).

Guilherme Rocha graduated in Mechanical-Aeronautical Engineering from the Aeronautics Institute of Technology (1998), Master's degree in Aeronautical and Mechanical Engineering from the Aeronautics Institute of Technology (2002) and PhD in Electronic and Computer Engineering from the Aeronautics Institute of Technology (2011). He was Product Development Engineer at Embraer and Technical and Operational Director of Konatus Soluções Inteligentes. Has experience in Mechanical Engineering, focusing on systems engineering, logistics, maintenance, modeling and control systems, working mainly in the fields of aviation, the automobile industry and the health sector. He is currently Associate Professor at Aeronautics Institute of Technology and also a researcher at ITA's Logistics Engineering Laboratory - AeroLogLab-ITA. Has worked in research and development of projects related to certification, maintenance, operation and risk management of critical systems.

João Castro Silva is a Portuguese student who is about to conclude his Master´s degree in Mechanical Engineering at the University of Aveiro, Portugal. As part of an Erasmus program, he spent six months in The Netherlands, learning more about Maintenance and Asset Management at the University of Twente, which led him and his colleague Gonçalo Soares to write a SLR on the progression of Maintenance in Industry 4.0 era.

Gurminder Singh has completed his PhD from Indian Institute of Technology Delhi in the field of additive manufacturing and having good SCI journals, conference papers, and book chapters in the leading publishing houses. He has three patents in the field of additive manufacturing. He has done his masters from Thapar University in CAD/CAM engineering and bachelors from Guru Nanak Dev Engineering College Ludhiana in Production Engineering.

Ravinder Pal Singh is working as Senior Research fellow in Department of Mechanical Engineering, Indian Institute of Technology Delhi.

Gonçalo Soares is a Portuguese Mechanical Engineering student at the University of Aveiro, about to finish his master's degree. He spent six months at the University of Twente learning more about Maintenance and Asset Management, which led him and his colleague João Silva to write an SLR on the progression of Maintenance in Industry 4.0 era.

Milos Tanasijevic graduated, received his master's and doctoral degrees in the Belgrade University – Faculty of Mining and Geology (FMG). A PhD thesis was received in 2007. The topic of the thesis is Dependability of the mechanical component of Bucket-wheel excavator. He is employed as a university professor. In a professional career, he is dealing with design, reliability, availability and dependability of mining machines. As author or co-author, he has published about 80 articles in monographs, scientific journals, proceedings and two books (Machine elements, Maintenance engineering in the mining industry (2015. and 2008, FMG). In scientific articles, he focused specifically on optimizing the system based on the application of fuzzy logic. Deals with the development of concepts of dependability and safety centred maintenance.

Vincenzo Tarsitano graduated in Environmental Engineering (Politecnico di Torino). He is working as quality, environment and safety consultant for a business company.

Adithya Thaduri works as Associate Senior Lecturer in the Division of Operation and Maintenance Engineering at Luleå University of Technology. He has experience in coordination of four European projects (IN2RAIL, INFRALERT, IN2SMART and FR8RAIL) and three national projects (InfraSweden, Mindi and SKF) in the area of Railways and have worked in collaboration in other seven projects. I recently got funding for one European project for Railways (IN2SMART2) and two national projects; one from Vinnova to Railway and other from Coal India Limited to Mining. He is part of over 35 deliverables/reports within above mentioned projects. He has over 40 research publications (28 after PhD) in journals, book chapters and conference proceedings. He is been teaching Maintenance Engineering course for master's programme for two years. His areas of research are machine learning and context-aware maintenance decision making within the framework of Maintenance 4.0 in Railways, asset maintenance analytics, prognostics and degradation modelling of railway infrastructure, reliability predictions, maintenance planning and optimization, RAMS, LCC and Risk assessment, predictive analytics of mining machines, and cybersecurity.

Phillip Tretten is working at Division of Operation and Maintenance, Luleå University of Technology, Sweden.

Leo A. M. van Dongen (1954) has worked for the Netherlands Railways (NS, 100% state owned) for 35 years. He retired as Chief Technology Officer (CTO), responsible for the asset management of the rolling stock fleet, workshops and maintenance equipment in August 2019. Since 2010 he is also professor in Maintenance Engineering at the faculty of Engineering Technology, in the Department of Design, Production and Management of the University of Twente. After his studies in mechanical engineering, he completed his doctoral research at Eindhoven University of Technology on energy efficiency of drive trains for electric vehicles. At DAF Trucks he was active in the development of diesel engines. His career within NS has concentrated mainly on technological functions. These include project manager of electric locomotives, secretary to the executive board, fleet manager at NS Reizigers, and, within NedTrain, he was responsible for technical fleet management, maintenance systems, spare parts purchase, maintenance management and construction of new capital goods. He promotes the academic development of the engineering profession and encourages further research into maintenance processes: from design methodologies for capital goods to the development of associated maintenance concepts, not only for the initial investment, but also for the management during the entire life cycle!

Pratima Verma is a Post Doctoral Fellow in the Department of Information Management at Chaoyang University of Technology, Taichung, Taiwan. She has worked as an Assistant Professor in Strategic Management area at Indian Institute of Management (IIM) Bodh Gaya, India. Prior to joining IIM, she was Postdoctoral Fellow in the Department of Management Studies at IIT Madras, India. She obtained her PhD from IIT Kanpur where she worked in the domain of Strategic Management and Horizontal Strategy in the Department of Industrial & Management Engineering, in the year 2017. She received her MBA in Finance and Human Resource Management from BBDNITM, Uttar Pradesh Technical University- Lucknow, India in the year 2011. She completed her graduation (B.Tech) in Information Technology in the year 2009 from BBNITM, Lucknow. She has one year of experience in teaching. She also awarded JRF/SRF in the area of human resource management. She has published ten articles in reputable international journals and presented fourteen papers at international conferences. Her research paper entitled "Time Table Scheduling for Educational Sector on an E-Governance Platform: A Solution from an Analytics Company" has been selected for best paper award in the International Conference on Industrial Engineering and Operations Management (IEOM) held in Bandung, Indonesia, March 6-8, 2018. She was invited to serve as session chair for Human Factors and Ergonomics Track at the International Conference on Industrial Engineering & Operations Management at Kuala Lumpur, Malaysia. She is a contributing author in journals including IJPMB, IJISE, IJBIS, and Benchmarking: An International Journal, etc.

Index

L

M

N

P

R

S

T

U

www.ingramcontent.com/pod-product-compliance
Lightning Source LLC
LaVergne TN
LVHW062141300926
849161LV00010B/2593
* 9 7 8 1 7 9 9 8 3 9 0 4 0 *